Developments in Soil Science 26

ENZYMOLOGY OF DISTURBED SOILS

Further Titles in this Series

5A. G.H. BOLT and M.G.M. BRUGGENWERT (Editors)
SOIL CHEMISTRY. A. BASIC ELEMENTS

8. M. SCHNITZER and S.U. KAHN (Editors)
SOIL ORGANIC MATTER

11A. L.P. WILDING, N.E. SMECK and G.F. HALL (Editors)
PEDOGENESIS AND SOIL TAXONOMY. I. CONCEPTS
AND INTERACTIONS

11B. L.P. WILDING, N.E. SMECK and G.F. HALL (Editors)
PEDOGENESIS AND SOIL TAXONOMY. II. THE SOIL ORDERS

13. P. KOOREVAAR, G. MENELIK and C. DIRKSEN
ELEMENTS OF SOIL PHYSICS

14. G.S. CAMPBELL
SOIL PHYSICS WITH BASIC-TRANSPORT MODELS
FOR SOIL-PLANT SYSTEMS

15. M.A. SMULDERS
REMOTE SENSING IN SOIL SCIENCE

17. K. KUMADA
CHEMISTRY OF SOIL ORGANIC MATTER

19. L.A. DOUGLAS (Editor)
SOIL MICROMORPHOLOGY: A BASIC AND APPLIED SCIENCE

20. H.W. SCHARPENSEEL, M. SCHOMAKER and A. AYOUB (Editors)
SOILS ON A WARMER EARTH

21. S. SHOJI, M. NANZYO and R. DAHGREN
VOLCANIC ASH SOILS

22. A.J. RINGROSE-VOASE and G.S. HUMPHREYS (Editors)
SOIL MICROMORPHOLOGY: STUDIES IN MANAGEMENT
AND GENESIS

23. J. DVORÁK and L. NOVÁK
SOIL CONSERVATION AND SILVICULTURE

24. N. AHMAD and A. MERMUT
VERTISOLS AND TECHNOLOGIES FOR THEIR MANAGEMENT

25. E.G. GREGORICH and M.R. CARTER (Editors)
SOIL QUALITY FOR CROP PRODUCTION
AND ECOSYSTEM HEALTH

Developments in Soil Science 26

ENZYMOLOGY OF DISTURBED SOILS

Edited by

S. KISS, D. PAŞCA and M. DRĂGAN-BULARDA

Laboratory for Environmental Enzymology and Microbiology,
Department of Plant Physiology,
Babeş-Bolyai University,
3400 Cluj, Romania

1998

ELSEVIER

Amsterdam - Lausanne - New York - Oxford - Shannon - Singapore - Tokyo

ELSEVIER SCIENCE B.V.
Sara Burgerhartstraat 25
P.O. Box 211, 1000 AE Amsterdam, The Netherlands

First edition 1998

Library of Congress Cataloging in Publication Data
A catalog record from the Library of Congress has been applied for.

ISBN: 0-444-50057-X

♾ The paper used in this publication meets the requirements of ANSI/NISO Z39.48-1992 (Permanence of Paper).

Transferred to digital printing 2006.

PREFACE

The disturbance of soils, like other phenomena of environmental pollution, encountered in so many areas all over the world, has become a subject of extensive concern and has led to a vast literature in the field of enzymology, too.

The present work deals with the enzymology of three great categories of disturbed soils, including their remediation, namely with I. Enzymology of oil-contaminated soils; II. Enzymology of soils affected by industrial emissions; and III. Enzymology of technogenic soils.

Studies of enzyme activities in disturbed soils and in those subjected to remediation are reviewed considering both western and eastern literature.

Within each section (or chapter), the reviewed studies are grouped by countries, the order of which was established based on the year in which the first paper on the topic of the section (or chapter) had appeared.

The present work is a revised, enlarged, and updated synthesis of our previous review papers on Enzymology of oil-contaminated soils (Kiss, 1995), soils affected by industrial emissions (Kiss, 1997), and technogenic soils (Kiss *et al.*, 1993, 1996).

ACKNOWLEDGEMENTS

The authors wish to express their gratitude to Ir. Jacques Kiebert of the Elsevier Science B.V. for his kind co-operation in publishing this book, to Professor Brian Johnston of the University-Degree Protestant Theological Institute in Cluj, and to our colleague Dr. Vasile Muntean of the Biological Research Institute in Cluj for his valuable technical assistance in giving the manuscript a camera-ready form. We also thank Professor Vasile Cristea, Dean of the Faculty of Biology and Geology of the Babeş-Bolyai University in Cluj, for allowing us to use the printing facilities of the Faculty.

CONTENTS

Preface V

Acknowledgements VI

PART I. ENZYMOLOGY OF OIL-CONTAMINATED SOILS 1

INTRODUCTION 3

Chapter 1. SOIL ENZYME ACTIVITIES AS AFFECTED BY ACCIDENTAL OIL CONTAMINATION 5

Contamination with crude oil 5

Enzymological research in the Russian Federation 5

Contamination with crude oil and oil products 7

Enzymological research in Georgia 7

Contamination with oil products 8

Enzymological research in Germany 8

Enzymological research in the United States of America 10

Contamination with oil field wastewaters 11

Enzymological research in the Russian Federation 11

Chapter 2. ENZYMOLOGICAL EVALUATION OF THE BIOLOGICAL EFFECTS OF OIL CONTAMINATION OF SOILS IN EXPERIMENTAL MODELS 13

FIELD EXPERIMENTS 13

Contamination with crude oil 13

Enzymological research in Azerbaijan 13

Enzymological research in the United States of America 15

Enzymological research in Belorussia 17

Enzymological research in the Russian Federation 17

Contamination with oil products 19

Enzymological research in Germany 19

LABORATORY EXPERIMENTS 22
Contamination with crude oil 22
Enzymological research in Ukraine 22
Enzymological research in the Russian Federation 22
Contamination with crude oil and oil products 22
Enzymological research in the United States of America 22
Enzymological research in Azerbaijan 23
Enzymological research in the Russian Federation 24
Contamination with oil products 24
Enzymological research in China 24
Enzymological research in Austria 25
Enzymological research in the Russian Federation 26
Enzymological research in Germany 27
Contamination with oil well-drilling fluid additives 28
Enzymological research in Ukraine 28
Chapter 3. ENZYMOLOGICAL EVALUATION OF THE BIOLOGICAL EFFECTS OF THE REMEDIATION OF OIL-CONTAMINATED SOILS IN EXPERIMENTAL MODELS 29
FIELD EXPERIMENTS 29
Contamination with crude oil 29
Enzymological research in the Russian Federation 29
Enzymological research in Azerbaijan 34
Contamination with oil products 36
Enzymological research in the United States of America 36
Enzymological research in Germany 39
Contamination with oil field wastewaters 41
Enzymological research in the Russian Federation 41

LABORATORY EXPERIMENTS 41
Contamination with crude oil 41
Enzymological research in the Russian Federation 41
Enzymological research in Azerbaijan 45
Enzymological research in Canada 45
Contamination with oil products 49
Enzymological research in the United States of America 49
Enzymological research in the Russian Federation 51
Enzymological research in Austria 52
CONCLUSIONS 53
REFERENCES 53
PART II. ENZYMOLOGY OF SOILS AFFECTED BY INDUSTRIAL EMISSIONS 63
INTRODUCTION 65
Chapter 4. STUDIES OF THE SOIL ENZYMOLOGICAL EFFECTS OF THE COMPONENTS OF INDUSTRIAL EMISSIONS, THROUGH EXPERIMENTS MODELLED IN THE LABORATORY OR IN *IN SITU* ARTIFICIAL MICROCOSMS 67
Chapter 5. STUDIES OF THE SOIL ENZYMOLOGICAL EFFECTS OF INDUSTRIAL EMISSIONS ORIGINATING FROM A POINT SOURCE (AN INDUSTRIAL PLANT) 69
Nonferrous metallurgical plants 69
Enzymological research in Sweden 69
Enzymological research in the United States of America 76
Enzymological research in Poland 77
Enzymological research in Romania 82
Enzymological research in Canada 83
Enzymological research in the Russian Federation 87
Enzymological research in Kazakhstan 91
Enzymological research in Switzerland 91
Enzymological research in Austria 92

Ironworks 94

Enzymological research in Sweden 94

Enzymological research in Poland 95

Ore enrichment works 95

Enzymological research in Sweden 95

Enzymological research in the Russian Federation 96

Coking plants 97

Enzymological research in Ukraine 97

Enzymological research in the United Kingdom 99

Refractory brickworks 100

Enzymological research in the United Kingdom 100

Pulp and paper mills 102

Enzymological research in the Russian Federation 102

Synthetic fibre factories 102

Enzymological research in Belorussia 102

Oil production plants 103

Enzymological research in Sweden 103

Other chemical factories 104

Enzymological research in the Russian Federation 104

Enzymological research in Austria 106

Enzymological research in Poland 106

Enzymological research in Ukraine 108

Biotechnological plants 109

Enzymological research in the Russian Federation 109

Oil- and coal-fired power plants 110

Enzymological research in Sweden 110

Enzymological research in the Russian Federation 111

Enzymological research in Poland 111

Atomic power plants 112

Enzymological research in the Russian Federation 112

ADDENDA 112
Military waste disposal operations 112
Enzymological research in the United States of America 112
Rocket destruction operations 114
Enzymological research in the Russian Federation 114
Chapter 6. STUDIES OF THE SOIL ENZYMOLOGICAL EFFECTS OF INDUSTRIAL EMISSIONS ORIGINATING FROM MULTIPLE SOURCES (MANY INDUSTRIAL PLANTS MANUFACTURING DIFFERENT PRODUCTS, BUT SITUATED IN THE SAME, INDUSTRIAL AREA) 117
Enzymological research in Sweden 117
Enzymological research in Poland 118
Enzymological research in the Russian Federation 119
Enzymological research in Germany 120
Enzymological research in the Czech Republic 120
Enzymological research in the United Kingdom 121
Enzymological research in the United States of America 122
Enzymological research in Austria 123
Enzymological research in Finland 127
ADDENDA 128
Urban soils 128
Enzymological research in Poland 128
Enzymological research in Germany 130
Enzymological research in Finland 131
Enzymological research in the Russian Federation 131
Roadside soils 131
Enzymological research in the United States of America 131
Enzymological research in the Russian Federation 133
Enzymological research in India 133
Enzymological research in Bulgaria 134
CONCLUSIONS 135
REFERENCES 135

PART III. ENZYMOLOGY OF TECHNOGENIC SOILS 153
INTRODUCTION 155
Chapter 7. TECHNOGENIC SOILS FROM COAL MINE SPOILS 157
Enzymological research in Ukraine 157
Enzymological research in the United States of America 162
Enzymological research in Poland 175
Enzymological research in the Russian Federation 181
Enzymological research in Hungary 186
Enzymological research in Germany 187
Enzymological research in New Zealand 193
Enzymological research in the United Kingdom 195
Enzymological research in Spain 198
Enzymological research in the Czech Republic 204
Chapter 8. TECHNOGENIC SOILS FROM POWER PLANT WASTES 207
Enzymological research in Poland 207
Enzymological research in Hungary 207
Enzymological research in Romania 208
Enzymological research in the United States of America 208
Chapter 9. TECHNOGENIC SOILS ON EXHAUSTED PEATLANDS 211
Enzymological research in the Russian Federation 211
Enzymological research in Belorussia 215
Chapter 10. TECHNOGENIC SOILS FROM RETORTED OIL SHALE 217
Enzymological research in the United States of America 217
Chapter 11. TECHNOGENIC SOILS FROM IRON MINE SPOILS 225
Enzymological research in Romania 225
Enzymological research in the Russian Federation 228
Chapter 12. TECHNOGENIC SOILS FROM MANGANESE MINE SPOILS 231
Enzymological research in Ukraine 231
Enzymological research in Georgia 232
Chapter 13.TECHNOGENIC SOILS FROM LEAD AND ZINC MINE WASTES 235
Enzymological research in the United Kingdom 235
Enzymological research in Romania 237

Chapter 14. TECHNOGENIC SOILS FROM GOLD MINE WASTES 245
Enzymological research in New Zealand 245
Chapter 15. TECHNOGENIC SOILS FROM APATITE AND PHOSPHORITE MINE WASTES 247
Enzymological research in the Russian Federation 247
Chapter 16. TECHNOGENIC SOILS FROM SULPHUR MINE SPOILS 251
Enzymological research in Ukraine 251
Chapter 17. TECHNOGENIC SOILS FROM LIME AND DOLOMITE MINE SPOILS 253
Enzymological research in Ukraine 253
Chapter 18. TECHNOGENIC SOILS FROM REFRACTORY CLAY MINE SOILS 255
Enzymological research in Ukraine 255
Chapter 19. TECHNOGENIC SOILS FROM BENTONITIC CLAY MINE SPOILS 259
Enzymological research in Georgia 259
Chapter 20. TECHNOGENIC SOILS FROM PUMICE MINE SPOILS 261
Enzymological research in Germany 261
Chapter 21. TECHNOGENIC SOILS ON SAND OPENCAST MINE FLOOR DRIFT AND SPOILS 263
Enzymological research in Poland 263
Enzymological research in Georgia 265
Chapter 22. TECHNOGENIC SOILS ON EXHAUSTED GRAVEL QUARRIES 267
Enzymological research in Lithuania 267
Chapter 23. TECHNOGENIC SOILS FROM OVERBURDENS REMAINING AFTER PIPELINE CONSTRUCTION 269
Enzymological research in the Russian Federation 269
Enzymological research in the United States of America 271
Enzymological research in Ukraine 272
Chapter 24. RECULTIVATION OF SOILS REMAINING AFTER TOPSOIL "MINING" 275
Enzymological research in New Zealand 275

Chapter 25. REVEGETATION OF SOILS USED FOR COVERING RUBBLE DUMPS 279
Enzymological research in Germany 279
Chapter 26. REVEGETATION OF LANDFILL SOILS 281
Enzymological research in the United Kingdom 281
Chapter 27. TRANSFORMATION OF HARBOUR MUDS INTO AGRICULTURAL SOILS 285
Enzymological research in Germany 285
Chapter 28. RECULTIVATION OF SOILS DISTURBED BY DIFFERENT INDUSTRIAL ACTIVITIES 287
Enzymological research in Canada 287
CONCLUSIONS 291
REFERENCES 292
SUBJECT INDEX 321
Index of the scientific names of microorganisms, plants, and animals 331
Index of geographic names 333

PART I. ENZYMOLOGY OF OIL-CONTAMINATED SOILS

INTRODUCTION

The modern society largely depends on petroleum products used as fuels or as raw materials for synthesis of plastics, surfactants, dye-stuffs, drugs, pesticides, etc.

Contamination of soils with crude oil, drilling muds, oil well waters, petroleum products and wastes can occur in many places (oil fields, refineries, chemical and other industrial plants, farms, highways, railroads, etc.), during processing, transportation, storage, and utilisation, due to spillage and leakage.

Oil contamination of soils negatively affects agricultural productivity, health of plants, animals, and humans. Oil from the contaminated soils can penetrate into ground- and surface waters, polluting them. Some components of oil from the contaminated soils can volatilise, causing air pollution.

Investigation of the soil biological effects of oil contamination and remediation of oil-contaminated soils are urgent tasks for environmental scientists all over the world.

Enzymological methods are used, besides other methods, to evaluate the soil biological effects of oil contamination and the efficiency of the remediation technologies applied. Part I of the book aims at reviewing these soil enzymological investigations.

Söhngen (1913) was the first to prove that there are microorganisms in soils capable of degrading oil hydrocarbons using them as carbon and energy sources. Later, it was established that many soil bacteria, actinomycetes, yeasts, and filamentous microfungi take part in the degradation of both aliphatic and aromatic hydrocarbons (e.g. Beerstecher, 1954; Davis, 1967; McGill *et al.*, 1981; Atlas, 1984; Atlas and Bartha, 1992). Bioremediation of oil-contaminated soils is based on this degrading activity of soil microbiota and, therefore, the remediation technologies should enhance the growth of the native and/or introduced hydrocarbon-degrading microorganisms in the contaminated soils (e.g. Terry, 1998).

The first data on soil enzyme activities as related to numbers of oil-degrading soil bacteria were published by Kiss (1957). It was found that in the four soils studied there was a parallelism between values of invertase activity and numbers of paraffin-degrading acid-fast mycobacteria.

The first enzymological investigation of a soil contaminated with oil following the accident of a heating oil-transporting vehicle was described by Steubing (1967). Ten years later, Aliev and Gadzhiev (1977) published the first enzymological results obtained in a soil contaminated experimentally with crude oil. The aim of this experiment was to better understand the soil biological effects of oil contamination which is necessary for elaborating efficient remediation technologies.

At present, a rich literature on enzyme activities in oil-contaminated soils is available from East, Central, and West Europe, North America and Asia.

In reviewing this literature, the chronological course of the investigations will be followed. First, the soil enzymological effects of accidental oil contamination were studied, then oil contamination of soils and remediation of oil-contaminated soils were modelled by field experiments and, for obtaining deeper knowledge, by laboratory experiments.

Chapter 1. SOIL ENZYME ACTIVITIES AS AFFECTED BY ACCIDENTAL OIL CONTAMINATION

Contamination with crude oil

Enzymological research in the Russian Federation

Mukatanov and Rivkin (1980) studied three soils contaminated with crude oil due to pipeline leaks and two adjacent, uncontaminated (control) plots in the Bashkirian Pre-Urals. The first two soils, belonging to the steppe zone, were contaminated with crude oil in 1974 and 1976, respectively; they were used as a pasture and as an arable land cultivated with oats, respectively. The third soil, located on the forest-steppe zone, was contaminated with crude oil in 1974; following contamination, the pine trees growing on this soil died out. In 1978, the soils were sampled for determining their dehydrogenase and catalase activities. Dehydrogenase activity in the first and second contaminated soils was a little higher than in the controls, whereas the reverse was true for the third soil. At the same time, catalase activity was lower in each contaminated soil than in the controls (see also pages 5-9 in Andreson and Khaziev, 1981).

The nine soil plots studied by Andreson and Khaziev (1981) were weakly or strongly contaminated with crude oil due to pipeline leaks in the areas of the Ishambai and Arlan oil fields located on the steppe and forest-steppe zones of the Bashkirian Pre-Urals, respectively. Uncontaminated plots served for comparison. The effects of oil contamination on dehydrogenase, catalase, and invertase activities greatly depended on the soil type: in chernozems exhibiting a high buffering capacity the activities remained stable, whereas in those with low buffering capacity (humid meadow dark-brown, soddy-podzolic and grey forest soils) the activities decreased after oil contamination (see also Khaziev and Fatkhiev, 1981).

Tishkina and Kireeva (1986), Kireeva *et al.* (1996) and Kireeva (1996b) carried out soil studies in the Kaltasin district (Bashkiria) where, following the rupture of a pipeline, a land of more than 3 ha became contaminated with crude oil. The oil content in the upper (0-20-cm) layer of the contaminated soil (grey forest soil of heavy loam texture) was 9.6% (weight/weight). Plots (>200 m^2) were set up on the contaminated soil. Plots on uncontaminated soil were the controls. The 0-20-cm layer of each plot was systematically sampled for different analyses.

The results obtained during the first two years showed that the contaminated soil, in comparison with the uncontaminated control, exhibited lower enzyme (dehydrogenase, catalase, invertase, urease, and phosphatase) activities, contained actinomycetes, nitrifying, and cellulose-degrading microorganisms in smaller numbers; only its respiration (CO_2 evolution) rate and the number of microfungi were higher. Number of organic N-utilising heterotrophic bacteria rather increased, whereas that of the inorganic N-utilising heterotrophic bacteria decreased as a result of oil contamination.

Samosova *et al.* (1982) have analysed the chernozem soil in a ryefield contaminated with Romashkino crude oil (Tataria) in March 1977. Degree of contamination was 7.8%. The oil accumulated in the surface soil layers. In the most contaminated places the rye plants died out. An adjacent, uncontaminated soil was the control. The soil samples, taken on May 15 and July 6, 1977, were analysed to determine their dehydrogenase, protease, and urease activities and numbers of bacteria, actinomycetes, microfungi, and cellulose-degrading microorganisms. At both sampling dates, the contaminated soil was less dehydrogenase- and protease-active and more urease-active than the control soil. Numbers of all microbial groups increased following contamination; the highest increase occurred in the number of bacteria growing on a nutrient medium containing crude oil. Comparative analysis of the crude oil and the residual oil extracted from the contaminated soil revealed that *n*-alkanes were the most easily biodegradable oil components which is in concordance with literature data (e.g. Beerstecher, 1954; Atlas and Bartha, 1992).

A similar soil enzymological and microbiological study was carried out by Fil'chenkova (1982). An arable land and a pasture, both on calcareous chernozems in the area of the Kama River (Tataria), were studied.The degree of contamination with crude oil was higher on the arable land (3.2%) than on the pasture (0.12%). Uncontaminated plots were the controls. In both soils, oil contamination resulted in a decrease of dehydrogenase and protease activities and of the numbers of nitrifying bacteria and actinomycetes and in an increase of the global number of bacteria. Due to the oil contamination, the number of cellulolytic microorganisms decreased in the arable soil and increased in the pasture soil. With time, the negative effects of oil contamination tended to diminish.

The seven soils studied by Antonenko and Zanina (1992) were mostly meadow-swampy alluvial soils located along the central zones of the Ob River (West Siberia). Four of the soils were contaminated with crude oil due to pipeline leaks. The 0-10-cm layer of the contaminated soils contained 1-4 g of hydrocarbons kg^{-1}. Dehydrogenase, invertase, protease, and urease activities were low in both contaminated and uncontaminated soils. But activities of polyphenol oxidase and peroxidase (enzymes playing a role in the formation of humic acids) were higher in the contaminated than in the uncontaminated soils. The activities showed a decreasing trend with increasing soil depth, excepting polyphenol oxidase activity in one contaminated soil and peroxidase activity in two contaminated soils, in which these activities manifested a reverse trend.

The spilled crude oil bog soils in the Tyumen region were studied, from enzymological and microbiological viewpoints, by Kaigorodova (1996). Contaminated soils containing in the upper layers less than 90 g of oil kg^{-1} were compared with more contaminated soils containing 200-460 g of oil kg^{-1}. The cellulase activity strongly decreased, whereas urease activity increased with increasing oil content. The hydrocarbon-oxidising microorganisms represented about 0.1 and 1% of the total microbial numbers in the less and heavily contaminated soils, respectively. The dominant hydrocarbon-oxidising microorganisms were spore-forming bacilli and actinomycetes in the less contaminated soils, and pseudomonads and flavobacteria in the heavily contaminated ones.

Contamination with crude oil and oil products

Enzymological research in Georgia

Two yellow peaty gley soil and two meadow swampy soil plots located on the Colchis Lowland in West Georgia and contaminated with crude oil and oil products were selected by Yashvili *et al.* (1982) for enzymological and microbiological analyses. In comparison with the adjacent uncontaminated soils, the contaminated ones, especially their surface layers, were more enzyme-active (dehydrogenase, catalase, and invertase) and richer in microorganisms.

The highest increase was registered in the numbers of bacteria utilising organic and inorganic nitrogen, and the lowest increase in the number of microfungi.

Contamination with oil products

Enzymological research in Germany

As Steubing (1967) specifies, the accident of a vehicle transporting heating oil on the Gießen-Lich highway on September 19, 1963 led to an 18,000-l oil spill. The oil as a stream contaminated the adjacent areas, including an about 2 by 12-m site in the neighbouring mixed forest. The soil on this site and that on an uncontaminated site at 15-m distance from the contaminated site were sampled from their 0-5-, 10-15-, 20-25-, and 45-50-cm depths in March 1964. Both contaminated and control soils were of pseudogley type. The samples were submitted to enzymological and microbiological analyses.

Figure 1 shows that dehydrogenase activity and respiration (CO_2 evolution) were higher and cellulose decomposition was lower in the heating oil-contaminated soil than in the control one, at each depth.

Ammonification was also more intense, whereas total numbers of bacteria, microfungi, and algae were smaller in the contaminated soil, especially in its 0-5-cm layer, than in the control soil. But the number of hydrocarbon-oxidising mycobacteria increased in the contaminated soil which explains the higher dehydrogenase activity, respiration, and ammonification in this soil.

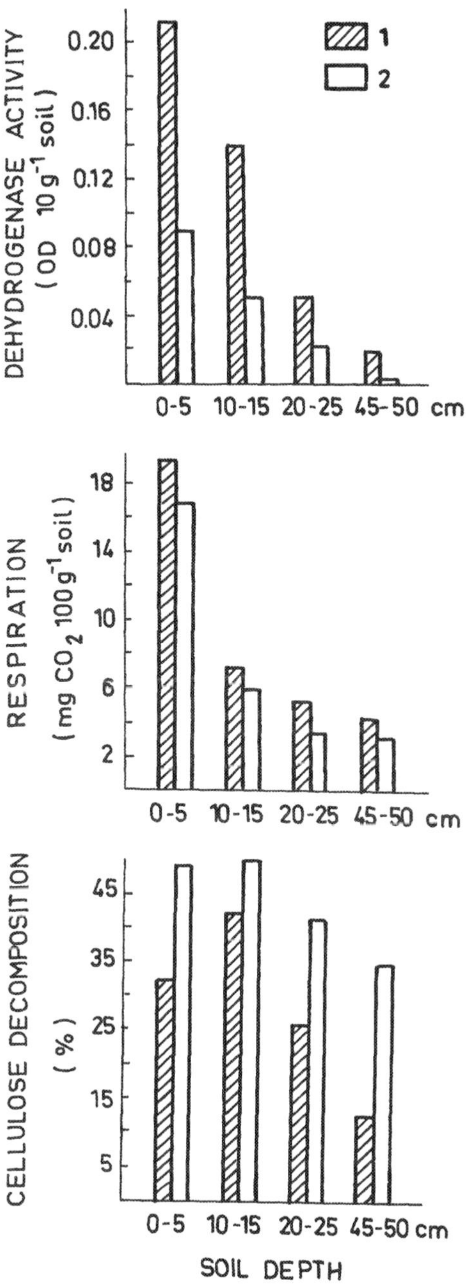

Figure 1. Dehydrogenase activity, respiration, and cellulose decomposition in heating oil-contaminated and control soil.

Redrawn from. L. Steubing, Angewandte Botanik, 1967, Vol. 40, p. 280.

1 – Heating oil-contaminated soil.

2 – Control soil.

Enzymological research in the United States of America

Frankenberger *et al.* (1989) performed chemical, microbiological, and enzymological studies at a site located in San Bernardino County, California, where, in the winter months of 1983, a diesel fuel spill in a paved parking area of a commercial establishment was discovered. It was estimated that approximately 4,500 l of diesel fuel leaked out of an underground storage tank. For investigating the areal extent of contamination (lateral and vertical movement), 18 exploratory borings were drilled (namely, four in June 1983, five in November 1983, and nine in March 1984) and used for soil sampling from depths of about 0.9-1.5 to 13-14 m.

The chemical analyses showed that diesel fuel was present in the parking area within the subsoil of both the near-surface (less than 3 m in depth) and capillary zone above the water table. In June 1983, the hydrocarbon quantities ranged from <10 to 1500 mg kg^{-1} soil. The migrating groundwater transported the contaminant plume in an eastward direction toward a creek and the leading edge of the plume migrated approximately 18-23 m down-gradient of the storage tank (source of contamination).

The samples taken in November 1983 were also used for determination of the numbers of diesel fuel-degrading and glucose-utilising bacteria, filamentous fungi, and yeasts and of dehydrogenase activity. The numbers of diesel oil-degrading bacteria were about two to three orders of magnitude higher than those of diesel oil-degrading filamentous fungi and yeasts. The same relationship was observed with the glucose-utilising microorganisms. Both diesel oil-degrading and glucose-utilising microorganisms showed wide variations with soil depth. Tha ratio of diesel fuel-degrading to glucose-utilising microorganisms was relatively high (0.917 for bacteria and 0.908 for filamentous fungi and yeasts), which indicates that the contaminated site was well enriched for hydrocarbon degradation.

Based on a regression analysis, insignificant positive correlations were found between oil content and microbial populations, but the correlation between dehydrogenase activity and number of diesel fuel-degrading microorganisms was significantly ($P<0.05$) positive. At the same time, dehydrogenase activity negatively, but insignificantly correlated with the oil content.

Contamination with oil field wastewaters

Enzymological research in the Russian Federation

Soil enzymological effects of contamination with oil field wastewaters were studied by Samosova *et al.* (1979). They determined dehydrogenase, urease, and protease activities in soils contaminated with wastewaters from the oil field areas of Tataria. The high salt content (especially Na^+, Ca^{2+}, Cl^-, SO_4^{2-}) of these waters caused salinisation of soils. Thus, Na^+ content was 150-200 times higher in strongly contaminated than in the control soils. Mobile Mn, Cu, and Zn contents increased 2-3 times in the contaminated soils which became phytotoxic. These changes brought about by contamination were accompanied by great diminutions of dehydrogenase and urease activities and disappearance of protease activity. Global numbers of soil microorganisms were also smaller in the contaminated than in the uncontaminated soils.

Chapter 2. ENZYMOLOGICAL EVALUATION OF THE BIOLOGICAL EFFECTS OF OIL CONTAMINATION OF SOILS IN EXPERIMENTAL MODELS

FIELD EXPERIMENTS

Contamination with crude oil

Enzymological research in Azerbaijan

Aliev and Gadzhiev (1977) added different amounts of crude oil (0, 10, 20, 40, and 100 kg m^{-2} to plots of a grey forest soil in the oil field area of the Apsheron Peninsula and determinaed periodically soil enzyme (catalase, polyphenol oxidase, invertase, protease, and phosphatase) activities and numbers of microorganisms during three years (1972-1975). The results obtained in enzymological analyses (Table 1) show that the enzyme activities decreased in parallel with the rate of crude oil addition. The only exception was catalase, the

Table 1
Effects of crude oil contamination on the enzyme activities in a grey-brown soil

Rate of crude oil addition (kg m^{-2})	Catalase	Polyphenol oxidase	Invertase	Protease	Phosphatase
No (control)	6.9	1.8	3.5	0.16	1.2
10	7.0	1.5	2.9	0.13	1.0
20	7.3	1.3	2.5	0.11	0.9
40	6.8	1.0	2.3	0.09	0.7
100	6.2	0.5	1.6	0.06	0.6

Reproduced from: S.A. Aliev and D.A. Gadzhiev, Izvestiya Akademii Nauk Az.SSR, Seriya Biologicheskikh Nauk, 1977, No. 2, p. 47.

Expression of enzyme activities: catalase in ml of O_2 g^{-1} soil 2 min^{-1}; polyphenol oxidase in mg of purpurogallin 100 g^{-1} soil 30 min^{-1}; invertase in mg of "glucose" g^{-1} soil 24 h^{-1}; protease in mg of NH_2-N g^{-1} soil 24 h^{-1}; and phosphatase in mg of P_2O_5 10 g^{-1} soil h^{-1}.

activity of which increased when the soil was treated with crude oil at the two low rates (10 and 20 kg m^{-2}); however, this activity also decreased at the rates of 40 and 100 kg of crude oil m^{-2}. The global numbers of microorganisms (bacteria, actinomycetes, and microfungi) and intensity of respiration (CO_2 evolution), like catalase activity, were highest in the plots treated with 10 and 20 kg of crude oil m^{-2} and exceeded to a large extent the values registered in the control soil. This findings indicate that small amounts of crude oil were not toxic to microorganisms of the soil studied; contrarily, the crude oil stimulated the microbial growth under these conditions.

In the area of the Apsheron oil fields, Ismailov (1982) has initiated a new experiment in July 1978. Plots of a grey-brown soil were contaminated with Binagadin crude oil (20 kg m^{-2}). Uncontaminated plots served for comparison. Soil samples were taken for analyses 3 days, 3, 9, 12, 15, and 22 months after the beginning of the experiment. Sampling depths were 0-3, 3-13, 13-23, and 23-48 cm. Crude oil contamination led to increased soil catalase and urease activities and respiration (CO_2 evolution), while dehydrogenase, invertase, and protease activities decreased. After 22 months, more than 60% of crude oil was degraded, but dehydrogenase activity remained low in contrast to invertase and protease activities which, expecially in the deeper soil layers, tended to increase to values registered in the control plots. At each soil depth, numbers of bacteria plus actinomycetes and separate numbers of bacteria were greater in the contaminated plots than in the control ones during the whole experimental period. The maximal numbers were obtained after 9 and 15 months, respectively. Numbers of yeasts (*Candida* and *Torulopsis*) increased only in the upper soil layers. Increase in the numbers of bacteria growing on nutrient media containing crude oil or oil products was remarkable in each soil layer. These bacteria represented an average of 46% of the total bacterial numbers in control plots and 76% in the contaminated plots. However, the soil in contaminated plots remained toxic for higher plants during the whole experimental period (see also Ismailov, 1988).

In this experiment, nitrate reductase and protease activities, numbers of denitrifying, ammonifying, N_2-fixing, and nitrifying microorganisms were also determined by Ismailov (1983). It has been found that nitrate reductase activity (expressed in mg of NO_3^- reduced by 10 g of soil in 24 h) gave much lower values in strongly and moderately contaminated soil samples (10 and 60, respectively) than in the uncontaminated control soil (150). The oil-caused decrease in the activity of accumulated soil nitrate reductase was compensated for by oil-stimulated proliferation of denitrifying microorganisms producing new nitrate reductase molecules and maintaining denitrification at a high level. Similarly, inhibition of accumulated soil protease activity by oil was compensated for by its stimulatory effect on the growth of ammonifying microorganisms producing new protease molecules. Oil contamination also led to higher numbers of aerobic and anaerobic N_2-fixers and to suppression of nitrifiers (see also Ismailov, 1988).

Enzymological research in the United States of America

Linkins *et al.* (1978) selected study sites in the polygonal coastal arctic tundra at Barrow, Alaska, in typical high centre and low centre polygon complexes. Ambient temperature Prudhoe Bay oil was uniformly surface-applied at 5 or 12 l m^{-2} rate on high centre polygon centres and low centre rim, trough and basin systems on June 30, 1975. Mid-season samples were taken between July 4 and 16, 1975, 1976, and 1977, and used for determination of exo- and endocellulase, and aryl hydrocarbon (benzo[a]pyrene) hydroxylase activities, soil and vascular plant (*Lezula confusa, Carex aquatilis*, and *Dupontia fischeri*) root respiration. Uncontaminated soils served as controls.

In the soils contaminated with 5 l of oil m^{-2}, exo- and endocellulase activities decreased, aryl hydrocarbon hydroxylase activity and soil respiration (O_2 consumption and CO_2 evolution) increased, indicating a shift in the catabolic activity of soil microbiota. These trends were paralleled in the soils contaminated with 12 l of oil m^{-2}, but usually after a lag period of one year, which was attributed to some toxic effect of the oil at high concentration. These results suggest that tundra soil microbiota can actively modify oil and can utilise it to support metabolism. Higher respiration rates in oiled soils than in control soils indicate that soil microbiota degrades and utilises oil faster than the normal residual plant material. Oil at each rate reduced root respiration in each soil during the first and second years, but in the third season root respiration was greater than or equal to that of the control roots. However, the plant root adaptation to oil was incomplete. This was shown by the annual decrease in viable root biomass after exposure to oil.

In another experiment, conducted by Linkins and Fetcher (1983), ambient temperature Prudhoe Bay crude oil was applied as a 20 l m^{-2} uniform surface spray on a 3 by 20-m plot of *Eriophorum vaginatum*-dominated tussock tundra at Eagle Creek, Alaska, in August 1979.

The 0-15-cm soil layer became visibly contaminated with the oil. A control plot (3 by 20 m) was delineated adjacent and upslope of the oil spill plot.

During the 1979-1982 period, exo- and endocellulase, and acid phosphatase activities were determined in the oil-contaminated layer and in the layer immediately below the contaminated layer of tussock and intertussock soils in the oil spill plot and in the corresponding layers of tussock and intertussock soils in the control plot.

First of all, a general finding should be pointed out: values of the same enzyme activity in the tussock soil and intertussock soil of the same plot were not significantly (P<0.05) different.

Excepting the first year, in the other years cellulase activities had significantly lower values in the oil-contaminated layer than in the corresponding layer of the control plot. The reduction was more pronounced in the exocellulase than in the endocellulase activity, i.e. exocellulase, which acting on cellulose catalyses the formation of products readily utilisable by microorganisms, was more sensitive to oil contamination than was endocellulase. Thus, in 1982, the exocellulase to endocellulase activity ratio shifted from the 0.50-0.62 range in control soils to the 0.27-0.16 range in the oiled soil layers of the contaminated plot.

Another finding was that the layer below the oil-contaminated layer and the corresponding layer of the control plot did not differ significantly in respect of their exocellulase or endocellulase activity.

Unlike cellulase activities, acid phosphatase activity did not show a significant reduction until the third year (1981). The reduction remained through the following year. But like cellulase activities, acid phosphatase activity in the layer below the oil-contaminated layer was not significantly different from that measured in the corresponding layer of the control plot.

Plant studies were also carried out. Of the results obtained, we cite the following: oil caused a significant reduction in mycorrhizal root numbers and root respiration rates in *Betula nana*, but not in *Eriophorum vaginatum* root tips which had grown through the oil-contaminated soil layer.

Based on the soil enzyme and plant studies, the conclusion was drawn that, in oil spills with limited vertical movement, removal of only the contaminated soil layer may promote an acceptable long-term cost-effective restoration of the site.

Enzymological research in Belorussia

Zimenka and Kartyzhova (1986a, b) installed microplots on a soddy-podzolic soil of light loam texture. The microplots were contaminated with crude oil (0.5 or 3 l m^{-2}) from the Rechitsa oil field. Uncontaminated plots served for comparison. The experiment lasted three years (1983-1985). The test plant was Kondor oat variety. Oil contamination at the low rate (0.5 l m^{-2}) had little effect on soil catalase and dehydrogenase activities, but affected urease activity which increased during the first two years of the experiment. At this oil rate, the nitrifying bacteria, actinomycetes, and microfungi were not affected significantly and some stimulation occurred in the growth of cellulolytic microorganisms, whereas growth of ammonifying and denitrifying microorganisms was inhibited and growth and development of oat plants were retarded. Soil respiration (CO_2 evolution) decreased in 1983 and showed increased values in 1984 and 1985. It was estimated that self-decontamination of these oil-polluted plots requires 30 months.

Oil contamination at the higher rate (3 l m^{-2}) resulted in decreased catalase and increased urease activity during the whole experimental period. Dehydrogenase activity decreased in 1983, had values close to those of the control soil in 1984 and greatly increased in 1985. The numbers of most microbial groups (ammonifiers, denitrifiers, cellulose-degraders, actinomycetes, microfungi) increased, but the numbers of nitrifiers decreased. Respiration decreased in 1983, increased in 1984 and was similar to that of the control plots in 1985. Due to the poor growth and development of oat plants, great losses occurred in grain yield in each year. The loss was minimal (13%) in 1985. For self-decontamination of these plots more than three years are necessary.

Enzymological research in the Russian Federation

The experiment described by Khaziev *et al.* (1988b) was carried out on a dark-grey forest soil of medium loam texture (Bashkiria). Rates of oil contamination were 0, 8, 16, and 25 l m^{-2}. For analyses, samples were taken from three soil depths (0-20, 20-30, and 40-50 cm) 3 days and one year after the beginning of the experiment. Catalase and dehydrogenase activities decreased at each crude oil rate on day 3, but after one year both activities showed increased values and, at the lowest oil rate, their values were close to those recorded in the uncontaminated soil. On day 3, invertase activity was not significantly affected by any of the soil rates, but after one year it exhibited a little higher values at the 8 and 16 l oil m^{-2} rates.

Decrease and increase in enzyme activities were always more evident in the 0-20-cm layer than in the deeper ones. After 3 days, no nitrate and mobile phosphorus were detectable in the oil-contaminated soil, but after one year they were present in each soil layer; their amounts were highest in the 0-20-cm layer.

In a similar experiment, conducted by Kireeva *et al.* (1997a), microplots were installed on another Bashkirian soil (grey forest soil of heavy loam texture) and contaminated with Tyumen crude oil at the same rates as those used in the previous experiment (i.e. 0, 8, 16, and 25 l m^{-2}). After contamination, the arable layer (0-20 cm) of the microplots was systematically sampled during 10 years for determination of phosphohydrolase activities. Some of the results obtained are summarised in Table 2, which shows that each activity decreased with both the

Table 2
Effect of crude oil contamination on phosphohydrolase activities in a grey forest soil of heavy loam texture

Phosphohydrolases	Rate of oil contamination (l m^{-2})	Activity values Time elapsed after contamination 1 month	6 months	12 months	10 years
Phosphomono-	0	4.23	3.15	5.29	6.40
esterase	8	2.12	2.02	1.56	9.00
(mg phenol-	16	1.76	1.26	1.34	4.00
phthalein)	25	0.18	0.27	0.95	3.40
Phytase	0	0.24	0.22	0.18	
(mg P_2O_5)	8	0.20	0.18	0.14	
	16	0.12	0.10	0.08	
	25	0.04	0.01	traces	
RNase	0	0.78	0.79	0.82	
(mg RNA)	8	0.51	0.44	0.40	
	16	0.25	0.22	0.18	
	25	0.11	0.08	0.04	
DNase	0	0.404	0.394	0.410	
(mg DNA)	8	0.200	0.180	0.120	
	16	0.050	0.020	0.014	
	25	0.013	0.006	0.002	
ATPase	0	0.086	0.095	0.092	
(mg P)	8	0.045	0.040	0.036	
	16	0.027	0.020	0.018	
	25	0.012	0.010	0.004	

Reproduced from: N.A. Kireeva, E.I. Novoselova and F.Kh. Khaziev, Pochvovedenie, 1997, No. 6, pp. 724-725.

contamination rate and time. The only exception was phosphomonoesterase activity which exhibited the highest value in year 10 after contamination with the smallest amount of crude oil (8 l m^{-2}).

Due to the decreased phosphohydrolase activities, the mobile P content also strongly decreased.

In this experiment, activities of soil enzymes participating in the N cycle were also determined (Kireeva *et al.*, 1997b). Protease, urease, asparaginase, glutaminase as well as nitrate, nitrite, and hydroxylamine reductase activities were measured. The results obtained 1, 6, and 12 months after oil contamination are presented for each activity.

Protease, asparaginase, glutaminase, and nitrate and nitrite reductase activities were inhibited by the crude oil. Degree of inhibition increased with rate of contamination and with time. But urease and hydroxylamine reductase activities were stimulated in the oil-contaminated soil. The stimulation increased with rate of contamination, but it did not show any evident trend to increase with time.

All these changes in enzyme activities were associated by a decrease in NO_3-N and free amino acid contents in parallel with contamination rate and time and by an increase in the NH_4-N content in parallel with only the contamination rate.

It has also been found that in year 10 protease activity in plots contaminated with 8 and 16 l of crude oil m^{-2} was of the same level as in the uncontaminated plot. The activity in the most contaminated plot also increased, during the last 4 years, but even in year 10 it remained lower than in the uncontaminated plot. The urease-stimulating effect of crude oil was discernible even in year 10.

Contamination with oil products

Enzymological research in Germany

Janke *et al.* (1992) installed 25 lysimeters on the experimental field of the Institute of Plant Ecology in Gießen. Each lysimeter was filled with about 10 l of topsoil (0-10 cm) taken from a fallow land on loamy sand soil. The soil was contaminated with 200 ml of heating oil (10 lysimeters) and with 400 ml of heating oil (10 lysimeters); the remaining 5 lysimeters served as controls. Soil samples were collected at 4-week intervals during 10 months in 1990 and 9 months in 1991. Dehydrogenase and catalase activities, respiration (CO_2 evolution), nitrification, and microbial numbers were determined. After an approximately 5-month-long lag period, dehydrogenase activity in the oil-contaminated soil increased and the increase was more pronounced in 1991 than in 1990. Beginning with the 9th month, dehydrogenase activity increased to a larger extent in the heavily rather than in the slightly contaminated soil. Catalase activity in the oil-contaminated soil exceeded the control values during the last 7 months. But catalase activity was influenced to a lesser extent by the rate of oil contamination than was dehydrogenase activity. Respiration behaved essentially like dehydrogenase activity. However, respiration of the contaminated soil became largely different from that of the control soil only in month 6 and remained nearly at this level up to the end of the experiment. The respiration-increasing effect of oil was more evident in 1991 than in 1990. Oil contamination at both rates resulted in strong inhibition of nitrification during the whole experimental period. Global number of heterotrophic bacteria and that of microfungi were practically not affected by oil contamination, but the numbers of streptomycetes, especially during the last months, were much smaller in the contaminated than in the control soil.

In parallel with these field experiments, "storage" experiments were also carried out (Weißmann and Kuntze, 1994). The same soil (loamy sand topsoil) and oil product (heating oil) were used in both types of experiments. In the storage experiments, 10-kg quantities of soil were left uncontaminated or contaminated with 200 and 400 ml of heating oil, respectively, then kept at 60% of maximum water-holding capacity at ambient temperature. The soils were sampled at 4-week intervals, during 18 months in the April 1990 – September 1991 period. The samples were analysed for catalase and dehydrogenase activities, nitrification capacity, and selected *n*-alkanes (chain lengths with 14-22 C atoms).

As shown in Figure 2, during the first 5 months, values of catalase and dehydrogenase activities in uncontaminated and contaminated field and storage soils did not significantly differ from each other. Thereafter, both activities in the contaminated soils increased continuously up to the end of the experimental period. The increase in catalase activity was higher and that in the dehydrogenase activity was lower in the field than in the storage soils. In each case, the increasing effect of oil was more pronounced at its higher rather than its lower rate.

Nitrification was almost completely depressed by both oil rates in both field and storage soils. A 25% recovery of nitrification capacity occurred only in the field soils towards the end of summer 1991.

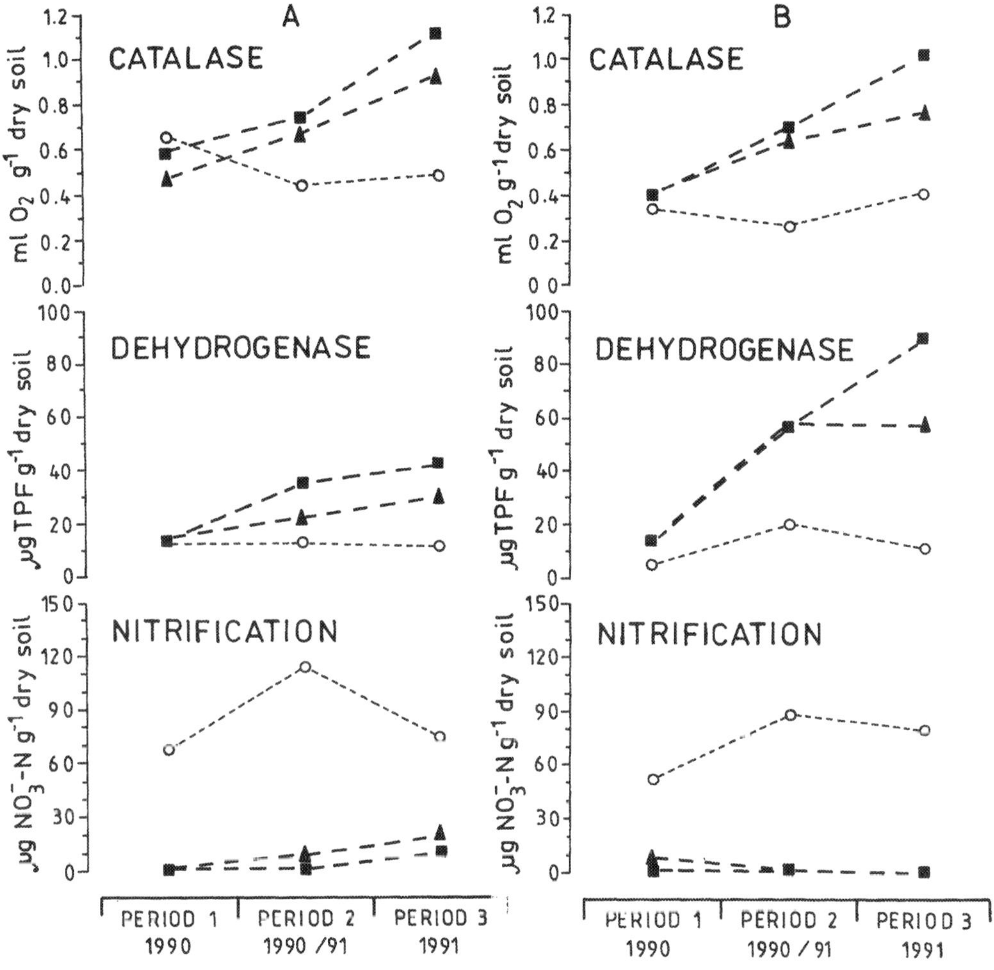

Figure 2. Catalase, dehydrogenase, and nitrification activities in field experiments (A) and storage experiments (B).

Redrawn from: S. Weißmann and C. Kunze, Angewandte Botanik, 1994, Vol. 68, p. 139.

Period 1 – April 1990 to August 1990.
Period 2 – September 1990 to March 1991
Period 3 – April 1991 to September 1991
o – Control soils. ▲ - Addition of 200 ml of heating oil. ■ - Addition of 400 ml of heating oil.
TPF – Triphenylformazan .

In the field soils, content of C_{14}-C_{22} *n*-alkanes strongly decreased, and the decrease correlated with the increase in dehydrogenase and catalase activities. Unexpectedly, no changes in contents of C_{14}-C_{22} *n*-alkanes were observed in the storage soils. However, the possibility that other oil components were degraded in these soils is not excluded.

Growth of vegetation, studied under field conditions, was severely retarded in the contaminated soils.

LABORATORY EXPERIMENTS

Contamination with crude oil

Enzymological research in Ukraine

Demidienko and Demurdzhan (1988) treated samples of an uncontaminated common chernozem with crude oil from the Bitkov oil field located in the Dnieper-Donets Depression. Rate of oil addition was 1% (weight/weight). The enzymological analysis of the soil-crude oil mixtures and control samples showed that, due to the crude oil, protease activity slightly and nitrite reductase activity strongly decreased, whereas invertase and nitrate reductase activities slightly and urease and hydroxylamine reductase activities strongly increased.

Enzymological research in the Russian Federation

For mathematical modelling of the microbial and enzymatic processes taking place in oil-contaminated soils (Kireeva and Vodop'yanov, 1996), samples of two Bashkirian soils (leached chernozem and grey forest soil) were contaminated with different amounts of crude oil, and analysed immediately and after 3 days, 1, 3, and 12 months of incubation. Numbers of soil heterotrophic bacteria, actinomycetes, nitrifying, and cellulose-degrading microorganisms, catalase and invertase activities continuously decreased, whereas number of microfungi rather increased with incubation time. Dehydrogenase and urease activities were highest on day 3, then decreased. The experimental and calculated data showed a high degree of correspondence.

Contamination with crude oil and oil products

Enzymological research in the United States of America

Frankenberger and Johanson (1982) used surface samples of three Californian soils, selected to obtain a wide range of pH, organic C content, and texture, for studying the effect of crude oil and refined oil products on soil dehydrogenase activity. Field-moist samples of 5 g each were contaminated with crude oil, leaded gasoline, kerosene, diesel fuel, and motor oil at 0, 20, 40, and 60% loading rates (weight/weight), then incubated under aerobic conditions at room temperature (24±2°C) for 30 days. Dehydrogenase activity was measured after 0, 3, 7, 14, and 30 days of incubation. The activity was found to increase with the rate of crude oil and refined oil application. Generally, the maximum value of dehydrogenase activity was registered after 30 days of incubation in the case of crude oil-contaminated soils and after 7 days in soils contaminated with refined oil products. When all soils and contaminations were considered, maximum values of dehydrogenase activity (expressed in μg of triphenylformazan produced by 1 g of soil in 24 h) were the following: crude oil 1,180; leaded gasoline 56; diesel fuel 56; motor oil 37; kerosene 32. In other words, the crude oil was more stimulatory than the refined oil products. These findings suggest that the additives present in refined oil products may have a major influence on the dehydrogenase reaction in soils.

Enzymological research in Azerbaijan

Ismailov (1982) contaminated samples of a garden soil with crude oil from the Apsheron oil fields or with oil components (*n*-hexadecane, cyclohexane, aromatic hydrocarbons, tar) at 1% rate (on field-moist soil weight basis). Four enzyme activities (catalase, dehydrogenase, invertase, and urease) and respiration (CO_2 evolution) were determined in the mixtures and in the uncontaminated, control soil. Crude oil stimulated dehydrogenase and urease activities, had little effect on catalase activity and inhibited invertase activity and respiration. *n*-Hexadecane had a stimulating effect on catalase, dehydrogenase, and urease activities and an inhibiting one on invertase activity and respiration. In the cyclohexane-contaminated samples, catalase activity increased, dehydrogenase and invertase activities decreased, whereas urease activity and respiration remained unchanged. The aromatic hydrocarbons decreased each enzyme activity and respiration. Contamination with tar led to decreased catalase and dehydrogenase activities and to increased respiration. (The effects of tar on invertase and urease activities were not studied.) One can deduce from these findings that the effects of *n*-alkanes and cycloalkanes are less inhibitory on soil biological activity than those of the aromatic hydrocarbons (see also Ismailov, 1988).

For explaining phytotoxicity of such oil-contaminated soils in which the amount of hydrocarbons greatly decreased due to their biodegradation, Ismailov *et al.* (1984) carried out an experiment using samples of an uncontaminated grey-brown soil from the Apsheron oil field area. The soil samples were contaminated with crude oil and *n*-hexadecane (at 1% rate, on weight/weight basis) and with four partial oxidative degradation products of oil hydrocarbons (at 0.5% rate), then analysed for determination of their dehydrogenase activity (expressed in mg of triphenylformazan produced by 10 g of soil). The following values were recorded in the uncontaminated and contaminated soil samples: 9.0 (uncontaminated); 9.5 (crude oil); 11.8 (*n*-hexadecane); 10.0 (1-hexadecanol); 8.6 (palmitic acid); 0.1 (benzoic acid); 0 (salicylic acid). It is evident from these values that the crude oil, *n*-hexadecane and 1-hexadecanol stimulated, whereas the organic acids inhibited dehydrogenase activity. The acids can accumulate in soil and exhibit phytotoxic effects (see also Ismailov, 1988).

Enzymological research in the Russian Federation

Kireeva (1995f) compared the effects of crude oil and oil products (e.g. benzine, tar, asphalt) on enzyme activities in a Bashkirian leached chernozem. Rates of contamination were 0, 0.5, 5, and 8% (weight/weight). After contamination (3 days, 1 and 12 months), samples, taken from the 0-20-cm soil depth, were analysed for determination of six enzyme activities.

The results obtained may be generalised as follows: the crude oil and oil products initially rather inhibited and later stimulated dehydrogenase, urease, and protease activities, and rather constantly inhibited catalase, peroxidase, and invertase activities. The effect increased or decreased with the rate of contamination, depending on the nature of contaminant and enzyme. The mean values of activities in the soil contaminated with crude oil (C), benzine (B), tar (T), and asphalt (A) and in the uncontaminated soil (U) presented the following order: U<B<C<A<T (dehydrogenase); C<B<A<U<T (catalase); C<A<B<T<U (peroxidase); A<C<T<B<U (invertase); U<B<T<A<C (urease); A<C<U<T<B (protease).

Contamination with oil products

Enzymological research in China

Zhou *et al.* (1990) contaminated soil samples with oil hydrocarbons at rates of 0, 100, 250, and 500 ppm, then incubated them under normal soil moisture or water-logged conditions for 90 days. The residual hydrocarbon content and the enzyme activities in the samples were determined after 0, 7, 21, 45, and 90 days of incubation. The results have shown that biodegradation of hydrocarbons was a faster process at their lower concentrations than at their higher ones, and under normal soil moisture than under water-logged conditions. For example, half-life times of the aliphatic hydrocarbons applied at 250 and 500 ppm were 18 and 20 days, respectively, under normal soil moisture conditions, and 51 and 70 days, respectively, under water-logged conditions.The corresponding half-life times of the aromatic hydrocarbons were greater, namely 22 and 25, and 90 and >90 days, respectively. The enzyme activities measured (invertase, protease, phosphatase, and peroxidase) were higher under normal soil moisture than under water-logged conditions. During the incubation period, the hydrocarbon content decreased parallelly with the decrease of invertase activity. The other activities rather increased during incubation. This is why Zhou *et al.* consider that by measuring invertase activity in hydrocarbon-contaminated soils useful data can be obtained for evaluating the capacity of these soils to degrade the contaminants.

Enzymological research in Austria

Bauer *et al.* (1991, 1992, 1994) and Pennerstorfer *et al.* (1992) compared the sensitivity of different soil microbiological and enzymological methods to asses the biological effects of oil contamination. Soil was collected from the top 50 cm of an agricultural field and passed through a 2-mm sieve. A part of the samples were contaminated with heating oil (2% volume/weight) and the contamination was repeated twice during the experiment after 3 and 6 weeks, respectively. The moisture content of the samples was kept constant during the 8-week incubation period. The following methods were applied weekly: soil respiration, glucose induced respiration, capacity to reduce 2,3,5-triphenyltetrazolium chloride (TTC-dehydrogenase activity), 2-(*p*-iodophenyl)-3-(*p*-nitrophenyl)-5-phenyltetrazolium chloride (INT-dehydrogenase activity), and dimethylsulphoxide (DMSO) reduction.

Excepting glucose-induced respiration, the other methods gave higher values in the oil-contaminated than in the control samples. The sensitivity indicator of the methods (i.e. ratio between the value in contaminated soil and that in the control soil taken as 1) increased in the order: glucose-induced respiration (0.965), DMSO reduction (1.228), soil respiration (1.425), INT-dehydrogenase activity (1.458), and TTC-dehydrogenase activity (2.464). This means that, of the methods applied, measurement of TTC-dehydrogenase activity is the most sensitive one for assessing biological effects of oil contamination of soil.

Bauer *et al.* (1992), Pennerstorfer *et al.* (1992) and Kandeler *et al.* (1994) studied five Lower-Austrian agricultural soils having different textures, chemical, and biological properties (pH, humus, respiration, TTC-dehydrogenase activity) (soils: 1 – calcareous chernozem; 2 – degraded chernozem; 3 – regorendzina; 4 – calcareous chernozem; 5 – gleyic chernozem). The sieved (2 mm) soil samples moistened to 40% of their water-holding capacity were contaminated with a mixture of *n*-tetradecane, 5-methyl-3-heptanone, and naphthalene (1:3.67:1.08) modelling the benzine. Rate of application was 3% (weight/weight). Uncontaminated samples were used as controls. During incubation (at 20°C) that lasted 20 weeks, TTC-dehydrogenase and alkaline phosphatase activities, respiration, glucose-induced respiration, nitrogen mineralisation, and residual hydrocarbon content were determined at 2- to 4-week intervals.

All soils reacted similarly to contamination, but with different intensities. Soil 1 (calcareous chernozem) was the least sensitive and soil 2 (degraded chernozem) was the most sensitive to contamination. In comparison with the control samples, the biological parameters decreased immediately after contamination, but later increased reaching very high values during weeks 4-8 (dehydrogenase activity, respiration) or during weeks 8-12 (phosphatase activity, N mineralisation) and, thereafter, showed, in general, a decreasing trend. Determination of the residual contaminants indicated that complete decontamination occurred in each soil. Average and maximum decontamination times were 13.6 and 20 weeks, respectively.

Enzymological research in the Russian Federation

Kireeva (1995b) contaminated samples taken from the 0-20-cm layer of a leached chernozem and a grey forest soil (Bashkiria) with *n*-hexadecane, cyclohexane, or benzene at rates of 0, 0.5, 1, and 2% (weight/weight). During incubation (3 days up to 24 months), the samples were periodically analysed enzymologically and microbiologically. The analytical data presented in Kireeva's paper were obtained after 1 and 6 months of incubation.

Contamination with *n*-hexadecane and cyclohexane resulted in increased soil dehydrogenase, catalase, invertase, and urease activities, the increase being highest at the lowest (0.5%) hydrocarbon rate. Benzene caused strong diminution of all activities, the diminution being greatest at the highest (2%) benzene rate.

Number of heterotrophic bacteria changed similarly to the enzyme activities. At the same time, soil respiration (CO_2 evolution), numbers of nitrifying and cellulose-degrading microorganisms decreased with increasing rate of each hydrocarbon. The decreasing effect followed the order: *n*-hexadecane ≈ cyclohexane < benzene.

In another experiment, described by Kireeva *et al.* (1997a, b), phosphatase and urease activities were determined in grey forest soil samples contaminated with 0, 0.5, 1, and 2% oil products (weight/weight) and incubated for 1, 6, and 12 months. Both activities were stimulated by *n*-hexadecane and cyclohexane and inhibited by the aromatic oil fraction. In the case of phosphatase activity, both positive and negative effects increased with the contamination rate and incubation time, but in the case of urease activity the stimulating effect of *n*-hexadecane and cyclohexane diminished with their increasing rate and did not change during the incubation. At the same time, the inhibitory effect of the aromatic oil fraction increased with contamination rate and incubation time.

Enzymological research in Germany

Koch and Wilke (1996) analysed samples of two regosols on a field which in the past had been exposed to overflow irrigation in Berlin-Buch. The field had been contaminated with numerous organic and inorganic compounds, including polycyclic aromatic hydrocarbons (PAHs) (which, as it is well known, comprise carcinogens), polychlorinated biphenyls (PCBs), and heavy metals. Soil 1 was less contaminated, whereas contamination of soil 2 was heavier, representing the medium contamination level. Thus, contents of the PAHs fluoranthene (FLA) and benzo[a]pyrene (BaP) and of the PCB-52 were 0.02, <0.01, and <0.2 $\mu g\ g^{-1}$ soil, respectively, in soil 1, and 0.18, 0.12, and 0.02 $\mu g\ g^{-1}$ soil, respectively, in soil 2. The Cu, Zn, and Cd contents were also higher in soil 2 than in soil 1.

Reaction mixtures were prepared for determination of soil dehydrogenase activity. Additions to the reaction mixtures were (per g of soil): up to 10 μg of FLA, up to 70 μg of BaP, up to 100 μg of PCB-52, 30-70 μg of Cu, 50-250 μg of Zn, and up to 70 μg of Cd, alone or in combination of one organic contaminant with one heavy metal. No contaminant was added to the control reaction mixtures. The incubation was carried out at 22°C and lasted 1, 8, and 15 days.

It was established that in the less contaminated soil dehydrogenase activity was not significantly affected by any FLA, BaP, or PCB-52 additions after 1, 8, or 15 days of incubation. The only exception was registered in reaction mixtures with PCB-52, in which a 12% inhibition occurred in dehydrogenase activity after 8 days of incubation (but no inhibition was evident in reaction mixtures incubated for 1 or 15 days). In the same less contaminated soil, the heavy metals inhibited dehydrogenase activity and the degree of inhibition increased with increasing heavy metal addition and was highest after one day of incubation. The effect of the combination of organic contaminant + heavy metal on dehydrogenase activity was not stronger than that of the heavy metal alone, but FLA + 60 μg of Cu g^{-1} soil caused a stronger inhibition than Cu alone.

In the more contaminated soil, 5 μg of FLA g^{-1} soil had no significant effect on dehydrogenase activity. Combination of 5 μg of FLA with 130 μg of Cu or 400 μg of Zn g^{-1} soil inhibited more strongly the dehydrogenase activity than did Cu or Zn alone.

Contamination with oil well-drilling fluid additives

Enzymological research in Ukraine

Effects of the oil well-drilling fluid additives on soil enzyme activities were studied by Demidienko and Demurdzhan (1988). Twenty inorganic and organic compounds and mixtures used for the preparation of oil well-drilling fluids were added to samples of an uncontaminated common chernozem from the Dnieper-Donets Depression, at a rate of 1% (weight/ weight). In comparison with the untreated control samples, each fluid additive exerted a negative or positive effect at least on one of the six enzyme activities studied (invertase; urease; protease; nitrate, nitrite, and hydroxylamine reductase). The majority of the fluid additives decreased invertase, nitrate and hydroxylamine reductase activities and increased the other activities. The increase was most pronounced in the case of urease activity.

Chapter 3. ENZYMOLOGICAL EVALUATION OF THE BIOLOGICAL EFFECTS OF THE REMEDIATION OF OIL-CONTAMINATED SOILS IN EXPERIMENTAL MODELS

FIELD EXPERIMENTS

Contamination with crude oil

Enzymological research in the Russian Federation

Tishkina and Kireeva (1986) dealt with the remediation of the *accidentally* crude oil-contaminated land in the Kaltasin district (Bashkiria) (see pages 5-6). For enhancing biodegradation of oil, plots were amended with mineral fertilisers without or with farmyard manure, in six variants:

1. uncontaminated (control) soil;
2. contaminated soil containing 9.6% oil in the 0-20-cm layer;
3. contaminated soil plus single NPK dose (60, 90, and 90 kg ha^{-1}, respectively);
4. contaminated soil plus double NPK dose;
5. contaminated soil plus single NPK dose plus farmyard manure (70 t ha^{-1}); and
6. contaminated soil plus double NPK dose plus farmyard manure (70 t ha^{-1}).

The test plant was barley. Enzyme activities were determined in the 0-20-cm layer.

The results showed that NPK and farmyard manure exerted positive effects on dehydrogenase, catalase, invertase, and urease activities, and a negative one on phosphatase activity in the contaminated soil. NPK was less efficient than farmyard manure. Dehydrogenase and urease activities became higher, whereas catalase and invertase activities remained lower in the amended contaminated soil than in the uncontaminated soil.

The grain yield of barley was the highest in the uncontaminated soil (2.33 t ha^{-1}), intermediary in the amended contaminated soil (0.24-1.04 t ha^{-1}) and lowest in the unamended contaminated soil (0.2 t ha^{-1}).

Khaziev *et al.* (1988b) used mineral fertilisers and farmyard manure for remediation of *artificially* contaminated plots on a grey forest soil of heavy loam texture. The rates of contamination were 0, 8, 16, and of 25 l crude oil m^{-2}. Mineral fertilisers (N, P, and K) were applied at rates of 40, 60, and 45 kg ha^{-1}, respectively, and farmyard manure at a rate of 40 t ha^{-1}. Unfertilised plots served for comparison. The test plants were oats. Five enzyme activities (dehydrogenase, catalase, invertase, urease, and phosphatase) were measured after one year (Table 3).

Table 3
Effects of mineral fertilisers (NPK) and farmyard manure (FYM) on enzyme activities in the arable layer of a heavy loamy grey forest soil one year after crude oil contamination

Fertilisation	Dehydrogenase	Catalase	Invertase	Urease	Phosphatase
		No oil contamination			
No	0.121	3.05	15.67	0.389	5.20
NPK	0.127	3.16	16.52	0.460	5.01
FYM	0.115	4.10	13.77	0.402	5.19
NPK + FYM	0.119	3.14	13.20	0.468	6.01
		Crude oil 8 l m^{-2}			
No	0.125	2.12	9.86	0.792	1.56
NPK	0.124	3.01	10.96	0.432	1.39
FYM	0.126	2.49	8.88	0.856	1.32
NPK + FYM	0.121	2.09	9.44	0.815	2.43
		Crude oil 16 l m^{-2}			
No	0.169	1.34	8.97	0.375	1.34
NPK	0.113	1.78	10.81	0.841	1.31
FYM	0.135	2.11	7.92	0.677	1.00
NPK + FYM	0.121	2.44	9.13	1.182	0.86
		Crude oil 25 l m^{-2}			
No	0.154	1.33	9.13	0.291	0.95
NPK	0.087	2.83	9.09	0.446	1.46
FYM	0.123	2.15	11.76	0.869	1.26
NPK + FYM	0.111	3.33	9.81	1.115	1.44

Reproduced from: F.Kh. Khaziev, E.I. Tishkina, N.A. Kireeva and G.G. Kuzyakhmatov, Agrokhimiya, 1988, No. 2, p. 58.

Expression of enzyme activities: dehydrogenase in mg of triphenylformazan g^{-1} soil 24 h^{-1}; catalase in mg of O_2 g^{-1} soil min^{-1}; invertase in mg of "glucose" g^{-1} soil 24 h^{-1}; urease in mg of NH_3 g^{-1} soil 24 h^{-1}; and phosphatase in mg of phenolphthalein g^{-1} soil h^{-1}.

The comparison of unfertilised plots showed that oil contamination at each rate decreased catalase, invertase, and phosphatase activities and increased dehydrogenase activity. Urease activity was stimulated at 8 l of crude oil m^{-2} and inhibited at the two higher rates. In the NPK-fertilised plots, the enzyme activities (excepting urease activity) were lower in the contaminated than in the uncontaminated soil. In farmyard-manured plots, the contaminated soil was less catalase-, invertase-, and phosphatase-active and more dehydrogenase- and urease-active than the uncontaminated soil. In the plots treated with both NPK and farmyard manure, two activities (invertase and phosphatase) were lower and three (catalase, dehydrogenase, and urease) were higher in the contaminated than in the uncontaminated soil.

The favourable effects of fertilisation on the numbers of different microbial groups in the oil-contaminated soil plots increased in the order: NPK < farmyard manure < NPK + farmyard manure.

It is evident from these findings that application of NPK fertilisers and farmyard manure was a more efficient measure for increasing the enzymatic and microbial potential of oil-contaminated soil than was NPK fertilisation or farmyard manuring alone. Nevertheless, the biological potential of the uncontaminated soil was never attained in the oil-contaminated soil during this one-year experiment. Similarly, the grain yield of oats increased in fertilised contaminated plots but never reached the yield obtained in the corresponding uncontaminated plots.

Besides NPK fertilisation and farmyard manuring, other technologies were also modelled in Bashkiria for enhancing remediation of crude oil-contaminated soils. Both field and laboratory experiments were carried out. The remediation-enhancing technologies comprised use of green manures (Kireeva, 1995d); a special compost called Biohumus (Kireeva, 1995e); activated sludge and sawdust (Kireeva *et al.*, 1996); wastewater (Kireeva, 1995c); hydrocarbon-oxidising microbial cultures (Zhdanova *et al.*, 1994; Kireeva, 1996a); a biopreparation called Paprin (Kireeva, 1996b); and a nonionic surfactant (Kireeva *et al.*, 1997b).

In the field experiment using green manures (Kireeva, 1995d), microplots (1.3 by 1.3 m) were installed on a leached chernozem and a grey forest soil and contaminated with Tyumen crude oil at rates of 0, 1, 2, 5, 10, and 25% (weight/weight of the 0-20-cm soil layer), then treated with green manure (sweetclover, vetch-barley mixture, or oats at a rate of 20 t green biomass ha^{-1}), thoroughly mixed into the 0-20-cm soil layer. Addition of green biomass was repeated systematically during several years.

Dehydrogenase and catalase activities decreased, whereas urease activity increased with increasing crude oil addition. At the same time, addition of each green biomass resulted in increased enzyme activities. In the first year, the enzyme activity-increasing effect of sweetclover biomass was stronger than that of the other two biomasses, but in the third year, the effects of the three green biomasses were similarly high, and the enzyme activities in the oil-contaminated soils attained the activity levels of the uncontaminated soils. Soil respiration (CO_2 evolution), numbers of heterotrophic bacteria, actinomycetes, and cellulose-degrading microorganisms also increased under the influence of green biomasses and, in parallel, the residual oil content in soils decreased.

The grain yield of two-row barley sown in microplots contaminated with 1, 2, 5, and 10% crude oil and treated with green biomass exceeded, in the third year, the yield recorded in the uncontaminated soils. There was no significant difference between the yield-increasing effects of the three green biomasses. In the microplots contaminated with 25% crude oil, the grain yield in the green biomass-treated microplots reached, in the third year, only about 80% of the yield obtained in the uncontaminated microplots.

In the field experiment using Biohumus (Kireeva, 1995e), microplots on a grey forest soil were contaminated with 0, 8, 16, and 25 l of crude oil m^{-2}, then treated with Biohumus. This compost was prepared from cattle, pig, and horse dungs mixed with sawdust and chopped straw, and inoculated with a Californian population of the earthworm *Eisenia foetida* adapted to farmyard manure, then subjected to maturation for 6 months. Biohumus was applied several times at rates of 4-8 t ha^{-1}. The test plants were oats and barley. Enzyme activities were determined in the 0-20-cm soil layer 1 and 12 months after the Biohumus treatment.

The results were similar to those registered in the previous experiment: dehydrogenase, catalase, polyphenol oxidase, peroxidase, invertase, and phosphatase activities decreased and urease activity increased in parallel with the rate of crude oil contamination; Biohumus attenuated the negative effects of crude oil, increasing each enzyme activity in each plot. The only exception was urease activity as in the contaminated plots Biohumus diminished the oil-induced high urease activity.

Biohumus enhanced oil degradation. Grain yields of oats and barley, in both the second and third years, were significantly greater in the Biohumus-treated contaminated plots than in the untreated control ones. The yields in the third year in plots contaminated with 8 and 16 l of crude oil m^{-2} and treated with Biohumus attained the yields obtained in the uncontaminated plots. But in the plots contaminated with 25 l of crude oil m^{-2} and treated with Biohumus, the yields were about 60% of those of the uncontaminated plots.

In the field experiment using activated sludge and sawdust (Kireeva *et al.*, 1996) and in that using Paprin (Kireeva, 1996b), plots (>200 m^2) set up on the land accidentally contaminated with crude oil in the Kaltasin district and uncontaminated plots were used (see pages 5-6).

A part of the contaminated plots were treated with activated sludge or with mixtures of activated sludge and birch sawdust at rates of 10 and 20 t of activated sludge ha^{-1} and 25 t of sawdust ha^{-1}. The plots were sown with two-row barley. During several years, the 0-20-cm soil layer was systematically analysed.

The negative effects of oil contamination on most of the studied enzymatic and microbial parameters are mentioned on pages 5-6. Both sludge and sludge + sawdust treatments brought about increases in all enzymatic and microbial parameters, especially in the second year, but the majority of them (dehydrogenase, catalase, invertase, and urease activities, numbers of actinomycetes, nitrifying, and cellulose-degrading microorganisms) remained lower in the treated contaminated plots than in the uncontaminated ones.

The treatments also resulted in the intensification of oil degradation and in increased barley grain yields which, in the third year, exceeded even the yields recorded in the uncontaminated plots.

There was no significant difference between the remediating effects of sludge and sludge-sawdust mixture.

Another part of the microplots were treated with Paprin. This biopreparation consists of processed biomass of hydrocarbon-oxidising *Candida* and *Rhodotorula* yeasts grown on *n*-alkanes. It is rich in proteins and vitamins. The treatment was carried out with suspensions containing 5 or 50 g of Paprin l^{-1} water. The test plants were oats (in the first year), two-row barley (in the second year), and vetch-barley mixture (in the third year).

Dehydrogenase activity determined 1, 3, 6, and 12 months after the Paprin treatment followed the order: Paprin-treated contaminated soil>uncontaminated soil>untreated contaminated soil, at each time. The activity increased with time in the Paprin-treated soil and manifested little time-dependent changes in the other two soils.

Respiration (CO_2 evolution) and the number of heterotrophic bacteria were also the highest in the Paprin-treated contaminated soil and increased with time.

The oil content in soil determined during 3 years continuously decreased and, after 3 years, the initial 9.6% oil content was reduced to 4.95 and 1.01% in the untreated and Paprin-treated contaminated soil, respectively.

The yield of test plants was highest in the Paprin-treated plots in each year. For example, the vetch-barley yield (t ha^{-1}) was 0.66 in the untreated contaminated plots, 2.80 in the uncontaminated plots, and 4.95 in the treated contaminated plots.

Nevertheless, the cost-effectiveness of this technology needs to be evaluated.

Most of these field experiments were referred to in a monograph elaborated in 1994 and published in 1995 (Kireeva, 1995a).

Gainutdinov *et al.* (1988) have dealt with the problems of remediation of the oil-contaminated soils in the forest-steppe zone of Tataria. They carried out a crop rotation experiment (bare fallow, winter rye, spring wheat, pea), in which plots were contaminated with crude oil (2.5, 6.3, 12.5, and 25 l m^{-2}) and submitted to remediation (mineral and organic fertilisation, liming, soil loosening) in the first year. At 2.5-12.5 l oil m^{-2} rates, mineral (N+P) fertilisation plus farmyard manuring resulted in increased soil dehydrogenase and protease activities and this increase correlated with the decrease of oil content and was accompanied by higher numbers of hydrocarbon-degrading microorganisms.

The spilled crude oil bog soils in the Tyumen region, studied by Kaigorodova (1996) (see also page 7), were submitted to recultivation which was based on mixing of the upper soil layers. Due to this measure, cellulase activity increased and urease activity remained at a high level. The hydrocabon-oxidising microorganisms became numerous, corresponding to about 10% of the total microbial number and being represented by corynebacteria, spore-forming bacilli, pseudomonads, and flavobacteria. All corynebacterium strains isolated were hydrocarbon-oxidisers and grew at both 20 and 5°C. For enhancing the bioremediation, the following recommendations were made: selection of the most active hydrocarbon-oxidising microbial strain from the isolated local strains, mass cultivation of this strain and utilisation of its culture for inoculation of the oil-contaminated soil under recultivation.

Enzymological research in Azerbaijan

In the spring of 1979, Ismailov *et al.* (1988a) installed microplots (1 m^2 each) on a grey-brown soil (located in the dry subtropical zone of the country) and contaminated them with

4.4 l of crude oil m^{-2}. The control microplots were not contaminated. Then, the microplots were submitted to remediation treatments in a series of variants. N and P fertilisers were used as aqueous solutions of $(NH_4)_2SO_4$ and KH_2PO_4 at rates of 12 g of N and 9 g of P_2O_5 m^{-2}, respectively. Cultures of three hydrocarbon-degrading microorganisms (*Candida guilliermondii* 916, *Pseudomonas aeruginosa* 30, and an unidentified bacterium, strain 7) were applied as inocula (0.8 g dry microbial biomass m^{-2}). Two biopreparations were also tested. They were obtained from grape marc cut into small pieces or sawdust and a suspension of *C.guilliermondii* (0.5 g dry yeast biomass to 200 g of grape marc or to 100 g of sawdust). During the experiment (at months 8, 19, and 32), soil enzyme activities (dehydrogenase, catalase, invertase, urease, and protease), respiration (O_2 consumption and CO_2 evolution), and phytomass were determined. Global numbers of bacteria, yeasts, and filamentous fungi and the residual oil content were also determined periodically.

All treatments applied to oil-contaminated soil led to increased activity of each enzyme, excepting protease activity which was very low in all variants. Four treatments, namely addition of biopreparation 1 (grape marc) or 2 (sawdust), mixture of the three hydrocarbon-degrading microorganisms and $(NH_4)_2SO_4$ plus KH_2PO_4, gave especially good results in increasing the enzymatic potential of oil-contaminated soil. The enzymatic activities correlated with respiration and microbial numbers. The phytomass obtained after 32 months was 3.3 g m^{-2} in oil-contaminated microplots and increased to 7.0-48.8 g m^{-2} in oil-contaminated and treated microplots, but the phytomass of uncontaminated microplots (65 g m^{-2}) was not achieved. Oil degradation was enhanced by each treatment. In 8 months, 56-76% of hydrocarbons were degraded in the treated microplots, but only 49% in those receiving no remediation treatment. After 8 months as after 20 months, the smallest amounts of residual oil were found in microplots treated with biopreparation 1 or inoculated with the mixture of the three microbial cultures. Biodegradation of aliphatic hydrocarbons was faster than that of the other oil components.

In another experiment, Ismailov *et al.* (1988a) used whey and baker's yeast factory wastewater for remediation of oil-contaminated microplots (0.5 × 0.5 m) on a grey-brown soil. The microplots contaminated with crude oil (2 l per microplot) were left untreated for 9 months, then treated with whey or wastewater (2 l per microplot) and this treatment was repeated after 4 months. Both whey and wastewater had beneficial effects on soil dehydrogenase activity, respiration (CO_2 evolution), and oil degradation. Thus, during 6 months after the first treatment, oil degradation increased from 16 to 28-29% (in microplots in which before oil contamination the vegetation was practically lacking) and from 11 to 42-44% (in microplots having herbaceous vegetation before oil contamination).

Contamination with oil products

Enzymological research in the United States of America

The soil studies performed by Frankenberger *et al.* (1989) at the diesel fuel spill site located in San Bernardino County were referred to on pages 10-11, where some details on the exploratory borings were also given.

Two of the five borings drilled in November 1983 and five of the nine borings drilled in March 1984 were used as injection wells of nutrients (NPK) and H_2O_2 (about 150 l of water to which 250 mg of NH_4NO_3, 100 mg of K_2HPO_4, and 100 mg of H_2O_2 (30%) were added per l) for on-site bioremediation of the contaminated area. Starting in April 1984, the injection of the solution of nutrients and H_2O_2 was repeated at 2-week intervals for 4 months. Chemical analyses for total hydrocarbons and volatile aromatics, carried out at 3-month intervals in the July 1984 – June 1985 period, indicated no contamination of soil, more precisely the residual contamination had attained background levels. In other words, the on-site bioremediation was efficient.

Although the remediated soil was not analysed enzymologically, the remediation has, however, a soil enzymological aspect: the O_2 necessary for the hydrocarbon-oxidising microorganisms was produced from the injected H_2O_2 by the catalase enzyme and nonenzymatic catalysts present and active in the soil under remediation.

In the investigations performed by Wang and Bartha (1990) at the New Jersey Agricultural Experiment Station, 10 lysimeters (each having a surface area of 90 by 90 cm and a depth of 60 cm) were used. The soil provided in bulk from nearby construction sites needed improvement to approach the water-holding capacity and organic matter content of topsoil. The improvement was achieved by adding to the soil at the approximate rate of 5% by volume a peat-sand-perlite mixture (equal parts). A 20-cm thick sand layer was placed on the bottom of lysimeters which were then filled with soil (thickness of soil layer was 35 cm). The free rim of lysimeters was 5 cm high.

Sets of three lysimeters were contaminated with jet fuel, heating oil, and diesel oil, respectively, at a rate of 2.3 ml cm^{-2}, corresponding to 50-70 mg of oil product g^{-1} soil. The soil in all lysimeters was watered weekly by hand at 2.6 ml cm^{-2}.

From the sets of three identically contaminated lysimeters, one was left untreated except for weekly watering. One was, in addition, tilled weekly by a hand shovel to a depth of 15 cm. The third received full bioremediation treatment consisting of liming (55 mg of powdered agricultural limestone cm^{-2}) to raise soil pH from 6.7 to 7.4, fertilization (10 mg of urea and 4.3 mg of superphosphate cm^{-2}), and weekly tilling and watering. The last (No. 10) lysimeter, receiving no oil product and no treatment except for watering, served as source of uncontaminated control soil.

The experiments lasted 20 weeks, from late April to late September 1988. Periodically, the lysimeter soils were sampled to a depth of 15 cm to determine their residual hydrocarbon content, fluorescein diacetate (FDA) hydrolysis activity (which is due to soil enzymes), and toxicity (Microtox test assessing acute toxicity by measuring reduction in light emission by a *Photobacterium phosphoreum* preparation; ryegrass and soybean seed germination, and plant growth).

The determinations have shown that the uncontaminated control soil did not contain hydrocarbons in measurable amounts. During the 20-week experimental period, the hydrocarbon content decreased in the contaminated soils. The decrease was the smallest in the untreated soil, intermediary in the tilled soil, and the greatest in the treated (limed, fertilized, and tilled) soil. Consequently, tilling was less efficient than the full bioremediation treatment. This order of hydrocarbon disappearance was recorded with each of the three oil products. But their persistance increased in the order: jet fuel<heating oil<diesel oil. However, 20 weeks of bioremediation treatment lowered hydrocarbon concentration to below 5 mg g^{-1} soil for even the most persistent diesel oil, while in the untreated soil diesel oil concentration was still at 21.8 mg g^{-1}.

FDA hydrolysis activity (Figure 3) showed only very slight fluctuations in the uncontaminated control soil. In the contaminated soils, after a 6-week lag, the activity increased in the same order as hydrocarbon biodegradation, i.e., untreated<tilled<treated soil. The activity was inversely correlated with oil persistance: jet fuel stimulated activity the most, heating oil and diesel oil to lesser degrees. The 6-week lag in activity was caused by hydrocarbon biodegradation intermediates which strongly inhibited the activity. Therefore, the stimulation

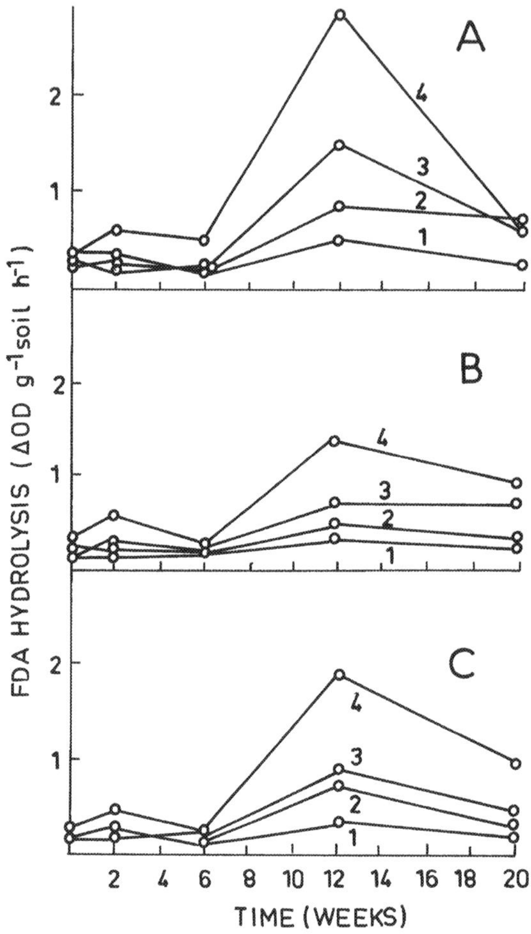

Figure 3. Changes in FDA hydrolysis activity with time in soil contaminated with jet fuel (A), heating oil (B), and diesel oil (C).

Redrawn from: X. Wang and R. Bartha, Soil Biology & Biochemistry, 1990, Vol. 22, pp. 503-504, with kind permission from Elsevier Science.

1 – Control soil.

2 – Untreated soil.

3 – Tilled soil.

4 – Treated soil.

became apparent only after most of the inhibitory intermediates had disappeared. This is why the activity peaked at week 12.

The Microtox measurements have revealed that at time 0 all three oil contaminants exhibited moderate toxicity, but their subsequent behaviours differed in terms of toxicity. Jet fuel was rapidly detoxified: toxicity values of the treated and untreated soils returned in 2 and 6 weeks, respectively, to background value of the control soil. Toxicity of heating oil increased during the initial phase of the experiment, but started to decrease after 6 and 12 weeks for the treated and untreated soils, respectively. By week 20, both soils returned to background toxicity value. Diesel oil exhibited a similar toxicity pattern to heating oil, but even in a more pronounced manner. Toxicity declined to background level by 20 weeks only in treated soil. Significant residual toxicity was still evident in the untreated soil, correlating with the detection of polycyclic aromatic residues in this but not in the treated soil.

Seed germination and plant growth data were consistent with the hydrocarbon residue and Microtox measurements.

The conclusion has been drawn that the full bioremediation treatment applied can restore oil-contaminated soils in 4-6 weeks to a degree that they can support plant cover; recovery of the soil is complete in 20 weeks.

In this experiment, soil dehydrogenase activity was also determined (Wang and Bartha, 1994), and it was found that in the soil contaminated by diesel oil (50 $\mu g\ g^{-1}$) and subjected to bioremediation treatment, dehydrogenase activity increased more than 10 times as compared with the contaminated but untreated soil. In contrast to FDA hydrolysis activity which peaked at week 12, the peak of dehydrogenase activity was reached at 6-7 weeks, at the time when over 80% of the contaminating hydrocarbon had been already degraded.

Enzymological research in Germany

The area studied by Kraatz and Schröder (1993) and Kraatz *et al.* (1993a, b) was used in the past as a coking and gas works in Saarbrücken. For more than 100 years the works polluted the soil with different wastes (sediments, ashes, slags, building rubbles) and chemical compounds, including polycyclic aromatic hydrocarbons (PAHs). One of the causes of the contamination with PAHs was the leakage of tar oil-holding tanks. Content of PAHs was very high, up to 1 $g\ kg^{-1}$ in the whole soil profile to depths of 8-10 m, near the groundwater. Consequently, there is a high risk that the PAHs will be transported into the Saar River flowing at only 500 m from the contaminated area. Besides PAHs, smaller amounts of benzene, toluene, ethyl benzene, xylenes (BTEX) were also present in the soil.

In April 1981, an on-site remediation experiment was initiated. The contaminated soil was excavated and used for forming two types of heaps. The soil in heaps 1 contained more clay, less PAHs and BTEX, and had a lower C-to-N ratio in comparison with heaps 2. Both heaps 1 and 2 were subjected to a dry and wet remediation procedure.

In the dry procedure, the homogenised soil was mixed with about 10% (weight/weight) bark mulch (C-to-N>40). At approximately 2-month intervals, the soil-mulch mixture was moistened and aerated by means of a special turning device. In the wet procedure, the soil slurry was transferred into a bioreactor where the conditions for biodegradation of contaminants were maintained optimal. Control soils 1 and 2 were contaminated soils from heaps 1 and 2, not subjected to either the dry or the wet remediation procedure.

After 8 months of remediation, the soil of all heaps were recultivated. For this, on each heap three plots were installed: two plots were sown for the development of pioneer and ruderal plant communities, and one plot was kept bare fallow.

During two years (April 1981 – April 1983), the soil was analysed 6-9 times for determining some enzyme activities, substrate-induced respiration (SIR), and some chemical parameters. Colonisation of soil by epigeic mesofauna was also evaluated.

Invertase activity was determined in all dry- or wet-remediated heaps 1 and 2 and plots installed on these heaps, whereas dehydrogenase and catalase activities, and SIR were assayed only in the dry-remediated heaps 1 and 2 and plots.

Invertase activity was higher in control soil 1 than in control soil 2, and in heaps 1 than in heaps 2, i.e. the less PAH-contaminated soil was more invertase-active than the more contaminated one. Dry remediation resulted in higher invertase activity than the wet remediation in both heaps 1 and 2 and plots. The plots covered by pioneer and ruderal vegetation exhibited greater soil invertase activity than did the plots kept bare fallow. This finding was valid for both heaps 1 and 2, dry- or wet-remediated.

Dehydrogenase and catalase activities, and SIR, determined – as mentioned above – only in the dry-remediated heaps, strongly increased, like invertase activity, after addition of bark mulch to the contaminated soil in April 1981. Later, the activities and SIR stabilised at lower level, which was, however, several times higher than that in the untreated contaminated (control) soil, but 3-5 times lower in comparison with the level of activities measured in less contaminated, herbaceous vegetation-covered zonal soils. It should be added that dehydrogenase and catalase activities, and SIR, like invertase activity, were favourably influenced by the pioneer and ruderal vegetation.

Dry remediation as compared to wet remediation, and vegetation-covered plots in comparison with plots kept bare fallow were also more favourable for colonisation of soil by epigeic mesofauna.

The humus dynamics of remediated soil remained unbalanced as the proportion of humic acid fraction in total C content was high and the content of fulvic acids and nonhumic substances (polysaccharides, proteins) was low, which explains the much lower metabolic activity in the remediated soil than in the less contaminated, herbaceous vegetation-covered zonal soils.

Contamination with oil field wastewaters

Enzymological research in the Russian Federation

For remediation of soils affected by salts due to their contamination with oil field wastewaters, Gainutdinov *et al.* (1988) selected contaminated and uncontaminated, control plots on a calcareous chernozem of heavy loam texture. Washing of contaminated plots with fresh-water (2,000-3,000 m^3 ha^{-1}, repeated three times), their treatment with gypsum (22 t ha^{-1}), or acidification (12.5 t of sulphuric acid ha^{-1}) were applied for remediation. Soil enzymatic activities (dehydrogenase, protease, urease) were determined during 2 years and plant productivity was estimated during 4 years. Washing had little effect on enzymatic activities, whereas gypsum and sulphuric acid rather depressed them. Washing alone did not improve plant productivity, but in combination with gypsum or sulphuric acid treatment had a beneficial effect on plant biomass production. In the fourth year, however, plant productivity in each contaminated and treated plot remained lower than that registered in the control plots.

LABORATORY EXPERIMENTS

Contamination with crude oil

Enzymological research in the Russian Federation

Andreson and Khaziev (1981) have contaminated samples of a chernozemic soil with West Siberian crude oil at rates of 0, 50, and 250 g kg^{-1} soil, then treated them with NH_4NO_3 or $(NH_4)_2SO_4$ or with superphosphate (0.1 g of N or P_2O_5 kg^{-1} soil). An emulsifier of pellicular oil (EPN-5) was also tested. EPN-5 was applied as a concentrated (100%) aqueous solution or as aqueous dilutions (corresponding to 20 to 80% concentrations). The soil humidity was adjusted to 40% of water-holding capacity. The incubation took place at room temperature; its duration was 4 weeks.

Dehydrogenase and catalase activities increased 1.5-2 times under the influence of EPN-5 at 20% concentration and decreased by 80 and 100% EPN-5, at both rates of crude oil addition. Similarly, 20-40% EPN-5 stimulated the growth of soil bacteria and microfungi, whereas EPN-5 at 60 to 100% concentrations caused reduction of microbial numbers, at both rates of oil contamination. Dehydrogenase activity showed 1.5-2-fold increases in samples treated with N fertilisers and lower increases in the superphosphate-treated samples. These increases occurred at both rates of oil contamination. NH_4NO_3 was a better N source than $(NH_4)_2SO_4$.

But in a similar experiment with a dark-grey forest soil, the negative effect of crude oil on phytase activity could not be removed by N or P fertilisation, or emulsifier application (Khaziev and Fatkhiev, 1981).

As mentioned on page 31, Kireeva's research group conducted field and laboratory experiments using different technologies for enhancing remediation of crude oil-contaminated soils.

The field experiment using activated sludge and activated sludge-birch sawdust mixture for remediation of the accidentally oil-contaminated soil in the Kaltasin district and its results were described on page 33.

Under similar conditions a laboratory experiment was also carried out (Kireeva *et al.*, 1996). The soil samples taken from the 0-20-cm layer containing 9.6% crude oil were treated with activated sludge (13.32 mg g^{-1} soil), moistened to 60% of water-holding capacity and incubated at 30°C. Uncontaminated and untreated soil samples were the controls.

After 1 and 12 months of incubation, dehydrogenase, catalase, and invertase activities, numbers of heterotrophic bacteria, nitrifying, and cellulose-degrading microorganisms showed the order: uncontaminated soil > treated contaminated soil > untreated contaminated soil. In other words, the negative effects of crude oil were partially removed by the sludge treatment. Oil contamination resulted in increases only in urease activity, numbers of microfungi and hydrocarbon-oxidising microorganisms. Residual oil content after 12 months of incubation was 5.92% in the untreated soil and 3.04% in the sludge-treated contaminated soil.

Raw wastewater from a farm breeding calves was introduced in sedimentation ponds in succession and after 6 months of purification the clarified wastewater was used by Kireeva

(1995c) for enhancing remediation of crude oil-contaminated samples of a grey forest soil of heavy loam texture.

The soil samples were contaminated with 0, 5, and 10% crude oil (weight/weight), then moistened to 60% of water-holding capacity with pure water (control), wastewater or its dilutions (1 part of wastewater to 1, 2, or 5 parts of pure water).

The enzymological analyses carried out after 3 days, 1, and 8 months of incubation showed that the crude oil decreased, while the wastewater and its dilutions increased soil dehydrogenase, catalase, and phosphatase activities. The negative effects of crude oil were attenuated or eliminated by the wastewater and its dilutions. Thus, after 8 months, the soil samples treated with 10% crude oil plus wastewater and its 1:1 and 1:5 dilutions exhibited increased enzyme activities in the order: 1:5 dilution > wastewater > 1:1 dilution (dehydrogenase activity); 1:1 dilution > wastewater > 1:5 dilution > (catalase activity); 1:5 dilution > 1:1 dilution > wastewater (phosphatase activity).

Under the influence of wastewater and its dilutions, numbers of actinomycetes, microfungi, nitrifying, and cellulose-degrading microorganisms became higher in both uncontaminated and contaminated soil samples.

The analysis of the residual oil content indicated that the treatment with wastewater and its dilutions resulted in intensification of oil degradation.

Similar results were obtained with samples containing 9.6% crude oil due to the accidental contamination in the Kaltasin district (Kireeva, 1995a).

For enhancing crude oil degradation in contaminated soils (leached chernozem and grey forest soil), two mixed and two pure cultures of hydrocarbon-oxidising microorganisms were tested by Zhdanova *et al.* (1994) and Kireeva (1996a). Both mixed cultures originated from oil industrial wastewaters. They contained *Pseudomonas* sp., *Candida* sp. (or *Arthrobacter* sp.), and an unidentified Gram-negative rod-shaped bacterium. In pure cultures, *Candida maltosa* and *Pseudomonas* sp. strain 6 were tested. A biopreparation consisting of buckwheat (*Fagopyrum*) chaff inoculated with one of the mixed cultures (*Pseudomonas* sp., *Candida* sp., unidentified bacterium) was also studied. Rates of oil contamination of soil samples ranged from 0.5 to 20% (weight/weight), and rates of inoculation varied between 10^6 and 10^9 microbial cells g^{-1} soil.

Each culture enhanced oil degradation, the mixed cultures being more efficient than the pure ones. The negative effects of crude oil on soil enzyme activities were partially or completely removed by the cultures. Thus, in soil samples contaminated with 5 or 10% crude oil, then treated with a mixed culture or the biopreparation mentioned above and incubated for 3 and 12 months, dehydrogenase, catalase, invertase, and urease activities exhibited significantly higher values after both incubation times, in comparison with the values of the non-inoculated contaminated samples. The activity-increasing effect of the cultures was more pronounced at 5 than at 10% oil contamination, and due to the biopreparation than due to the mixed culture. Invertase and urease activities of the less contaminated samples treated with the biopreparation even exceeded after 12 months the activity levels of the uncontaminated samples.

Kireeva *et al.*(1997c) added 0, 0.5, 1, 2, 5, and 10% crude oil to samples of a leached chernozem and a grey forest soil and then, for emulsifying the oil and thus making it more available for microbial degradation, the samples were treated with aqueous solution of the nonionic surfactant Neonol AF-14 at a rate of 1 mg surfactant g^{-1} soil.

After 3 days and 1 month of incubation, it was found that the surfactant treatment led to attenuation of the decreasing effects of crude oil on soil dehydrogenase and catalase activities and of its increasing effect on soil urease activity. The effects of surfactant on the growth of heterotrophic bacteria, actinomycetes, and cellulose-degrading bacteria were similar to those exerted on dehydrogenase and catalase activities. The oil content significantly decreased in the surfactant-treated samples. Thus, in soil samples contaminated with 10% crude oil, the oil content was reduced to 5.04% after 1 month of incubation.

Most of these laboratory experiments were referred to in Kireeva's (1995a) monograph.

Fomenko *et al.* (1996) described laboratory experiments, in which some biopreparations (Firezyme, Degradoil, Bacispecin) containing live bacteria and enzymes were tested as agents for enhancing biodegradation of oil in and recultivation of soils contaminated with crude oil at the enterprises of the "Yuganskneftegaz" Company (Nefteyugansk city, Tyumen region).

Contamination degree of the studied soils was different. The biopreparations were applied at different rates. Incubation was carried out at laboratory temperature (20-25°C) and lasted 33 days. The residual oil content was determined periodically.

The results have shown that activity of biopreparations increased with increasing degree of oil contamination, and during 1 month of incubation 90% of the oil was degraded, irrespective of the rate of biopreparation addition.

Enzymological research in Azerbaijan

Ismailov *et al.* (1988b) have verified the effect of the hydrocarbon-degrading yeast strain *Candida guilliermondii* 916 on the remediation of oil-contaminated soils (see page 35). 0.5-kg samples of a grey-brown soil from the Apsheron Peninsula were contaminated with 10% (weight/weight) Binagadin crude oil and inoculated with a yeast suspension (10^6 or 10^9 cells g^{-1} soil) using grape marc (cut into small pieces) as a carrier of yeast cells. Rates of grape marc application were 0.5 and 2% (on soil weight basis). N and P fertilisers (12 g of N and 9 g of P_2O_5 m^{-2}) were also added to the samples. Oil-contaminated but untreated samples served as controls. Incubation took place at optimum humidity and aeration conditions at 28°C and lasted 60 days, during which the samples were periodically submitted to enzymological, microbiological, and chemical analyses.

Dehydrogenase activity, respiration (CO_2 evolution), and oil degradation were most intense in the sample treated with 10^9 yeast cells g^{-1} soil + 0.5% grape marc and their values in this sample were especially high during the first 25 days of incubation and decreased later. During the 60-day incubation, numbers of the inoculated yeast cells continuously decreased from 10^9 to $\sim 10^3$ g^{-1} soil and from 10^6 to $\sim 10^2$ g^{-1} soil, respectively, but numbers of the heterotrophic bacteria became higher in the yeast-treated soil samples than in the untreated control. Finally, oil degradation was 30% in the treatment with 10^9 yeast cells g^{-1} soil + 0.5% grape marc, whereas it was only 8% in the untreated control.

Enzymological research in Canada

Although the contamination of the soil studied by Xu and Johnson (1995) resulted from a crude oil spill due to pipeline break, their experiment could not be dealt with in Chapter 1, because the oil-contaminated soil used for enzymological and other analyses had previously been submitted to different remediation treatments. Therefore, the direct effects of oil contamination on soil properties as compared to those of the uncontaminated soil could not be evaluated. See also the footnote on page 46.

The crude oil spill occurred in 1990 on a black chernozem located near Erskine, Alberta. The spill site sloped into slough; the upper portion was planted to grain crops, while the lower area was seeded to forage. At the spill site, the surface soil (0-15 cm) was excavated, homogenised, and stockpiled. The oil-contaminated soil studied by Xu and

Johnson was sampled from the stockpile. It contained 55 g of total petroleum hydrocarbons kg^{-1} soil. Previously bioremediated and solvent-treated samples of the oil-contaminated soil were also studied. The bioremediated soil had been treated in a bioremediation facility for 15 months. This treatment reduced the oil content to 25 g hydrocarbons kg^{-1} soil. The solvent-treated soil contained oil and solvent (naphtha) residues of 5 g hydrocarbons kg^{-1} soil. The uncontaminated soil used as a control was collected adjacent to the spill site. This soil contained no detectable petroleum hydrocarbons.

Samples of the four soils (oil-contaminated, bioremediated, solvent-treated, and uncontaminated) were sieved (2-mm screen) and packed in plastic cylinders (25 cm in height and 20 cm in diameter). Three seeds of barley (*Hordeum vulgare*) or field pea (*Pisum arvense*) were planted per cylinder. Urea, superphosphate, and potassium sulphate (300 mg of N, 65 mg of P, and 125 mg of K kg^{-1} soil, respectively) were placed 3 cm below the seeds at the time of seeding. The plants were grown in a growth chamber. The sampling was conducted four times on each plant during the growing period: tillering stage (day 25), stem extension stage (day 40), heading stage (day 60), and ripening stage (day 80) for barley. The field pea was sampled at the same time. On each sampling date, the shoots were excised at the surface of the soil, and the roots were separated from the soil by hand, washed and then dried (at 75°C for 12 h). The soils were homogenised and used for enzymological and other analyses*.

The root mass of barley was significantly greater than that of the field pea over the growing period in each of the four soils studied. At the end of the growing period, the barley root mass had values significantly decreasing in the order: uncontaminated soil > solvent-treated soil > bioremediated soil > oil-contaminated soil; the root mass of field pea was also significantly greater in the uncontaminated soil than in the other three soils, in which the field pea root mass decreased, but insignificantly, in the same order as the barley root mass. The shoot mass of both plants had a similar trend to the root mass.

In comparison with the root mass, the microbial biomass-C content in soils under both plants presented, during the growing period, a rather reversed order: oil-contaminated soil >

* It should be emphasised that at time 0, i.e. before fertilising and seeding no enzymological analyses were carried out. Thus, the direct soil-enzymological effects of crude oil contamination could not be assessed. The soils, their enzymological and other properties analysed after 25, 40, 60, and 80 days of plant growth had already been exposed to the complex influence of their hydrocarbon content, the fertilisers applied, and the growing plants.

or < bioremediated soil > solvent-treated soil > or < uncontaminated soil, the differences being significant only between the oil-contaminated or bioremediated soil and the solvent-treated or uncontaminated soil.

The microbial respired C corresponding to respiration of soils, from which the barley and field pea plants were removed, was significantly higher in the oil-contaminated than in the other soils during the whole growing period: oil-contaminated soil > bioremediated soil > or < solvent-treated soil > or < uncontaminated soil.

The average acid phosphatase activity in the four soils under barley was significantly greater than under the field pea (Figure 4). Under barley, the activity was highest in the oil-

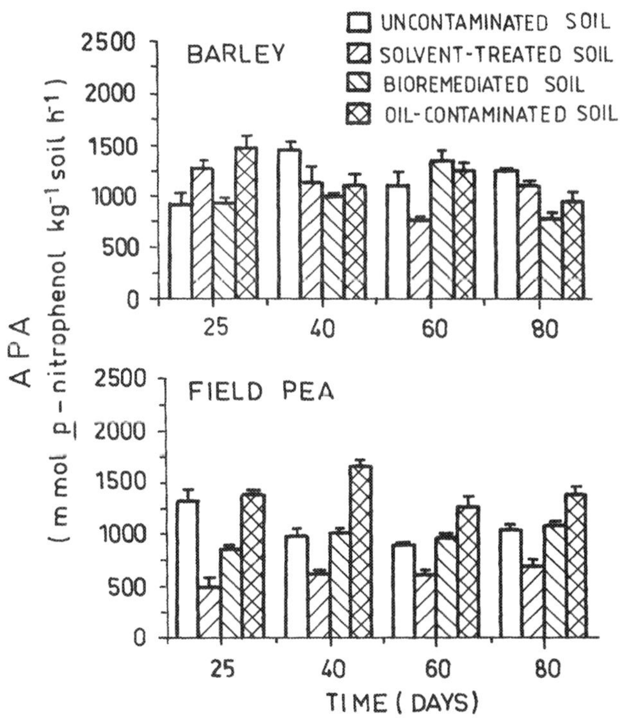

Figure 4. Acid phosphatase activity (APA) under barley and field pea in four soil materials over the growing season.

Redrawn from: J.G. Xu and R.L. Johnson, Plant and Soil, 1995, Vol. 173, p. 6, with kind permission from Kluwer Academic Publishers.

The error bars represent the standard deviation.

contaminated soil at tillering stage, in the uncontaminated soil at stem extension and ripening stages, and in the bioremediated soil at heading stage. The activity was lowest in the bioremediated soil during the growing period except at the heading stage.

The pattern of acid phosphatase activity in soils under the field pea was not as complicated as under barley. The activity was always highest in the oil-contaminated soil and lowest in the solvent-treated soil. Acid phosphatase activity in the four soils under the two plants significantly correlated with the microbial respired C: the linear correlation coefficient between the two measurements was 0.3373 ($P<0.05$) under barley and 0.6282 ($P<0.001$) under field pea.

It is evident from these findings that oil contamination had a negative effect on the growth of both plants (the effect being stronger on barley than on the field pea) and a positive effect on soil microorganisms and phosphatase production. This means that, under the conditions of this experiment, the soil microorganisms, although stimulated by contaminating oil, did not detoxify the soil or, at least, the detoxification was not complete.

The crude oil-contaminated black chernozem at the spill site near Erskine, Alberta was used for another experiment, too (Xu *et al.*, 1995). Three soil materials were sampled. Material 1 was taken from the 3-10-cm layer of the oil-contaminated soil. Material 2 was the same oil-contaminated soil but subjected to an artificial aggregation process in a field aggregator. Material 3 was sampled from the 3-10-cm layer of an uncontaminated soil adjacent to the spill site. Materials 1 and 2 contained 65 g of total petroleum hydrocarbons kg^{-1} soil. Material 3 was free of such hydrocarbons.

Pore structure analysis revealed that artificial aggregation significantly increased the inter-aggregate porosity and insignificantly the intra-aggregate porosity.

For biological analyses, the soil samples were preincubated at 60% of their water-holding capacity for 14 days, then assayed for their microbial biomass-N and FDA hydrolysis activity. Both parameters were found to show the following order in the three soil materials: aggregated oil-contaminated soil > untreated oil-contaminated soil > uncontaminated soil.

From the comparison of the two oil-contaminated soil materials with each other one can deduce that aggregation enhanced the microbial and enzymatic potential of the soil by improving the physical environment for aerobic microorganisms and, thus, facilitating biodegradation of hydrocarbons (i.e. use of hydrocarbons by microorganisms as C and energy sources).

Comparison of the two oil-contaminated soil materials with the uncontaminated soil indicates that for the microorganisms in the studied black chernozem the effect of hydrocarbons as C and energy sources was greater than their toxic effect.

Contamination with oil products

Enzymological research in the United States of America

Frankenberger (1988) has dealt with the use of urea as a nitrogen fertiliser in remediation of soils contaminated with refined oil products. Surface samples (0-15 cm) of three Californian soils of different pH, organic C content, and texture were studied. Field-moist soil samples (5 g each) were treated with 0, 5, 10, and 25% (weight/weight) of kerosene, diesel oil, motor oil, and leaded gasoline, and incubated at 25°C for 14 days, then submitted to urease activity determination.

The results have shown (Table 4) that the oil products inhibited soil urease activity. The degree of inhibition ranged from 23 to 67%. In each soil leaded gasoline was most inhibitory

Table 4
Effects of refined oils on urease activity in soils

Refined oils	Loading rates (%)	Percentage inhibition of urease activity in three soils studied		
		Altamont	Kitchen Creek	Pico
Kerosene	5	24	23	27
	10	28	31	28
	25	32	35	33
Diesel oil	5	28	24	27
	10	36	35	31
	25	47	39	44
Motor oil	5	23	31	29
	10	27	38	34
	25	38	45	48
Leaded gasoline	5	34	48	40
	10	46	51	46
	25	55	67	61

Reproduced from: W.T. Frankenberger, Jr., Bulletin of Environmental Contamination and Toxicology, 1988, Vol. 40, p. 67.

at each loading rate. The inhibitory effect of the other refined oils varied in the three soils studied and increased in the following orders: motor oil < kerosene < diesel oil (Altamont); kerosene < diesel oil < motor oil (Kitchen Creek); and kerosene ≈ diesel oil < motor oil (Pico).

The greater the loading rate, the less effective urea becomes as a nitrogen fertiliser. Because inhibition of urease activity is so high, availability of NH_4^+ derived from urea would be greatly delayed and, thus, growth of hydrocarbon-degrading microorganisms would be hindered. The conclusion is that urea should not be recommended as a nitrogen fertiliser to remediate soils contaminated with refined oil products, because it would do little benefit in bioremediation of such soils.

In the remediation experiments described by Song *et al.* (1990) and Song and Bartha (1990), fluorescein diacetate (FDA) hydrolysis activity was also determined together with some soil microbial parameters. A loam soil from the grounds of the Bayway Refinery, New Jersey (Exxon USA) previously not exposed to oil spill was studied. Soil samples (60 g dry weight) were placed in 600-ml beakers in a layer that was less than 2-cm thick for free access of oxygen. They were contaminated with jet fuel at a low (50 mg g^{-1} soil) and a high (135 mg g^{-1} soil) concentration. For bioremediation, some of the fuel-contaminated samples were limed (10 mg of $CaCO_3$ g^{-1} soil), fertilised with 60 μmol of N as NH_4NO_3 and 5 μmol of P as K_2HPO_4 g^{-1} soil, and "tilled" by a stainless steel wire loop. Moisture content of the soil was adjusted to 50% of water-holding capacity. For inactivating microorganisms and enzymes, some samples were poisoned with 2% $HgCl_2$. All samples were then incubated at 27°C for 18 weeks. During incubation, the samples were aerated and their moisture was maintained at a constant level.

FDA hydrolysis activity was determined at 2-week intervals. In uncontaminated soil the activity remained practically unchanged during 18 weeks of incubation. $HgCl_2$ poisoning resulted in a 90% inhibition of the activity. Jet fuel contamination, without bioremediation, led to a lasting depression of the activity at both fuel concentrations which was attributed to inhibition of the FDA assay by jet fuel degradation products. In soils submitted to bioremediation, jet fuel contamination first depressed the activity, but this brief depression was followed by a strong increase. The increase was of a 3-week duration at the low contamination level, but the activity kept increasing for 14 weeks in the case of the higher contamination level. Thus, the benefit of bioremediation in terms of FDA hydrolysis activity was obvious.

Although jet fuel contamination alone depressed FDA hydrolysis activity, it increased bacterial numbers and fungal mycelial length by 2 to 2.5 orders of magnitude. At the same time, the combination of jet fuel contamination with bioremediation increased the microbial parameters by 3 to 4 orders of magnitude. Interestingly, the contamination level (50 or 135 mg of jet fuel g^{-1} soil) made little difference in microbial parameters. During incubation a great part of the added jet fuel disappeared.

Enzymological research in the Russian Federation

Remediation of a grey forest soil contaminated with oil products was the objective of the investigations described by Khaziev *et al.* (1988a). Light gasoline (LG), extract of the second oil fraction (EOF) and mixture of EOF and tar (T) in equal proportion were the contaminants. They were tested at the following rates: LG and EOF 0.5, 1, 4, and 6%; EOF+T 0.5+0.5, 4+4, and 6+6% (percentage means ml of oil product per 100 g of air-dry soil). The control samples received no contaminants. For remediation, mineral fertilisers (N 120, P 180, and K 180 kg ha^{-1}) plus farmyard manure (140 t ha^{-1}) were used. The soil samples moistened to 60% of water-holding capacity and incubated at room temperature were analysed after 24 h and 3 months.

After 24 h of incubation, oil product additions to soil samples resulted in reduced enzyme (dehydrogenase, catalase, invertase, and urease) activities and in smaller numbers of bacteria, actinomycetes, and microfungi. The effects were dependent on contaminant concentration. LG was more toxic to enzymes than were the other oil products. Invertase was found to be the most sensitive enzyme to contamination. The only parameter that increased under the influence of contaminants was soil respiration (CO_2 evolution). Treatment with NPK plus farmyard manure did not remove the negative effects of oil products on enzyme activities, but respiration was further enhanced.

After 3 months, the oil products at 0.5 and 1% rates stimulated dehydrogenase and catalase activities, but the higher rates remained inhibitory. Oil contamination also resulted in higher global microbial numbers. In contaminated soil samples treated with NPK plus farmyard manure, urease activity, global microbial numbers, and respiration showed increased values.

Kireeva *et al.* (1996), based on the investigations of Khaziev *et al.* (1988a), used LG, EOF, and EOF+T each at a rate of 6% for contamination of soil samples. For enhancing remediation, activated sludge and birch sawdust were tested. The analyses carried out after 3 days, 1, 3, and 12 months of incubation gave the following principal results: the sludge and the sludge-sawdust mixture had some stabilising effect on the enzymatic and microbial potential of the contaminated soil; under the influence of sludge, the mineralisation coefficient (which is proportional to the amount of energy used in the microbial respiratory processes) increased 4.7 times with LG, 4.3 times with EOF, and 5.2 times with EOF+T.

Enzymological research in Austria

Bauer *et al.* (1992) and Pennerstorfer *et al.* (1992) have studied the effect of organic amendments (bark compost, bark mulch, and garbage compost) on bioremediation of heating oil-contaminated samples of two soils (calcareous chernozem and degraded chernozem). In other experiments (see page 26), these soils were found to be the least and the most sensitive, respectively, to oil contamination. In the bioremediation experiments, the soil samples were treated with organic amendments in a proportion of 4 parts of soil to 1 part of organic amendment, and contaminated with heating oil at a rate of 3.75% (weight/weight). The control samples received no organic amendment and/or no oil. The mixtures were incubated for 8 weeks. Soil respiration and TTC-dehydrogenase activity were determined weekly. The residual oil content was analysed at the end of the incubation period.

The two soils studied behaved differently. In the oil-contaminated calcareous chernozem, bark and garbage composts increased and bark mulch decreased respiration and dehydrogenase activity; these effects were remarkable only during the first two weeks of incubation. Degradation of oil increased in the order: no organic amendment < bark mulch < garbage compost ≈ bark compost. In other words, the organic amendments enhanced bioremediation of the oil-contaminated calcareous chernozem.

In the oil-contaminated degraded chernozem, the organic amendments decreased respiration and dehydrogenase activity, especially during the first 4 weeks of incubation. Degradation of oil increased in the order: bark mulch < bark compost < garbage compost < no organic amendment. This means that the organic amendments inhibited bioremediation of the oil-contaminated degraded chernozem.

CONCLUSIONS

The investigations reviewed show that the soil enzymological and microbiological effects of oil contamination are dependent on the site (soil, vegetation, climate, relief, etc.) and on chemical composition and the amount of oil contaminants. These factors also determine the rate and extent of bioremediation of oil-contaminated soils, i.e. degradation of oil contaminants by the native and/or introduced microorganisms in soils. Therefore, it is difficult to generalise from one case to another (Blackburn and Hafker, 1993). However, the soil enzymatic activities, that are easily measurable, may indicate toxicity of oil contaminants on soil life and also the capacity of the soil microbiota to catalyse self-decontamination of soil. The multidisciplinary investigation of oil contamination of soils and their remediation should always comprise enzymological measurements, too.

Development of a new branch of the enzymology of oil-contaminated soils can also be envisaged. Fan and Krishnamurthy (1995), based on investigations aiming at using isolated, then immobilised enzymes instead of live microbial cells for remediation of soils and waters contaminated with xenobiotics (e.g. Bollag, 1992), have suggested the "enzyme-accelerated remediation of petroleum-contaminated sites" and have expressed their hope that the enzymatic remediation technology will be efficient and cost-effective.

REFERENCES

Aliev, S.A. and Gadzhiev, D.A. (1977) Vliyanie zagryazneniya neftyanym organicheskim veshchestvom na aktivnost' biologicheskikh protsessov pochv (Effect of pollution by petroleum organic matter on the activity of biological processes of soils). *Izvestiya Akademii Nauk Az.SSR, Seriya Biologicheskikh Nauk*, No. 2, 46-49.

Andreson, R.K. and Khaziev, F.Kh. (1981) Bor'ba s zagryazneniem pochvogruntov neft'yu (Fighting against petroleum pollution of soils). *Korroziya i Zashchita v Neftegazovoi Promyshlennosti* (Moscow), No. 7, 1-46.

Antonenko, A.M. and Zanina, O.V. (1992) Vliyanie nefti na fermentativnuyu aktivnost' alluvial'nykh pochv Zapadnoi Sibiri (Effect of petroleum on the enzymatic activity of alluvial soils in West Siberia). *Pochvovedenie*, No. 1, 38-43.

Atlas, R.M. (Ed.) (1984) *Petroleum Microbiology*. Macmillan, New York.

Atlas, R.M. and Bartha, R. (1992) Hydrocarbon biodegradation and oil spill bioremediation. *Advances in Microbial Ecology*, **12**, 287-338.

Bauer, E., Pennerstorfer, C., Holubar, P., Plas, C. and Braun, R. (1991) Microbial activity measurement in soil – a comparison of methods. *Journal of Microbiological Methods*, **14**, 109-117.

Bauer, E., Pennerstorfer, C., Kandeler, E. and Braun, R. (1992) Beurteilung biologischer Bodendekontaminationsverfahren – Versuche im Labormaßstab (Evaluation of biological soil decontamination procedures – experiments on laboratory scale). In *ALVA-Tagung 1992. Fachgruppe Boden. Vorträge*, pp. 47-56. Bundesanstalt für Bodenwirtschaft, Vienna.

Bauer, E., Pennerstorfer, C., Kandeler, E. and Braun, R. (1994) Biologische Bodenreinigung (Biological soil decontamination). *Mitteilungen der Österreichischen Bodenkundlicher Gesellschaft*, No. 48/49, 193-206.

Beerstecher, E. (1954) *Petroleum Microbiology*. Elsevier, Amsterdam.

Blackburn, J.W. and Hafker, W.R. (1993) The impact of biochemistry, bioavailability and bioactivity on the selection of bioremediation techniques. *Trends in Biotechnology*, **11**, 328-333.

Bollag, J.-M. (1992) Decontaminating soil with enzymes. *Environmental Science & Technology*, **26**, 1876-1881.

Davis, J.B. (1967) *Petroleum Microbiology*. Elsevier, Amsterdam.

Demidienko, A.Ya. and Demurdzhan, V.M. (1988) Puti vosstanovleniya plodorodiya neftezagryaznennykh pochv chernozemnoi zony Ukrainy (Ways to restoration of the fertility of petroleum-polluted soils in the chernozem zone of Ukraine). In *Vosstanovlenie Neftezagryaznennykh Pochvennykh Ekosistem* (M.A. Glazovskaya, Ed.), pp. 197-206. Izdatel'stvo Nauka, Moscow.

Fan, C.-Y. and Krishnamurthy, S. (1995) Enzymes for enhancing bioremediation of petroleum-contaminated soils: a brief review. *Journal of the Air & Waste Management Association*, **45**, 453-460.

Fil'chenkova, V.I. (1982) Biologicheskaya aktivnost' karbonatnogo chernozema pri neftyanom zagryaznenii (Biological activity of a petroleum-polluted calcareous chernozem). In *Zashchita Rastenii i Okhrana Okruzhayushchei Sredy v Tatarskoi ASSR*, pp. 168-170. Oblastnoe Pravlenie NTO Sel'skogo Khozyaistva, Kazan.

Fomenko, V.V., Bormasheva, N.G., Omel'chenko, E.B., Khmarinov, L.K., Shevchuk, N.P. and Shimkevich, S.V. (1996) Analiz effektivnosti metodov likvidatsii neftyanykh zagryaznenii pochv s ispol'zovaniem ekologicheski chistykh biopreparatov na predpriyatiyakh AO "YUNG" (Analysis of the efficacy of methods for liquidation of petroleum pollution of soils by using ecologically clean biopreparations, at enterprises of the "Yugansneftegaz" Company). *Tezisy Dokladov 3. Mezhdunarodnoi Konferentsii "Osvoenie Severa i Problemy Rekul'tivatsii"* (Sankt-Petersburg, 1996), pp. 216-217.

Frankenberger, W.T., Jr. (1988) Use of urea as a nitrogen fertilizer in bioreclamation of petroleum hydrocarbons in soil. *Bulletin of Environmental Contamination and Toxicology*, **40**, 66-68.

Frankenberger, W.T., Jr. and Johanson, J.B. (1982) Influence of crude oil and refined petroleum products on soil dehydrogenase activity. *Journal of Environmental Quality*, **11**, 602-607.

Frankenberger, W.T., Jr., Emerson, K.D. and Turner, D.W. (1989) *In situ* bioremediation of an underground diesel fuel spill: a case history. *Environmental Management*, **13**, 325-332.

Gainutdinov, M.Z., Samosova, S.M., Artem'eva, T.I., Gilyazov, M.Yu., Khramov, I.T., Gaisin, I.A., Fil'chenkova, V.I. and Zherebtsov, A.K. (1988) Rekul'tivatsiya neftezagryaznennykh zemel' lesostepnoi zony Tatarii (Recultivation of petroleum-polluted soils in the forest-steppe zone of Tataria). In *Vosstanovlenie Neftezagryaznennykh Pochvennykh Ekosistem* (M.A. Glazovskaya, Ed.), pp. 177-197. Izdatel'stvo Nauka, Moscow.

Ismailov, N.M. (1982) Neftyanoe zagryaznenie i biologicheskaya aktivnost' pochv (Petroleum pollution and biological activity of soils). In *Dobycha Poleznykh Iskopaemykh i Geokhimiya Prirodnykh Ekosistem*, pp. 227-235. Izdatel'stvo Nauka, Moscow.

Ismailov, N.M. (1983) Vliyanie neftyanogo zagryazneniya na krugovorot azota v pochve (Effect of petroleum pollution on the nitrogen cycle in soil). *Mikrobiologiya*, **52**, 1003-1007.

Ismailov, N.M. (1988) Mikrobiologiya i fermentativnaya aktivnost' neftezagryaznennykh pochv (Microbiology and enzymatic activity of petroleum-polluted soils). In *Vosstanovlenie Neftezagryaznennykh Pochvennykh Ekosistem* (M.A. Glazovskaya, Ed.), pp. 42-56. Izdatel'stvo Nauka, Moscow.

Ismailov, N.M., Gadzhieva, V.I. and Gasanova, M.G. (1984) Koeffitsient mineralizatsii uglevodorodov kak pokazatel' samoochishchayushchei sposobnosti neftezagryaznennykh pochv i effektivnosti primenyaemykh metodov ikh rekul'tivatsii (Coefficient of hydrocarbon mineralisation as an indicator of the self-purification capacity of petroleum-polluted soils and of the efficiency of the methods applied for their recultivation). *Izvestiya Akademii Nauk Az.SSR, Seriya Biologicheskikh Nauk*, No. 6, 76-85.

Ismailov, N.M., Akhmedov, A.G. and Akhmedov, V.A. (1988a) Rekul'tivatsiya neftezagryaznennykh zemel' sukhikh subtropikov Azerbaidzhana (Recultivation of petroleum-polluted soils in the arid subtropics of Azerbaijan). In *Vosstanovlenie Neftezagryaznennykh Pochvennykh Ekosistem* (M.A. Glazovskaya, Ed.), pp. 206-221. Izdatel'stvo Nauka, Moscow.

Ismailov, N.M., Gadzhieva, V.I., Bairamova, E.M. and Serebryakova, O.B. (1988b) Zondirovanie mikrobiologicheskikh protsessov v neftezagryaznennoi sero-buroi pochve uglevodorodokislyayushchimi drozhzhami (Monitoring of the microbiological processes with hydrocarbon-oxidising yeasts in a petroleum-polluted grey-brown soil). *Izvestiya Akademii Nauk Az.SSR, Seriya Biologicheskikh Nauk*, No. 1, 97-102.

Janke, S., Schamber, H. and Kunze, C. (1992) Beeinflussung der bodenbiologischen Aktivität durch Heizöl (Effects of heating oil on the soil biological activity). *Angewandte Botanik*, **66**, 42-45.

Kaigorodova, S.Yu. (1996) Biologicheskaya aktivnost' neftezagryaznennykh bolotnykh pochv Tyumenskoi oblasti (Biological activity of the petroleum-polluted bog soils in the Tyumen region). *Programma i Tezisy 4. Nauchnoi Konferentsii Instituta Biologii "Aktual'nye Problemy Biologii"* (Syktyvkar, 1996), p. 56.

Kandeler, E., Pennerstorfer, C., Bauer, E. and Braun, R. (1994) Mikrobiologische Beurteilung biologischer Bodendekontaminationsverfahren im Modellversuch (Microbiological evaluation of the biological soil decontamination procedures in model experiment). *Zeitschrift für Pflanzenernährung und Bodenkunde*, **157**, 345-350.

Khaziev, F.Kh. and Fatkhiev, F.F. (1981) Izmenenie biokhimicheskikh protsessov v pochvakh pri neftyanom zagryaznenii i aktivatsiya razlozheniya nefti (Changes in biochemical processes in petroleum-polluted soils and activation of petroleum degradation). *Agrokhimiya*, No. 10, 102-111.

Khaziev, F.Kh., Tishkina, E.I. and Kireeva, N.A. (1988a) Vliyanie nefteproduktov na biologicheskuyu aktivnost' pochvy (Effect of petroleum products on the biological activity of soil). *Biologicheskie Nauki* (Moscow), No. 10, 93-99.

Khaziev, F.Kh., Tishkina, E.I., Kireeva, N.A. and Kuzyakhmetov, G.G. (1988b) Vliyanie neftyanogo zagryazneniya na nekotorye komponenty agroekosistemy (Effect of petroleum pollution on some components of the agro-ecosystem). *Agrokhimiya*, No. 2, 56-61.

Kireeva, N.A. (1995a) *Mikrobiologicheskie Protsessy v Neftezagryaznennykh Pochvakh* (Microbiological Processes in Petroleum-Polluted Soils). Bashkirskii Gosudarstvennyi Universitet, Ufa.

Kireeva, N.A. (1995b) Mikrobiologicheskaya otsenka pochvy, zagryaznennoi neftyanymi uglevodorodami (Microbiological evaluation of the soil polluted by petroleum hydrocarbons). *Bashkirskii Khimicheskii Zhurnal*, **2** (3-4), 65-68.

Kireeva, N.A. (1995c) Ispol'zovanie otkhodov sel'skokhozyaistvennogo proizvodstva v rekul'tivatsii neftezagryaznennykh pochv (Use of wastes of agricultural production for recultivation of petroleum-polluted soils). In *Voprosy Biotekhnologii*, pp. 15-21. Bashkirskii Gosudarstvennyi Universitet, Ufa.

Kireeva, N.A. (1995d) Intensifikatsiya biodestruktsii neftyanykh uglevodorodov v pochve sideratami (Intensification of the biodestruction of petroleum hydrocarbons in soil, with green manures). In *Voprosy Biotekhnologii*, pp. 22-27. Bashkirskii Gosudarstvennyi Universitet, Ufa.

Kireeva, N.A. (1995e) Ispol'zovanie biogumusa dlya uskoreniya destruktsii nefti v pochve (Use of biohumus for acceleration of the destruction of petroleum in soil). *Biotekhnologiya*, No. 5-6, 32-35.

Kireeva, N.A. (1995f) Pochvennye mikroorganizmy kak indikatory zagryazneniya uglevodorodami (Soil microorganisms as indicators of pollution by hydrocarbons). In *Reaktsiya Organizmov na Antropogennoe Vozdeistvie* (E.N. Khisamov, Ed.), pp. 17-29. Bashkirskii Gosudarstvennyi Pedagogicheskii Institut, Ufa.

Kireeva, N.A. (1996a) Biodestruktsiya nefti v pochve kul'turami uglevodorodokislyayushchikh mikroorganizmov (Biodestruction of petroleum in soil by cultures of hydrocarbon-oxidising microorganisms). *Biotekhnologiya*, No. 1, 51-54.

Kireeva, N.A. (1996b) Vosstanovlenie plodorodiya neftezagryaznennykh pochv (Restoration of the fertility of petroleum-polluted soils). *Neftyanaya i Gazovaya Promyshlennost'*, No. 7, 21-22.

Kireeva, N.A. and Vodop'yanov, V.V. (1996) Matematicheskoe modelirovanie mikrobiologicheskikh protsessov v neftezagryaznennykh pochvakh (Mathematical modelling of the microbiological processes in petroleum-polluted soils). *Pochvovedenie*, No. 10, 1222-1226.

Kireeva, N.A., Novoselova, E.I. and Khaziev, F.Kh. (1996) Ispol'zovanie aktivnogo ila dlya rekul'tivatsii pochv, zagryaznennykh neft'yu (Use of activated sludge for the recultivation of petroleum-polluted soils). *Pochvovedenie*, No. 11, 1399-1403.

Kireeva, N.A., Novoselova, E.I. and Khaziev, F.Kh. (1997a) Fosfogidrolaznaya aktivnost' neftezagryaznennykh pochv (Phosphohydrolase activity of petroleum-polluted soils). *Pochvovedenie*, No. 6, 723-725.

Kireeva, N.A., Novoselova, E.I. and Khaziev, F.Kh. (1997b) Fermenty azotnogo obmena v neftezagryaznennykh pochvakh (Enzymes of the nitrogen metabolism in petroleum-polluted soils). *Izvestiya Akademii Nauk, Seriya Biologicheskaya*, No. 6, 755-759.

Kireeva, N.A., Zhdanova, N.V. and Gorina, T.S. (1997c) Sposob vosstanovleniya plodorodiya zagryaznennykh neft'yu pochv (Method for restoration of the fertility of petroleum-polluted soils). *Neftyanoe Khozyaistvo*, No. 4, 51-52.

Kiss, S. (1957) A flotációs módszer alkalmazása a talaj saválló mükobaktériumainak közvetlen minőségi és mennyiségi vizsgálatára (Application of the flotation method for direct qualitative and quantitative study of the acid-fast mycobacteria in soil). *Kolozsvári Babeş és Bolyai Egyetemek Közleményei, Természettudományi Sorozat*, **2** (1-2), 329-344.

Kiss, S. (1995) Enzymology of oil-contaminated soils. *Studia Universitatis Babeş-Bolyai, Biologia*, **40** (1-2), 3-25.

Koch, C. and Wilke, B.-M. (1996) Einfluß von PAK, PCB und Schwermetallen auf die mikrobielle Aktivität in Rieselfeldböden (Effect of PAH, PCB and heavy metals on the microbial activity in irrigated field soils). *Mitteilungen der Deutschen Bodenkundlichen Gesellschaft*, **81**, 295-298.

Kraatz, M. and Schröder, D. (1993) Bodenmikrobiologische und humuschemische Untersuchungen während der mikrobiologischen on-site-Sanierung eines PAH-belasteten Substrates (Soil microbiological and humus chemical investigations during microbiological on-site decontamination of a PAH-contaminated substrate). *Mitteilungen der Deutschen Bodenkundlichen Gesellschaft*, **72**, 575-578.

Kraatz, M., Emmerling, C. and Schröder, D. (1993a) Bodenbiologische und humuschemische Untersuchungen während der mikrobiologischen on-site-Sanierung eines PAK-belasteten Substrates (Soil biological and humus chemical investigations during microbiological on-site decontamination of a PAH-contaminated substrate). In *Altlastensanierung '93* (F. Arendt, G.J. Annokkée, R. Bosman and W.J. van den Brink, Eds.), pp. 961-962. Kluwer, Dordrecht.

Kraatz, M., Schumacher, B., Emmerling, C. and Schröder, D. (1993b) Development of soil microbial activities, soil fauna, and humic matter during remediation and recultivation of PAH-contaminated soil. *II. Euro-Forum Altlasten* (Saarbrücken, 1993), **1**, 255-270.

Linkins, A.E. and Fetcher, N. (1983) Effect of surface-applied Prudhoe Bay crude oil on vegetation and soil processes in tussock tundra. In *Proceedings, 4th International Conference on Permafrost* (Fairbanks, 1983), pp. 723-728.

Linkins, A.E., Atlas, R.M. and Gustin, P. (1978) Effect of surface applied crude oil on soil and vascular plant root respiration, soil cellulase, and hydrocarbon hydroxylase at Barrow, Alaska. *Arctic*, **31**, 355-365.

McGill, W.B., Rowell, M.J. and Westlake, D.W.S. (1981) Biochemistry, ecology, and microbiology of petroleum components in soil. In *Soil Biology, Vol. 5* (E.A. Paul and J.N. Ladd, Eds.), pp. 229-296. Dekker, New York.

Mukatanov, A.Kh. and Rivkin, P.R. (1980) Vliyanie nefti na svoistva pochv (Effect of petroleum on properties of soils). *Neftyanoe Khozyaistvo*, No. 4, 53-54.

Pennerstorfer, C., Bauer, E., Kandeler, E. and Braun, R. (1992) Beurteilung biologischer Bodensanierungsverfahren. Versuche im Labormaßstab (Evaluation of soil decontamination procedures. Experiments on laboratory scale). In *Industrial Waste Management. Waste Reduction and Treatment, Site Remediation and Water Purification* (W.

Pillmann, Ed.), pp. 670-677. International Society for Environmental Protection, Vienna.

Samosova, S.M., Minibaev, V.G. and Artem'eva, T.I. (1979) Vliyanie zasoleniya pochvy neftepromyslovymi stochnymi vodami na svoistva pochvy i ee biologicheskuyu aktivnost' (Effect of the soil salinisation with wastewaters of petroleum industry on the properties of soil and on its biological activity). In *Sezonnaya Dinamika Pochvennykh Protsessov*, pp. 183-185. Akademiya Nauk Est.SSR, Tallinn.

Samosova, S.M., Kurbskii, G.P., Usacheva, G.M., Gubaidullina, T.S., Fil'chenkova, V.I. and Abushaeva, V.V. (1982) Izmenenie mikroflory i sostava nefti v chernozemnoi pochve Tatarii v pervyi period posle zagryazneniya (Changes of microflora and petroleum composition in a Tatarian chernozemic soil during the first period after pollution). In *Dobycha Poleznykh Iskopaemykh i Geokhimiya Prirodnykh Ekosistem*, pp. 235-245. Izdatel'stvo Nauka, Moscow.

Söhngen, N.L. (1913) Benzin, Petroleum, Paraffinöl und Paraffin als Kohlenstoff- und Energiequelle für Mikroben (Benzine, petroleum, paraffin oil and paraffin as carbon and energy source for microbes). *Zentralblatt für Mikrobiologie, Parasitenkunde, Infektionskrankheiten und Hygiene, 2. Abteilung*, **37**, 595-609.

Song, H.-G. and Bartha, R. (1990) Effects of jet fuel spills on the microbial community of soil. *Applied and Environmental Microbiology*, **56**, 646-651.

Song, H.-G., Wang, X. and Bartha, R. (1990) Bioremediation potential of terrestrial fuel spills. *Applied and Environmental Microbiology*, **56**, 652-656.

Steubing, L. (1967) Untersuchungen über die Veränderung der Mikroflora eines Waldbodens durch eingedrungenes Heizöl (Studies on the change in microflora of a forest soil by penetrated heating oil). *Angewandte Botanik*, **40**, 275-286.

Terry, R.E. (1998) Use of soil conditioners to enhance and speed bioremediation of contaminated soils. In *Handbook of Soil Conditioners* (A. Wallace and R.E. Terry, Eds.), pp. 551-571. Dekker, New York.

Tishkina, E.I. and Kireeva, N.A. (1986) Okul'turivanie neftezagryaznennykh serykh lesnykh pochv (Cultivation of petroleum-polluted grey forest soils). In *Povyshenie Plodorodiya Pochv v Usloviyakh Intensivnoi Sistemy Zemledeliya* (F.Kh. Khaziev, Ed.), pp. 145-151. Bashkirskii Filial Akademii Nauk SSSR, Institut Biologii, Ufa.

Wang, X. and Bartha, R. (1990) Effects of bioremediation on residues, activity and toxicity in soil contaminated by fuel spills. *Soil Biology & Biochemistry*, **22**, 501-505.

Wang, X. and Bartha, R. (1994) Effects of bioremediation on toxicity, mutagenesis, and microbiota in hydrocarbon-polluted soils. In *Remediation of Hazardous Waste Contaminated Soils* (D.L. Wise and D.J. Trantolo, Eds.), pp. 175-197. Dekker, New York.

Weißmann, S. and Kunze, C. (1994) Microbial activity in heating oil contaminated soil under field and controlled conditions. *Angewandte Botanik*, **68**, 137-142.

Xu, J.G. and Johnson, R.L. (1995) Root growth, microbial activity and phosphatase activity in oil-contaminated, remediated and uncontaminated soils planted to barley and field pea. *Plant and Soil*, **173**, 3-10.

Xu, J.G., Feng, Y., Johnson, R.L. and McNabb, D.H. (1995) Pore structures of oil-contaminated, aggregated oil-contaminated and uncontaminated soils in relation to microbial activities. *Environmental Technology*, **16**, 587-594.

Yashvili, N.N., Beradze, I.A., Dumbadze, T.K., Mardaleishvili, R.K. and Kobakhidze, N.V. (1982) Vliyanie zagryazneniya neft'yu i nefteproduktami na biologicheskuyu aktivnost' pochv Kolkhidskoi nizmennosti (Effect of pollution with petroleum and petroleum products on the biological activity of soils in the Colchis Lowland). *Izvestiya Akademii Nauk Gruz.SSR, Seriya Biologicheskaya*, **8**, 413-418.

Zhdanova, N.V., Kireeva, N.A. and Matytsina, O.I. (1994) Nekotorye puti intensifikatsii biodestruktsii nefti v pochve (Some ways for the intensification of biodestruction of petroleum in soil). *Neftyanaya i Gazovaya Promyshlennost'*, No. 5, 31-32.

Zhou, L.-K., Zheng, Q.-Y. and Song, M. (1990) Biodegradation of petroleum hydrocarbons and phenol substances in soil and soil enzyme activities (in Chinese with English summary). *Journal of Applied Ecology* (Shenyang), **1** (2), 149-155.

Zimenka, T.G. and Kartyzhova, L.Ya. (1986a) Uplyv naftavaga zabrudzhvannya na biyalagichnuyu aktyvnasts' dzyarnova-padzolistai legkasuglinkavai gleby (Effect of petroleum pollution on the biological activity of a soddy-podzolic light loam soil) *Vestsi Akademii Navuk BSSR, Seryya Biyalagichnykh Navuk*, No. 6, 52-55.

Zimenko, T.G. and Kartyzhova, L.E. (1986b) Vliyanie Rechitskoi nefti na biogennost' dernovo-podzolistoi pochvy (Effect of the Rechitsa petroleum on the abundance of microorganisms in a soddy-podzolic soil). In *Mikrobiologicheskie Protsessy v Pochvakh i*

Urozhainost' Sel'skokhozyaistvennykh Kul'tur, p. 125. Institut Botaniki Akademii Nauk Lit.SSR, Vilnius.

PART II. ENZYMOLOGY OF SOILS AFFECTED BY INDUSTRIAL EMISSIONS

INTRODUCTION

Many industrial plants emit gaseous, liquid, and solid, particulate pollutants to the atmosphere. The air pollutants are then dispersed over smaller or larger areas and deposited in wet or dry form on the landscape affecting soils and waters, in which they can accumulate. Most of the pollutants, e.g. heavy metals may be very harmful to natural and man-made ecosystems, to health of humans, animals, and plants, may exert detrimental effects on soil life.

The effects of heavy metals on the enzyme activities in soils were first studied in the '40s and '50s by the founders of soil enzymology (J.P. CONRAD, Ed. HOFMANN, V.F. KUPREVICH) (see e.g. Kiss, 1958; Skujinš, 1978). These first studies did not aim at enzymological indication of soil pollution; their purpose was to gather additional data proving that the reactions attributed to soil enzymes are really catalysed by enzymes, as they are inhibited or inactivated by heavy metals.

The soil enzymological effects of industrial emissions as environmental pollutants are studied under laboratory and natural conditions.

In laboratory experiments, the effects of some components of industrial emissions on enzyme activities in soils are estimated. These components (heavy metals, mineral acids, fluoride, phenol, etc.) are added in form of pure compounds to soil samples and the changes induced by the additions in soil enzyme activities are determined. Thus, the emissions from nonferrous metallurgical plants are *modelled* by using Cu, Zn, Pb, Cd, etc. salts with which soil samples are treated. The acid rains are *simulated* by treating soil samples with mineral acids (H_2SO_4, HNO_3, or HCl) or SO_2. The *in situ*, artificial microcosm studies are, conceptually, similar to the laboratory ones.

In the studies carried out under natural conditions, enzyme activities in soil samples collected in the vicinity of the pollution source and in those taken at different distances from this source are compared, i.e. soil enzyme activities are determined along a pollution gradient.

The pollutants originate from a point source (an industrial plant) or from multiple sources (many industrial plants manufacturing different products, but situated in the same, industrial area).

Consequently, three directions can be delineated in the studies dealing with enzymology of soils affected by industrial emissions:

I. studies of the soil enzymological effects of the components of industrial emissions through experiments modelled in the laboratory or in *in situ* artificial microcosms;

II. studies of the soil enzymological effects of industrial emissions originating from a point source (an industrial plant); and

III. studies of the soil enzymological effects of industrial emissions originating from multiple sources (many industrial plants manufacturing different products, but situated in the same, industrial area).

The studies in the first direction as laboratory or artificial microcosm studies contribute to the understanding of the soil enzymological effects of pollutants from industrial emissions, but their results can not be extrapolated as perfectly valid for the natural conditions.

The studies in the second and third directions, evidencing the soil enzymological effects of industrial emissions from point or multiple sources, make it possible to establish the responsibility of industries in environmental pollution and to evaluate the efficiency of the decontamination technologies applied.

Chapter 4. STUDIES OF THE SOIL ENZYMOLOGICAL EFFECTS OF THE COMPONENTS OF INDUSTRIAL EMISSIONS, THROUGH EXPERIMENTS MODELLED IN THE LABORATORY OR IN *IN SITU* ARTIFICIAL MICROCOSMS

Such studies are numerous, being described in hundreds of papers. Many synthesis papers were also published. They review the studies dealing with the soil biological and biochemical (including soil enzymological) effects of heavy metals (Doelman, 1978; Tyler, 1981; Bååth, 1989; Francis, 1989; Galiulin, 1989; Kobzev, 1989; Tiller, 1989; Gianfreda and Bollag, 1996); simulated acid rain (Tabatabai, 1985; Francis, 1986, 1989; Gianfreda and Bollag, 1996); sulphur dioxide (Kulhavý, 1987); pollutant inorganic and organic compounds (Nannipieri *et al.*, 1997). Due to existence of these excellent reviews, the studies in the first direction of the enzymology of soils affected by industrial emissions will not be dealt with in our present work.

Chapter 5. STUDIES OF THE SOIL ENZYMOLOGICAL EFFECTS OF INDUSTRIAL EMISSIONS ORIGINATING FROM A POINT SOURCE (AN INDUSTRIAL PLANT)

Reviewing these studies representing the second direction of the enzymology of soils affected by industrial emissions, twelve sections could be delineated, according to the type of industries emitting pollutants to the atmosphere: nonferrous metallurgical plants, ironworks, ore enrichment works, coking plants, refractory brickworks, pulp and paper mills, synthetic fibre factories, oil production plants, other chemical factories, biotechnological plants, oil- or coal-fired power plants, and atomic power plants. In additional sections, two activities causing point source pollution (military waste disposal and rocket destruction operations) are also considered.

Nonferrous metallurgical plants

Enzymological research in Sweden

The first soil enzymological studies in areas, where the unique, point source of pollution is a nonferrous metallurgical plant, were performed in Sweden. Rühling and Tyler (1973) have studied the area around Fliseryd. This village in south-eastern Sweden has only one industry processing metal oxides for production of batteries. The village is surrounded by pine (*Pinus sylvestris*) and spruce (*Picea abies*) forests. In April 1972, the F_1 needle mor layer (usually 3-5 cm deep) was sampled in 100 sites situated along a transect extending 12.5 km from the industrial plant. The elements emitted in large amounts from the plant are lead, nickel, and cadmium. The Pb concentrations in mor samples from sites situated within 0.5 km from the plant were, on average, 26 times higher than in samples from sites situated more than 7 km away. The corresponding factors for Ni and Cd were 54 and 22, respectively. Small gradients were found in vanadium (factor ~7), copper (~2), iron (~1.4), and chromium (~1.3).

The mor samples were also analysed for determining their dehydrogenase activity and respiration (CO_2 evolution) rate. Highly significant ($P<0.001$) negative correlations were found between dehydrogenase activity and concentration of Pb, Ni, Cd, and V, respectively or the sum of Pb+Ni+Cd concentrations (Figure 5).

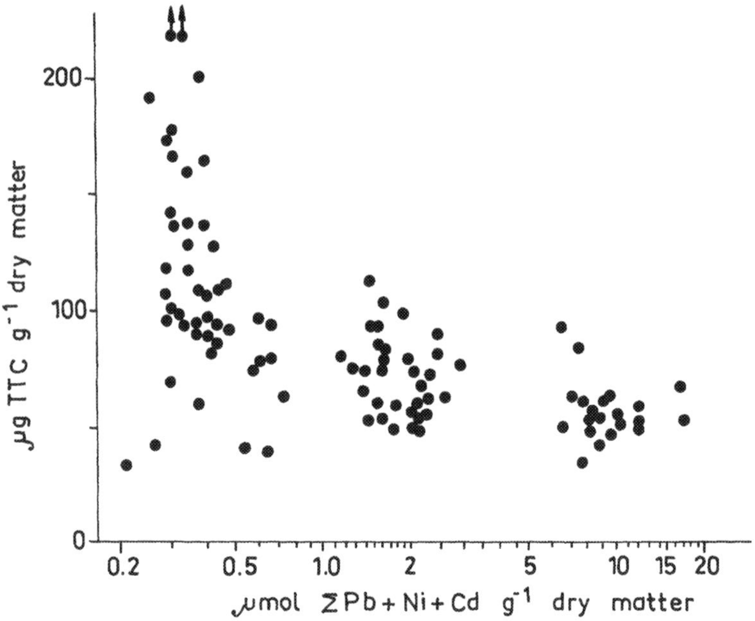

Figure 5. Dehydrogenase activity in the Fliseryd needle mor, related to heavy metal concentration.

Redrawn from: Å. Rühling and G. Tyler, Oikos, 1973, Vol. 24, p. 413.

TTC – 2,3,5-Triphenyltetrazolium chloride reduced at 18°C in 48 h.

Tyler (1973, 1974, 1975a, b, c, 1976, 1977, 1981, 1984) has performed detailed chemical, microbiological, and biochemical (including enzymological) studies in the Gusum area, too. Ebregt and Boldewijn (1977) and Nordgren *et al.* (1983) also carried out studies in this area. In Gusum, a small town in south-eastern Östergötland, a brass foundry has been operating since 1661. In 1966, the old foundry in central Gusum was replaced by a modern one outside the town, at 1.5 km from the old foundry. The new foundry, having no smoke-cleaning equipment, remained the unique, point source of heavy metal pollution. About 98% of total emissions of heavy metals as alkaline oxides are copper and zinc, and there are only small emissions of lead and cadmium. The vegetation in the Gusum area, mainly coniferous woodland dominated by spruce (*Picea abies*), developed on podzolic soils. Close to the foundry the vegetation was severely injured, most spruces being dead or dying. The organic topsoil (the F_1 layer of spruce needle mor) was sampled in many sites.

In July 1973, Tyler (1974) collected mor samples from 40 sites, distributed in all directions within a radius of 3 km from the foundry. Distinct metal concentration gradients were recorded in the samples. Thus, the samples taken close to the foundry (at 0.2-0.3 km from the main outlet) were exceedingly high in Cu and Zn (11,000-17,000 ppm and 16,000-22,000 ppm dry weight, respectively) compared with those taken about 3 km away (40-140 ppm and 200-600 ppm, respectively).

Urease and acid phosphatase activities, like respiration (CO_2 evolution) rate, correlated negatively and highly significantly with the Cu+Zn concentration of the mor samples. Within the recorded concentration range of Cu+Zn (240-39,000 ppm), with increasing Cu+Zn concentration the urease activity was reduced by a factor of 10*, acid phosphatase activity and respiration rate by a factor of 5. At the same time, no negative effect of high concentrations of Cu and Zn on the β-glucosidase activity has been established.

The spruce mor samples taken in March-April 1974 from 80 sites, distributed in a stratified random way in all directions within 7 km of the foundry, were used for determination of nitrogen mineralisation rate (increase of NH_4^+ + NO_3^- + NO_2^- content) during aerobic storage of mor samples in undisturbed stratification in a humid chamber at 22°C and constant water content corresponding to 47-50% of the water-holding capacity, for 10 weeks. It was found (Tyler, 1975a) that heavy metal pollution, especially Cu is a limiting factor for N mineralisation rate, and even moderate amounts of Cu and Zn, exceeding only three times the background levels registered at 7-9 km from Gusum (Cu: 12-20 ppm; Zn: 100-200 ppm), are sufficient to reduce the available N to plants.

Forty spruce mor samples collected in July 1973 and 150 samples taken in March-April 1974 were used by Tyler (1976) for determination of the rates of decomposition (dry weight loss) and phosphorus mineralisation (increase in acetic acid-soluble *o*-phosphate content)

---

* Urease activity was also determined in 150 mor samples collected in March-April 1974 (Tyler, 1975b). In these samples, the concentration range comprised only the low and middle parts of the metal gradient (corresponding only to 15-2,000 ppm Cu). Within this concentration range, the highly significant correlation between urease activity and heavy metals decreased in the order: Cd (-0.633) > Zn (-0.571) > Cu (-0.556). In other words, when the most metal-polluted mor samples are excluded, urease activity is reduced to the largest extent not by the most abundant Cu and Zn, but by the less abundant Cd.

during 5-month aerobic storage of the samples in a humid chamber at 22°C, keeping their water content at 72-75% of the water-holding capacity. There were highly significant negative correlations between the Cu or Zn or Cu+Zn concentration of samples and the decomposition rate, as well as between the Cu+Zn concentration and the P mineralisation rate. In five samples containing >20,000 ppm Cu+Zn, no organic carbon decomposition could be recorded. In stored samples, very close inverse relationships were observed between heavy metal concentration and acid phosphatase activity and high positive correlation coefficients were obtained between phosphatase activity, decomposition, and P mineralisation rates.

It was evidenced by statistical analyses that Cu is more responsible for reducing phosphatase activity than Zn at about equal concentrations. Thus, even in the range of 30-200 ppm Cu there is an inverse relationship ($P<0.05$) between Cu concentration and phosphatase activity of samples. The relationship is indicated even in the range of 15-80 ppm Cu. The Cu concentration in sites beyond the influence of pollution is 12-20 ppm. In other words, the negative effect of Cu on phosphatase activity is measurable already at a low degree of pollution. Other important findings: a higher pH of the mor sample may, to some extent, counteract the activity decrease; edaphically poor sites are more sensitive to heavy metal pollution as indicated by stronger decrease in phosphatase activity.

Tyler's report to the National Swedish Environment Protection Board published in Swedish in December 1974 and in English in February 1975 (Tyler, 1975b) is a synthesis of the studies we have referred to above. "The phosphatase map" presented by Tyler in this report (see Figure 6) shows that a certain reduction of acid phosphatase activity is noticeable as far as 3-5 km from the old foundry, that is within an area of about 5,000 ha. In the vicinity of foundry, as in the residential area, phosphatase activity is very much reduced.

Tyler's communication at the International Conference on Heavy Metals in the Environment, held in Toronto in 1975, is also a review of the investigations performed mainly in the Gusum area (Tyler, 1975c). In the communication, mention was also made of a reactivation experiment described in detail two years later (Tyler, 1977).

A metal-chelating agent, namely Na-EDTA (0.05 M) was used to reactivate acid phosphatase and urease in 10 + 10 spruce needle mor samples collected in the surroundings of the old and new brass foundries, respectively, at Gusum in the autumn of 1975. Ten needle mor samples from locally unpolluted areas in south-eastern Östergötland and, thus, containing only background levels of Cu and Zn served as controls.

Reaction mixtures were prepared, with and without Na-EDTA addition, for determination of acid phosphatase activity (at pH 5.0) and urease activity (at pH 6.7), and incubated at 22°C for 3 h. The difference between activity values measured in the presence and absence of Na-EDTA was calculated and expressed in percentage, too, for each mor sample.

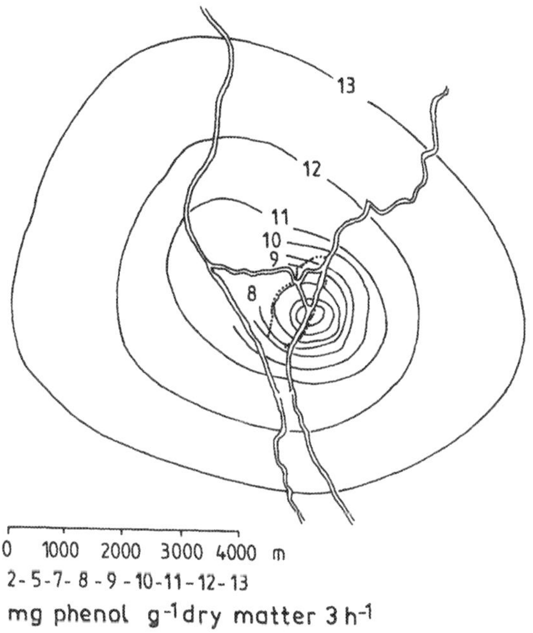

Figure 6. "The phosphatase map" of the spruce needle mor samples in the Gusum area.

Redrawn from: G. Tyler, Statens Naturvårdsverk, 1975, PM 542E, p. 37.

The 30 mor samples were divided into four groups according to their Cu+Zn concentration: <0.3, 3.4-11.9, 16.2-23.3, and 28.5-45.8‰ of the mor dry weight. The mean percent differences in enzyme activities measured in reaction mixtures with and without Na-EDTA addition were the following in the four mor groups: +<1, +17, +55, and +36% (acid phosphatase activity), and –8, +26, +37, and +5% (urease activity). In other words, Na-EDTA had no effect on phosphatase activity and slightly inhibited urease activity in the unpolluted mor samples, but brought about a partial reactivation of both enzymes in the polluted mor samples. The reactivating effect of Na-EDTA was most pronounced in the mor samples containing 16.2-23.3‰ Cu+Zn.

Ebregt and Boldewijn (1977) have determined amylase activity and rates of basal respiration (no organic substance added) and starch-induced respiration in spruce needle mor samples from 50 sites selected in a stratified random way to secure an approximately logarithmic distribution with regard to distance from the foundry. Amylase activity significantly decreased with the sum of Cu+Zn+Pb+Cd concentration increasing from 200 to 50,000 ppm. Both basal and starch-induced respirations behaved like amylase activity.

Nordgren *et al.* (1983) emphasise that the organic topsoil (spruce needle mor) close to the new foundry replacing – as already mentioned – the old one in 1966, reached in 15 years the same levels of heavy metals, about 20,000 μg of Cu and 20,000 μg of Zn g^{-1} oven-dry soil as those around the old foundry. By studying the microfungi in topsoil samples collected in October 1979 and May 1980 these authors have established that soil moisture and organic matter content had little influence on the microfungal community compared with the heavy metal pollution, along both the whole gradient and its most polluted part, whereas at the less polluted places soil moisture and organic matter accounted for more of the observed changes than did the Cu and Zn concentrations.

Tyler and Westman (1979) and Nordgren *et al.* (1986) performed soil studies in the area of the primary smelter in Rönnskär (Skellefteå region, northern Sweden), the strongest single point source of heavy metals in this country. The smelter has been emitting a series of heavy metals and other pollutants for 50 years, the dominant metals being Zn, Pb, Cu, and As, whereas Cd, Hg, V, Ni, Cr, Se are present in smaller amounts. The emissions also contain SO_2 in large amounts.

The vegetation of the area is dominated by spruce forests, frequently mixed with pine and deciduous trees.

In October 1978, Tyler and Westman (1979) have collected spruce needle mor samples from 68 sites located at 10-km distances from each other along 11 transects long of up to 80 km from the smelter. In such a way, the sites along the transects are situated circularly around the smelter.

Concentrations of 5 metals (Zn, Pb, Cu, Cd, and As), pH_{KCl}, phosphatase and urease activities, and respiration rate were determined in each sample. Then, the mean values of these variables in samples taken at sites located in 10-km intervals (0-10, 10-20..... 60-70, and 70-80 km) from the smelter were calculated. We quote only the results obtained in samples from the distances of 0-10, 20-30, and 70-80 km: metals (μg g^{-1} dry mor), i.e. Zn 352, 71, and

51; Pb 290, 78, and 39; Cu 115, 35, and 22*; Cd 9.60, 1.56, and 0.84; As 107, 53, and 32*; pH_{KCl} 3.30, 3.30, and 3.49; phosphatase activity (mg phenol g^{-1} organic dry matter of mor 2 h^{-1}) 4.69, 6.68, and 9.32; urease activity (μg NH_4^+-N g^{-1} organic dry matter 2 h^{-1}) 113, 146, and 206; respiration rate (μg CO_2-C g^{-1} organic dry matter 24 h^{-1}) 65, 94, and 138.

It is evident from these data that the metal concentrations decreased, pH increased with 0.2 units, enzyme activities and respiration rate considerably increased with increasing distance from the smelter.

The metal concentrations positively correlated with each other and negatively with phosphatase and urease activities and respiration rate. The correlation between these three indices and metal concentration was most significant in the case of Cd and Cu. The correlation of pH_{KCl} was positive and significant with phosphatase activity and respiration rate and positive but insignificant with urease activity.

As Tyler and Westman (1979) point out, elevated levels of several heavy metals in the topsoil of forested areas are measurable at great distances (at least 70-80 km) from the smelter, and the pollution has a measurable effect on enzyme activities and respiration rate in coniferous forest soils over a territory of at least 2,500-4,000 km^2.

These investigations were continued by Nordgren *et al.* (1986). In September 1981, they collected mor samples at 12 sites from a northern transect (2.4-40 km north of the smelter) and at 9 sites from a southern transect (3-55 km south of the smelter). Several chemical, microbiological, and biochemical indices were determined in the samples. The different heavy metals were 5 to 75 times higher at the site closest to the smelter compared with background levels. Despite emission of sulphur, no decrease in pH was recorded. Due to the pollutants, of which arsenic was found to be the representative one, urease activity decreased by about a factor of 4. Figure 7 shows that the lowest urease activity values were registered at the highest As concentrations on the northern transect and at the lowest As concentrations on the southern transect. At the same time, the phosphatase activity did not decrease significantly with the pollution level. The respiration rate, like urease activity, decreased by about a factor of 4, whereas the pollution-induced decrease in numbers of bacteria producing acid from or

* The Cu and As concentrations in mors even at 70-80 km from the Rönnskär smelter are higher than the normal values (Cu 8-12 ppm; As<10 ppm) registered in mors in the middle part of Norrland (northern Sweden) (Tyler and Westman, 1979).

hydrolysing various substrates was usually 8-11-fold. The mycelial lengths, like the phosphatase activity, were unaffected by the pollutants.

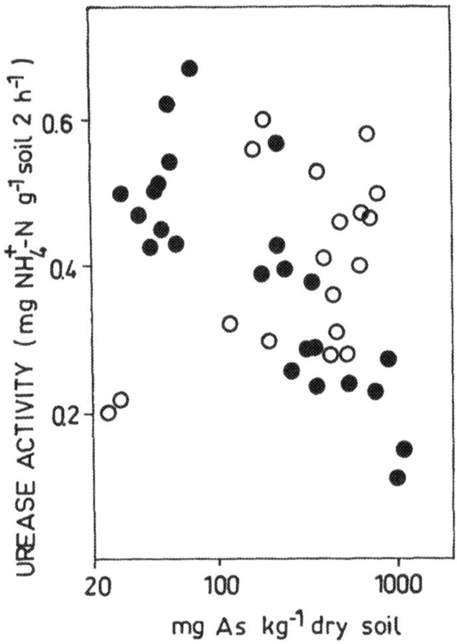

Figure 7. The effect of metal pollutants, represented by arsenic, on soil urease activity.

Redrawn from: A. Nordgren, T. Kauri, E. Bååth and B. Söderström, Environmental Pollution, Series A, 1986, Vol. 41, p. 94, with kind permission from Elsevier Science.

The full and empty circles show data from the northern and southern transect, respectively.

One can state that some results obtained by Nordgren *et al.* (1986) are in good agreement with and some other results differ from the results registered by Tyler and Westman (1979).

Enzymological research in the United States of America

Pancholy *et al.* (1975) have studied the soil factors preventing revegetation of a denuded area near an abandoned zinc smelter. First, they present a brief history of the studied area. A zinc roaster and smelter were operated at Henryetta in Okmulgee County, Oklahoma from 1916 to 1953, and the smelter was operated alone from 1953 to 1968. The area adjacent to the roaster and smelter was originally covered with post oak-blackjack oak (*Quercus stellata-Q. marilandica*) forest, but an area approximately 400 ha in extent was completely denuded downwind from the plant. Pancholy *et al.* selected a test plot of several hectares in the bare area and a control plot in a post oak-blackjack oak forest at about one mile from the test plot. Eight soil samples were taken in the test plot and eight in the control plot from the 0-15-cm depth in December 1972 and again in March 1973.

Soil dehydrogenase and urease activities were significantly lower ($P<0.01$) in the bare test plot than in the control plot, the mean values being 0.51 and 7.37 μl H 10 g^{-1} soil 24 h^{-1}, and 0.57 and 4.69 mg NH_3 g^{-1} soil 24 h^{-1}, respectively. Mean numbers of bacteria, actinomycetes, and microfungi and those of *Nitrosomonas* and *Nitrobacter* were also considerably lower in the bare than in the control soil. *Azotobacter* was lacking in both soils at both sampling periods. The weak biological potential of the bare soil is related to the high concentrations of Cd and Zn combined with the secondary effects of low pH.

Enzymological research in Poland

A seminar paper, presented by Kobus (1975), contains some data on soil investigations conducted in the area of the zinc smelter in Miasteczko Slaskie (Katowice voivodship). For analyses soil samples were taken along a transect, at 0, 0.5, 1.3, and 2.1 km from the smelter during 1972-1975. In this period, Zn concentration in soil increased from 80 to 1250 ppm at 0 km and to a lesser extent around the smelter, thus to 1/6 of these values at 0.5 km. The soil dehydrogenase and amylase activities did not exhibit considerable changes along the transect, but invertase activity increased with increasing distance from the smelter. In the total numbers of microorganisms, the annual variations were broader than the changes along the transect. Exceptionally, the number of *Azotobacter* cells was highest at 0 km.

In the area of this smelter, Badura *et al.* (1984) also performed soil chemical, microbiological, and enzymological investigations. Soil samples were collected from three pine (*Pinus sylvestris*)-dominated forest stands located at 1.4, 2.6, and 5.5-8 km from the smelter. For determination of Zn content the soils were analysed seasonally in 1979; the total numbers of microorganisms were determined monthly in the period of January 1979 – September 1980; the ammonium-oxidising capacity and cellulase activity were determined monthly in the November 1979 – May 1980 period.

The total Zn content (extractable with 8 N HNO_3) showed considerable differences in the soils of the three forest stands, the highest content being always found in the soil of the stand closest to the smelter. Contrarily, the water-extractable Zn content was of the same level in the three soils.

The total numbers of bacteria, actinomycetes, and microfungi varied to a larger extent depending on season than on the distance from the smelter. Similar results were obtained in the estimation of ammonium-oxidising capacity and cellulase activity. All these findings were explained by the uniformity of the mobile, biologically available Zn content in soils situated along a pollution gradient determined mainly by the total Zn content emitted from the smelter, which means that in these soils a large part of pollutants were immobilised and, thus, became biologically unavailable.

Greszta *et al.* (1979) performed field experiments by using dusts, collected from electrofilters operating in zinc and copper smelting works. The dusts, in which the metals are present as oxides, were mixed several times with the 0-30-cm (mellow sandy or slightly clayey sandy) soil layer of experimental plots installed in an only slightly polluted forest nursery at Klaj, in the south-eastern part of the Niepolomice Forest. Three types of dusts, namely a Pb-Zn dust (containing 6.30% Pb, 39.42% Zn), a Pb-Cu dust (7.41% Pb, 16.50% Cu), and a Cd-Pb-Zn dust (2.44% Cd, 15.29% Pb, 55.70% Zn), and a Pb-Cu sludge – wet dust extract from a copper smelter (8.68% Pb, 0.57% Cu, 3.75% Zn) were used in the following percent proportions to the soil weight: 50, 25, 10, and 1%. Plots to which no dust or sludge was added served as controls. Onto the plots, 2-year-old forest tree seedlings were planted. The plots were set up in the autumn of 1974. For enzymological and microbiological analyses soil samples were taken from the 3-5-cm depth, in June, August, and October 1975.

As expected, the soil enzyme activities studied (urease, invertase, and catalase) decreased with increasing amounts of dust or sludge additions. The only exception was the urease activity in the Cd-Pb-Zn dust-treated soil: the urease activity decreased in the soil treated with this dust in 1 and 10% proportions, but, when the soil contained 25 and 50% dust, this activity increased and even exceeded the activity value registered in the control soil. The urease activity-decreasing effect of the other heavy metal materials showed the order: Pb-Cu sludge ≈ Pb-Cu dust < Pb-Zn dust.

The reducing effect of the Cd-Pb-Zn dust on invertase activity was a little weaker than that of the other heavy metal materials.

The order in which these materials reduced catalase activity was: Cd-Pb-Zn dust < Pb-Zn dust ≈ Pb-Cu sludge < Pb-Cu dust.

The finding that the Cd-Pb-Zn dust acted less inhibitorily on the enzyme activities contrasts other findings: this dust had the most detrimental effect on soil microorganisms (excepting the ammonifying bacteria, whose number, like urease activity, increased at the highest Cd-Pb-Zn dust addition) and on the growth of tree seedlings.

In August 1980, a new experiment was initiated in the Niepolomice Forest, at 25 km north-east of Cracow (Greszta, 1988; Greszta *et al.* 1988). The plots of 240 m^2 each were installed in a mixed pine-oak forest, dominated by about 40-year-old pines (*Pinus sylvestris*) and were treated with dusts from electrofilters of six industrial plants: zinc, cadmium, and aluminium smelters, ironworks, electric power plant, and cement plant. The main components and their amounts in dusts were the following: zinc dust 22.06% ZnO, 43.74% SiO_2; cadmium dust 3.02% CdO, 4.07% PbO, 21.83% Al_2O_3, 45.34% SiO_2; aluminium dust 21.42% Al_2O_3, 43.18% SiO_2; ironworks dust 40.94% Fe_2O_3, 12.44% CaO, 20.92% SiO_2; power plant dust 10.75% CaO, 68.21% SiO_2; cement plant dust 47.04% CaO, 28.42% Al_2O_3, 16.26% SiO_2. The rate of dust application was: 0, 100, 500, 1000, 2000, and 5000 t km^{-2} $year^{-1}$ during 1 ½ years. The first quarter of the annual dose was applied on August 20-25, 1980 and the other quarters were administered at 3-month intervals.

For enzymological and microbiological analyses Zwoliński *et al.* (1988) took soil samples from the 0-5-cm layer (A_1 horizon) of the control (untreated) plot and of those treated with dusts from the cadmium and aluminium smelters, ironworks, and power plant. The samplings were carried out twice prior to the dust treatment (in May and June 1980) and 4-5 times annually in the 1980-1983 period.

The enzyme activities measured are specified in Table 5 together with the mean of the activity values recorded in the 1980-1983 period. Besides the mean values (x), the standard deviations (s), correlation coefficients (r) between the dust rate and enzyme activity, significance of correlations and number of analyses (n) are also given in Table 5.

One can see from this table that the effects of the four dusts on the six enzyme activities studied varied with the rate and source of dusts and with kind of enzymes.

At the largest rate applied, all dusts reduced activity of each enzyme. The correlation coefficients between the rates of Cd and Al dusts and the enzyme activities were always negative and significant. Asparaginase, phosphatase, and dehydrogenase activities were the most sensitive to the Cd dust, and again phosphatase and dehydrogenase activities manifested the highest sensitivity to the Al dust. The correlations of ironworks dust rates were insig-

Table 5
Soil enzyme activities in plots not treated and treated with industrial dusts

Source of dusts	Rate of dusts (t km^{-2} year^{-1})	Asparaginase* x + s	Asparaginase* r	Urease* x + s	Urease* r	Phosphatase* x + s	Phosphatase* r	β-Glucosidase* x + s	β-Glucosidase* r	Invertase* x + s	Invertase* r	Dehydrogenase* x + s	Dehydrogenase* r
Control	0	2.37±0.20		6.18±0.75		5.64±0.11		3.1±0.3		2.3±0.4		1.37±0.33	
Cadmi-	500	2.46±0.16	-0.699	5.71±0.84	-0.537	5.73±0.24	-0.868	3.0±0.5	-0.569	2.2±0.4	-0.540	1.14±0.19	-0.723
um	1000	2.26±0.31	$P<0.001$	4.31±0.96	$P=0.02$	4.97±0.36	$P<0.001$	2.7±0.6	$P=0.01$	1.8±0.4	$P=0.02$	0.76±0.25	$P<0.001$
smelter	2000	2.01±0.12	n=20	3.76±0.42	n=20	4.63±0.11	n=20	2.4±0.4	n=20	1.5±0.3	n=20	0.59±0.16	n=20
	5000	1.75±0.17		2.88±0.71		4.18±0.20		1.9±0.4		1.2±0.1		0.46±0.21	
Alumi-	500	2.41±0.10	-0.495	4.31±0.43	-0.472	5.13±0.30	-0.902	2.7±0.1	-0.405	2.2±0.3	-0.558	1.34±0.33	-0.675
nium	1000	2.27±0.21	$P=0.02$	3.55±0.38	$P=0.05$	4.66±0.28	$P<0.001$	2.2±0.1	$P=0.05$	1.9±0.2	$P=0.01$	0.95±0.18	$P=0.001$
smelter	2000	2.03±0.11	n=20	2.77±0.49	n=20	4.13±0.17	n=20	1.9±0.2	n=20	1.7±0.1	n=20	0.96±0.39	n=20
	5000	1.69±0.31		2.51±0.31		3.54±0.10		1.8±0.3		1.4±0.2		0.43±0.22	
Iron-	500	2.54±0.17	-0.352	5.71±0.25	-0.358	5.77±0.25	-0.327	3.2±0.5	-0.295	2.4±0.6	-0.380	1.53±0.37	0.336
works	1000	2.58±0.29	$P>0.05$	6.28±0.29	$P>0.05$	6.06±0.85	$P>0.05$	3.2±0.6	$P>0.05$	2.5±0.5	$P>0.05$	1.85±0.64	$P>0.05$
	2000	2.41±0.23	n=20	6.65±0.76	n=20	5.72±0.38	n=20	3.4±0.6	n=20	2.3±0.4	n=20	1.71±0.35	n=20
	5000	2.16±0.25		4.08±0.92		5.10±0.61		2.8±0.6		2.0±0.4		1.30±0.41	
Power	500	2.18±0.07	0.209	5.15±0.43	-0.361	5.66±0.49	-0.460	2.8±0.3	-0.272	2.1±0.4	0.465	1.40±0.17	0.562
plant	1000	2.45±0.16	$P>0.05$	6.08±0.34	$P>0.05$	5.88±0.34	$P=0.05$	3.3±0.7	$P>0.05$	2.6±0.4	$P>0.05$	1.54±0.31	$P=0.02$
	2000	2.57±0.17	n=16	6.12±0.89	n=20	5.63±0.58	n=20	3.7±0.7	n=16	2.8±0.6	n=16	1.74±0.17	n=16
	5000	1.80±0.15		3.80±0.57		4.39±0.24		2.1±0.3		1.8±0.3		1.17±0.10	

Reproduced from: J. Zwoliński, J. Olszowski, G. Olszowska and B. Zwolińska, Zeszyty Naukowe Akademii Rolniczej w Krakowie, Leśnictwo, 1988, No. 18, p. 120.

* Expression of enzyme activities: asparaginase and urease in mg NH_3 10 g^{-1} soil; phosphatase in mg P_2O_5 100 g^{-1} soil; β-glucosidase and invertase in ml 0.1 N $Na_2S_2O_3$ 10 g^{-1} soil; and dehydrogenase in mg triphenylformazan 10 g^{-1} soil.

nificant with all enzyme activities, whereas the correlations of power plant dust rates were significantly negative with phosphatase activity and significantly positive with dehydrogenase activity.

The mean values for the 1980-1983 period also showed that the Cd dust reduced significantly the total numbers of soil bacteria and actinomycetes and insignificantly the number of microfungi. The Al dust exerted significant effect only on the number of actinomycetes, the effect being positive. The dust from the ironworks influenced negatively and significantly only the number of microfungi. None of the microbial groups was significantly affected by the power plant dust.

The intensity of ammonification was not significantly influenced by any of the four dusts, while nitrification intensity significantly decreased under the influence of Cd dust. Intensity of cellulose decomposition was significantly reduced by the Cd, Al, and ironworks dusts, and respiration (CO_2 evolution) was significantly decreased by the Cd and Al dusts.

It is evident from these results that the effects of Cd dust were very detrimental to the soil life.

In continuation of the studies reviewed above, Starzecka (1989) has determined protease, C_x cellulase, acid and alkaline phosphatase activities in the surface soil layer (more precisely, in the 5-6-cm thick fermentation-humus layer – AoFH) of three types of experimental plots (A, B, and C) installed in the southern part of the Niepolomice Forest. Plots A (of 240-m^2 each) were installed in 1981 when they were treated with Cd dust at rates of 0, 100, 500, 1000, and 2000 t km^{-2} $year^{-1}$. In 1986-1987, half of each plot A was fertilised (720 kg NPK ha^{-1}) and limed (3000 kg magnesium-dolomite lime ha^{-1}); the fertilised and limed halves of plots A were designated plots B. In the spring of 1987, one of the 240-m^2 control plots (not treated with Cd dust in 1981) was divided into 1-m^2 "mini" plots (plots C), to which then Cd dust was added at rates proportional to those used in 1981 for plots A. All plots were sampled for soil enzymological and microbiological analyses in the period of June-October 1987. Thus, plots A and B served for assessing the long-term effects and plots C for estimating the short-term effects of Cd dust.

The analytical data have indicated that protease activity was nearly the same in all plots. This means that the Cd dust exhibited neither long- nor short-term effects on this activity which remained at an approximately constant level also in the fertilised and limed plots.

Contrarily, C_x cellulase activity sharply decreased with increasing Cd dust rate in plots A. The decrease was not so sharp in plots B treated with the two highest rates of Cd dust, which may be attributed to the fertilisers and lime. In plots C, the activity showed an increasing trend with the amount of Cd dust.

The activity of acid phosphatase was much higher than that of the alkaline phosphatase. In plots A, both phosphatase activities tended to decrease with the rate of Cd dust addition. In plots B, fertilisation and liming weakened the decreasing effect of the largest amount of Cd dust on both phosphatase activities. In plots C, the Cd dust-induced decrease of acid phosphatase activity was sharper than that of the alkaline phosphatase activity.

Due to the long-term effects of Cd dust in plots A, a considerable reduction also occurred in the total numbers of soil microorganisms (heterotrophic bacteria, actinomycetes, and microfungi) and in the intensity of soil respiration (CO_2 evolution), but the short-term effects of Cd dust in plots C were partly different: the total number of actinomycetes slightly decreased, whereas the total numbers of heterotrophic bacteria and microfungi and intensity of respiration increased.

The fertilisation and liming of plots B, as compared with plots A, resulted in increases in both microbial numbers and respiration intensity.

In conclusion one can state that the soil enzymological and microbiological effects of Cd dust are persistent as the enzymatic and microbial indices measured in 1987 showed more negative effects of Cd dust in plots A treated with Cd dust in 1981 than in plots C to which the same amounts of Cd dust were added in 1987. Some indices were even higher in the Cd dust-treated plots C than in the untreated, control plot C.

Enzymological research in Romania

The effects of the emissions from the aluminium works in Slatina (Olt county) on soil dehydrogenase activity and some microbiological indices were dealt with by Eliade *et al.* (1977) and Ionescu *et al.* (1977). The main pollutants in these emissions are fluoride, SO_2 and dust. Soil was sampled at 50, 500, and 700-1200 m from the works. Soil dehydrogenase activity, total numbers of bacteria, actinomycetes, microfungi, endosporogenic bacteria, N mineralisation and nitrification capacities were lowest in the vicinity of the pollution source.

Soreanu (1982) carried out soil chemical and enzymological studies in mixed, deciduous-coniferous forest stands heavily polluted by emissions from the lead smelter in Baia Mare (Maramureş county). Four places were selected for soil sampling: they are located at 0.1, 1, 2, and 10 km from the source of pollution. The place at 10 km is the unpolluted control.

On the polluted area, there were experimental plots installed. They were limed (5 t ha^{-1}) and fertilised with N (50 kg ha^{-1}) or P (70 kg ha^{-1}) or with both (50 kg N and 70 kg P ha^{-1}) by using ammonium nitrate or/and superphosphate. Unlimed and nonfertilised plots served for comparison.

The soils sampled in June and September 1976 and 1977 were analysed to determine their Pb content and dehydrogenase, invertase, urease, and phosphatase activities. As expected, the Pb content was much higher in the polluted than in the unpolluted area. Thus, the 0-4-cm soil layer at 0.1 km from the smelter contained 8,000 ppm Pb which was about 30 times higher than the Pb content in the 0-4-cm layer at 10 km. Of the four enzyme activities determined, dehydrogenase and urease activities were found to clearly differentiate the unpolluted soils from the polluted ones, and urease activity was also found to be a good indicator of the beneficial effects of liming and fertilisation on Pb-polluted soils in forest stands.

During 1985, Răuţă *et al.* (1988) conducted complex pedological investigations on an area of about 36 km^2 around the Zn-Pb smelter in Copşa Mică (Sibiu county). Nine profiles were described and 42 litter and soil samples were taken for analyses. In the most polluted zone, the concentrations of heavy metals in samples of the litter and the 0-3-cm soil layer were the following: Zn $>$ 1800 and 370 ppm; Pb $>$ 1300 and 200 ppm; Cu $>$ 100 and 25 ppm; Cd $>$ 50 and 10 ppm, respectively. No dehydrogenase activity could be detected in these heavily polluted samples. In both litter and soil, numbers of bacteria decreased, whereas those of cellulolytic microfungi increased with increasing concentrations of heavy metals.

Enzymological research in Canada

Freedman and Hutchinson (1980a, b) have reported on extensive studies of the pollutants and their effects on soils and vegetation in the area of a large Ni-Cu smelter at Sudbury, Ontario. In this area mining and smelting activity began about 1885. Initially, ore was roasted using open roastbeds which, by 1928, was forbidden, and all further roasting was carried out at three smelter facilities located near Sudbury (at Copper Cliff, Coniston, and Falconbridge). Since 1972, all smelting activity is centred either at Copper Cliff, where pollutants are vented through a 380-m "superstack" (the world's tallest smokestack) or at Falconbridge with a 93-m stack. The Coniston smelter was closed in 1972, with all production transferred to Copper Cliff. The stacks emite huge amounts of SO_2 and fine particulates, mostly iron oxides, although Cu and Ni emissions are also significant.

Along a 60-km transect originating at the Copper Cliff smelter, the aerial deposition of smelter-derived pollutants was estimated, foliar analyses as well as soil chemical, microbiological, enzymological, and faunal studies were carried out in the 1975-1977 period.

It was found, for example, that the foliage of the 17 plant species studied had consistently higher concentrations of Cu and Ni at sites closer to the smelter, relative to sites further away. Differences of 10- to 50-fold occurred between the 60-km site and those at 3.0 or 1.6 km for both foliar Cu and Ni. Although the within-site variation was high, Cu and Ni and (to a lesser extent) sulphate were also markedly elevated in surface soil horizons close to the smelter. Differences of up to 60-, 30-, and 11-fold for Cu, Ni, and sulphate, respectively, occurred between sites at 76.5 km from the smelter and those at <5 km.

Table 6 comprises data on total Cu and Ni concentrations, respiration rate, and acid phosphatase activity in duff-humus soil samples collected from sites at various distances from the Copper Cliff smelter, on June 2, 1977. Duff is defined as the material consisting of recognizable leaf fragments, and humus as the amorphous organic matter. The mean of the values recorded for 10 replicate collections per site and standard deviation are presented.

It is evident from this table that both respiration rate and acid phosphatase activity were lowest in the mineral soil at the denuded site and lower in the duff-humus soils at the contaminated than at the control forest site. The correlation coefficients calculated using the data of the table showed a stronger inverse relationship of both respiration rate and acid phosphatase activity with the total copper than with the total nickel concentration. The respiration rate and phosphatase activity also significantly correlated.

Chemical, microbiological, and enzymological studies in the area of another Canadian smelter were performed by Dumontet *et al.* (1992). This, Cu-Zn smelter is located on the north-east side of the city of Rouyn-Noranda (south-western Québec). The soil sampling was carried out in May 1986 along two transects oriented in the direction of the prevailing winds; one transect is in north-north-east (NNE) and the other in east-south-east (ESE) direction,

Table 6
Total copper and nickel concentrations, respiration rate, and acid phosphatase activity in duff-humus soil samples collected in sites at different distances from the Copper Cliff smelter

Nature of site	Distance of site from the smelter (km)	Total ppm dry weight		Respiration rate*		Acid phosphatase activity**	
		Copper	Nickel	per fresh weight	per dry weight	per fresh weight	per dry weight
Denuded (devegetated) hilltop	3.0	210±110	220±200	0.06±0.02	0.07±0.02	8.5±10.0	9.2±10.8
Contaminated forest site	3.0	2400±700	1400±400	0.18±0.09	0.26±0.14	41.6±16.2	63.9±19.3
Contaminated forest site	3.7	2600±800	1900±1400	0.31±0.16	0.58±0.38	46.2±19.3	78.5±44.7
Contaminated forest site	5.2	1400±400	1200±400	0.38±0.17	0.72±0.40	73.9±58.5	144.8±114.0
Control forest site	34.3	470±170	590±190	0.59±0.19	0.98±0.38	191.7±47.7	316.5±82.4
Control forest site	47.8	450±100	460±130	0.51±0.15	0.94±0.32	105.5±49.3	205.6±114.7
Control forest site	50.2	230±70	270±100	0.39±0.15	0.62±0.28	180.2±75.5	248.7±122.4

Reproduced from: B. Freedman and T.C. Hutchinson, Canadian Journal of Botany, 1980, Vol. 58, p. 1728.

* Expressed in µl of CO_2 min^{-1} g^{-1} fresh or dry weight soil.
** Expressed in µg of *p*-nitrophenol 20 min^{-1} g^{-1} fresh or dry weight soil.

from 1 to 27.5 and from 2 to 42.5 km away from the smelter. Nine sites (four on the NNE and five on the ESE transect) were selected. The samples were taken from the surface litter and the 0-15- and 15-30-cm mineral soil layers. Concentrations of Cu, Zn, Ni, Cd, Pb, and total heavy metals (TM) were determined in the litter and both soil layers, whereas respiration (CO_2 evolution) rate and acid phosphatase activity were measured only in the 0-15-cm layer, and expressed in mg CO_2-C kg^{-1} dry soil day^{-1} and μmol *p*-nitrophenol g^{-1} dry soil h^{-1}, respectively.

At each site, the concentration of TM decreased in the order: litter > 0-15-cm layer > 15-30-cm layer. The maximum amount of TM (29.5 meq 100 g^{-1} dry matter or 13,000 ppm) was recorded in the litter of the site at 1 km away from the smelter on the NNE transect, whereas the less polluted litter (0.38 meq TM 100 g^{-1} dry matter) was collected at 27.5 km from the smelter on the ESE transect. However, little correlation was found between heavy metal concentrations in the litter and the distance from the smelter. At the same time, in the 0-15-cm soil layer the log-concentration of individual heavy metals and TM decreased significantly ($P < 0.05$ or 0.01) with the log-distance from the smelter; the only exception was Ni as its decrease was not significant.

Table 7 shows that the correlations among heavy metal concentrations were always

Table 7
Correlations between heavy metal concentrations, respiration rate, and acid phosphatase activiy in the 0-15-cm soil layer

	Zn	Cd	Pb	Total metals	Respiration rate	Acid phosphatase activity
Cu	0.449	0.856**	0.929**	0.938**	-0.790*	-0.223
Zn		0.519	0.726*	0.725*	0.076	-0.524
Cd			0.895*	0.874**	-0.856**	-0.114
Pb				0.993**	-0.690	0.041
Total metals					-0.647	0.041
Respiration rate						0.388

Reproduced from: S. Dumontet, H. Dinel and P.E.N. Lévesque, The Science of the Total Environment, 1992, Vol. 121, p. 242, with kind permission from Elsevier Science NL, Amsterdam.

Significance: * $P < 0.05$; ** $P < 0.01$.

positive and, usually, significant. The respiration rate was negatively affected by the heavy metals with the exception of Zn, but only the effects of Cu and Cd were significant. The acid phosphatase activity was negatively, but insignificantly influenced by Cu, Zn, and Cd concentrations and was not affected at all by Pb and total metals. The correlation between respiration rate and acid phosphatase activity was positive, but insignificant. This activity correlated significantly ($P < 0.05$) only with the content of both soil C and N. One can deduce from these findings that in the area of the Rouyn-Noranda smelter the soil respiration rate indicated more sensitively the heavy metal pollution than did the acid phosphatase activity.

Enzymological research in the Russian Federation

Skvortsova *et al.* (1980) have determined asparaginase activity in brown forest soil samples collected at 0.6, 2.6, and 3.4 km from a lead smelter located in the Primorye area. The samples at the three distances contained 300, 295, and 100 mg Pb, 215, 175, and 95 mg Zn, 17, 0.9, and 0.5 mg Cd kg^{-1} soil. The asparaginase activity was 5-6 times lower at 0.6 km than at 2.6 or 3.4 km from the smelter. Microbial numbers showed a trend similar to that of asparaginase activity.

In the 1976-1993 period, Evdokimova (1982, 1989, 1992, 1995) and Evdokimova *et al.* (1984) have performed extensive investigations in the area of the "Severonikel'" Ni-Cu smelter complex, located in the northern taiga, near the town of Monchegorsk (Kola Peninsula). The area is dominated by adverse climatic conditions. The emissions contain mainly Cu, Ni, and Co (in form of sulphates, chlorides, sulphides, oxides) and SO_2. Due to their huge amounts, these pollutants affect large territories, within which four zones were selected for the investigations: the epicentre, the impact zone ("technogenic desert"), the buffer zone, and the background zone.

In each zone, not only the virgin podzolic forest soils (already affected by the smelter emissions for many decades) were investigated, but experimental plots were also installed.

In one of the experiments, the surface (0-30-cm) layer of the soil plots installed in the four zones mentioned above, namely at 0.2-1 km (epicentre) and at 5, 15, and 60 km from the smelter, was removed and replaced with cultivated mineral soil, at the end of May 1976. After 10 years, more precisely on June 3, 1986, 1-m^2 30-cm deep microplots taken from the plots were reintroduced to a "pure", unpolluted area located at 60 km from the smelter. The 0-10-cm layer of the microplots was sampled on the day of reintroduction and, thereafter, periodically, for determination of the Cu, Ni, and Co contents, invertase and urease activities, nitrification capacity, and other chemical and microbiological parameters. Table 8 comprises some of the results obtained during the first 15 months following reintroduction of microplots to the unpolluted area.

Table 8
Dynamics of the heavy metal contents, enzyme activities, and nitrification capacity in the 0-10-cm layer of 30-deep microplots of initially unpolluted podzolic soils after their exposure to the effects of emissions at different distances from the smelter for 10 years, followed by their reintroduction to an unpolluted area

Sampling date of the reintroduced soil	Distance at which the initially unpolluted soil was exposed to the effects of smelter emissions for 10 years			
	Epicentre	5 km	15 km	60 km
	Copper content (mg kg^{-1} soil)			
3.VI.1986	6232	86	70	49
23.IX.1986	3347	62	51	42
4.VI.1987	3603	84	57	34
3.IX.1987	3176	81	52	35
	Nickel content (mg kg^{-1} soil)			
3.VI.1986	2106	126	97	62
23.IX.1986	1004	137	83	53
4.VI.1987	1201	147	79	47
3.IX.1987	1003	139	95	43
	Cobalt content (mg kg^{-1} soil)			
3.VI.1986	84	18	18	9
23.IX.1986	50	13	13	13
4.VI.1987	62	17	18	13
3.IX.1987	46	15	16	17
	Invertase activity (mg glucose g^{-1} soil)			
3.VI.1986	0.6	12.0	9.6	10.3
23.IX.1986	0.9	11.3	10.5	14.4
4.VI.1987	0.8	17.4	13.1	12.5
3.IX.1987	1.2	10.2	9.8	8.5
	Urease activity (mg NH_4^+ 2 g^{-1} soil)			
3.VI.1986	0	2.9	2.5	2.2
23.IX.1986	0.2	2.2	2.2	1.8
4.VI.1987	0.3	2.3	3.2	2.3
3.IX.1987	0.2	3.3	2.8	1.9
	Nitrification capacity (mg NO_3-N kg^{-1} soil)			
3.VI.1986	0	35	41	31
23.IX.1986	0	40	53	46
4.VI.1987	0	50	55	56
3.IX.1987	0.5	40	40	40

Reproduced from: G.A. Evdokimova, Plodorodie Pochv Murmanskoi Oblasti, Kol'skii Nauchnyi Tsentr, Apatity, 1989, pp. 76 and 79.

It is evident from this table that during 10 years high amounts of Cu, Ni, and Co accumulated in the soil of epicentre and the accumulation decreased with increasing distance from the smelter. After reintroduction of soils to the unpolluted area, the heavy metal contents in the soil from the epicentre decreased cosederably in 3.5 months, but during the next ~12 months, decrease of heavy metal contents was much smaller*. These findings are explained by the migration of mobile forms of heavy metals from the surface to the deeper soil layers.

In the soils from the 5-, 15-, and 60-km distance, very little changes occurred in their metal contents during the whole 15-month period.

Table 8 also shows that in the soils exposed to the effects of smelter emissions in the epicentre, the invertase activity was very low and the urease activity and nitrification capacity were not demonstrable. In the epicentre soil reintroduced to the unpolluted area, invertase activity increased during the 15-month period, urease activity became detectable after 3.5 months, whereas a weak nitrification capacity was demonstrable after 15 months. The soils from the 5-, 15-, and 60-km distance manifested both enzyme activities and nitrification capacity which remained nearly at the same level after reintroduction of these soils to the unpolluted area.

In the 1988 1993 period, the invertase and urease activities, and nitrification capacity showed a trend to further increase in the soil reintroduced from the epicentre and exhibited only small changes in the soils reintroduced from the 5-, 15-, and 60-km distance (Evdokimova, 1995).

Other soil microbiological indices such as the numbers of bacteria and actinomycetes, bacterial and fungal biomass, species diversity of endosporogenic bacteria, microfungi, and algae, respiration (CO_2 evolution), decomposition of cellulose and plant residues, production of free amino acids, N_2 fixation (acetylene reduction), like enzyme activities and nitrification capacity, were negatively affected by the smelter emissions (Evdokimova, 1982, 1989, 1992, 1995; Evdokimova *et al.* 1984).

It has also been proved that farmyard manuring of soils affected by smelter emissions led to some diminution of the heavy metal contents in the surface layer, which was attributed to the formation of soluble heavy metal compounds with organic components of the farmyard manure and to the migration of these compounds from the surface to the deeper soil layers (Evdokimova, 1982; Evdokimova *et al.* 1984).

---

* The decrease remained extremely low in the 1988-1992 period, too (Evdokimova, 1995). It was deduced from other studies (Evdokimova, 1992, 1995) that complete decontamination from Cu and Ni would take place in 42-700 and 21-190 years, respectively.

Kovalenko (1996) and Kovalenko *et al.* (1996) have analysed, enzymologically and microbiologically, a series of litter and soil samples taken in pine plantations around the Cu smelter in Sredne-Uralsk. Three experimental plots were selected for sampling litter (01 and 02 horizons) and soil (A_1 horizon). Two of them, installed at 2 and 6 km from the smelter, are affected by the pollutants (S-containing compounds and heavy metals) present in the smelter emissions. The third plot, at 24 km from the smelter, served as a control.

Cellulase and catalase activities were lower, while polyphenol oxidase activity was higher in the contaminated plots than in the control plot. Thus, the sum of activities in the three horizons analysed (01 + 02 + A_1) was 1.3 and 6.5 times lower (cellulase), 3.2 and 5 times lower (catalase), and 1.3 and 1.5 times higher (polyphenol oxidase) in the plots at 6 and 2 km from the smelter, respectively, than in the control plot.

Cellulase activity strongly correlated with the global number of microorganisms, whereas polyphenol oxidase activity correlated with the phenol content in litter and soil.

The specific case of sulphate reductase activity should also be mentioned. In summer, the pollutants reach even the control plot, increasing its sulphate content. The sulphate induces the microbial synthesis of sulphate reductase and, thus, the activity of this enzyme increases in the litter and soil. Due to the increased activity, the sulphate content is continuously diminished and in the autumn no sulphate is detectable in the litter and soil. As in the absence of sulphate, i.e. of the inducing substrate, new sulphate reductase molecules are not synthesised by the microorganisms, the enzyme activity decreases in autumn. This finding shows that the sulphate reductase enzyme molecules produced by the microorganisms in summer did not accumulate, did not become resistant to inactivation and degradation.

The polluted plots always contain much sulphate, but their sulphate reductase activity is low. This means that some components of the smelter emissions inhibit the sulphate-inducible microbial synthesis of sulphate reductase.

There was a significant correlation found between sulphate reductase activity and number of the anaerobic bacteria belonging to the *Clostridium* genus.

In a similar study, Kovalenko *et al.* (1996) have determined catalase activity in the litter (01 and 02) and soil (A_1) samples taken from plots in pine plantations at 4, 5, 6, and 11.5 km from the Blagoslovo aluminium works emitting fluoride and S-containing compounds to the atmosphere. Pollution led to decreased catalase activity in these samples, too.

Enzymological research in Kazakhstan

Skvortsova *et al.* (1980) have established that in close vicinity of a lead smelter (Chimkent district, southern Kazakhstan) asparaginase activity in the polluted common serozem (containing 1000 mg Pb, 400 mg Zn, 50 mg Cu, 10 mg Cd kg^{-1} soil) suffered a 3.25-fold diminution compared with the unpolluted soil. Numbers of microorganisms were also lower in polluted rather than unpolluted soil.

The area of a metallurgical plant located in northern Kazakhstan was also studied from soil enzymological and microbiological viewpoints (Aristovskaya and Chugunova, 1989). Samples were taken from chernozemic soils under wheat, along a 30-km transect. The soil urease activity was twice and the respiration (CO_2 evolution) was 1.5 times as high at 30 km from the plant as in its vicinity. Intermediary values were registered at distances of 2.5, 5, and 10 km from the plant.

Enzymological reseach in Switzerland

Wallis (Rhonetal), where Polomski (1985) has performed soil studies, is a meteo-rologically nearly closed inner alpine valley which during many decades had been affected by fluoride-containing emissions originating from an aluminium smelter. The fluoride emission has been controlled efficiently only since 1981. In 1983, Polomski has sampled the 0-6-cm layer of soils from eight spruce forest and ten grassland sites located at different distances from the smelter. All samples were analysed chemically, but only the forest soil samples were submitted to enzymological analyses. The coefficients of correlations between the analysed parameters were calculated.

The content of water-soluble fluoride decreased with increasing distance from the smelter and correlated significantly and negatively with pH and $CaCO_3$ content and positively with ammonium acetate-extractable Al content. The phosphatase and dehydrogenase activities increased and β-glucosidase activity decreased with the distance from the smelter; more precisely, there were negative significant correlations between the fluoride content and phos-phatase and dehydrogenase activities (r = -0.90 and r = -0.83, respectively), whereas the fluoride – β-glucosidase correlation was positive, but less significant (r = +0.61).

Enzymological research in Austria

Pohla *et al.* (1991) and Kandeler *et al.* (1992) described multidisciplinary investigations in the area of a copper smelter in Brixlegg (located on the Inntal Lowland, Federal Region of Tirol). Along a pollution gradient, four grassland sites (each of about 1000 m^2) were studied. They are situated at 300 m (site A, most polluted), 1125 m (site B), 2425 m (site C), and 5900 m (site D, unpolluted) from the smelter. Samples were taken from the 0-5- and 5-10-cm soil depths in September 1990 and May 1991. The results obtained at the two sampling periods were similar.

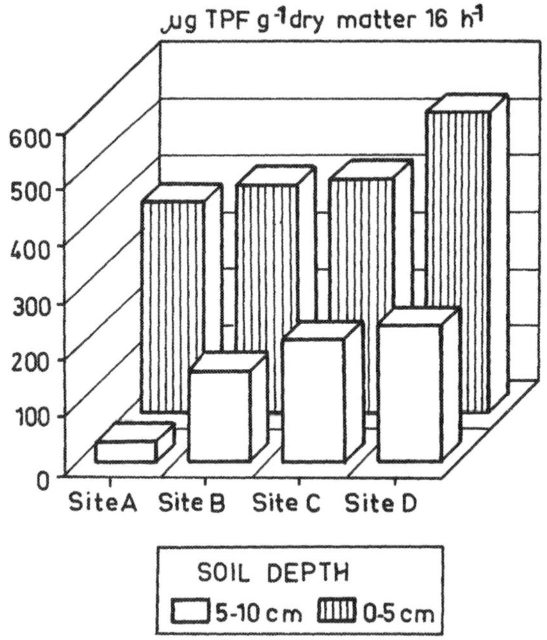

Figure 8. Influence of heavy metal pollution on soil dehydrogenase activity at four grassland sites.

Redrawn from E. Kandeler, G. Lüftenegger and S. Schwarz, Zeitschrift für Pflanzenernährung und Bodenkunde, 1992, Vol. 155, p. 320.

TPF – Triphenylformazan.

The concentration of each heavy metal determined (Zn, Cu, Pb, Ni, Cd) decreased with increasing distance from the smelter. At site A, the 5-10-cm layer contained more heavy metals (excepting Cd) than the 0-5-cm one, whereas at the other sites both layers contained nearly the same amounts of heavy metals. In the two layers of the four sites the mean heavy metal concentrations (mg kg^{-1} dry soil) were the following: Zn 3138, 856, 228, and 96.7; Cu 2746, 664, 131, and 45.7; Pb 1530, 225, 50.8, and 36.8; Ni 72.4, 48.4, 23.8, and 35.3; Cd 15.9, 3.8, 1.2, and 0.4, respectively.

As shown in Figure 8, dehydrogenase activity was higher in the 0-5-cm than in the 5-10-cm layer at each site and increased with decreasing heavy metal pollution (from site A to site D). The increase was insignificant in the 0-5-cm layer and significant ($P < 0.01$) in the 5-10-cm one.

At site A, as compared with site D, other enzyme activities (protease, alkaline phosphatase, and arylsulphatase) as well as basal and substrate (glucose)-induced respirations and nitrification were also significantly ($P < 0.001$) inhibited. At the same time, other enzyme activities (cellulase, xylanase, and β-glucosidase) were significantly higher, whereas the urease activity and N mineralisation were insignificantly higher at site A than at site D.

It was also found that organic fertilisation of site A led to the diminution or even to the elimination of the inhibitory effects of heavy metals in the 0-5-cm layer. Heavy metal pollution exerted negative effects on the soil fauna, too.

The conclusion was drawn that the microbial parameters and enzyme activities measured in the 5-10-cm soil layer are suitable for bioindication of heavy metal pollution.

Finkernagel and Schinner (1986) have compared a forest site in the Brixlegg area with a less polluted forest site near another locality in Tirol, Terfens. The enzyme activities (dehydrogenase, urease, phosphatase, and xylanase) and the respiration (CO_2 evolution) rate were found to be significantly lower in the polluted than in the less polluted forest soil.

On both sites, four 4-m^2 experimental plots were installed for comparing the effects of liming and acid treatment on the enzyme activities and the respiration in the two soils. The plots received the following treatments: 1 – control (no treatment); 2 – 1 kg lime; 3 – 20 l sulphuric acid solution of pH 3.2; 4 – 1 kg lime + 20 l sulphuric acid solution of pH 3.2. During the 7-month experimental period, lime was applied only once (at the beginning of the experiment), whereas the acid treatment was carried out at the beginning and still 3 times at regular intervals.

The liming resulted in increased pH, dehydrogenase activity, respiration rate, in decreased phosphatase activity and in little changes in urease activity in both soils. The very low xylanase activity in the polluted soil remained at the same low level after liming, too. In contrast, the liming of the less polluted soil led to the decreased xylanase activity during the first two months, then the activity tended to increase.

The acid treatment brought about minor changes in pH, dehydrogenase and urease activities in both soils. The respiration was stimulated in both soils. The phosphatase and xylanase activities remained unchanged in the polluted soil, whereas in the less polluted soil phosphatase activity reached, after an initial decrease, the level of the control, and xylanase activity increased after the first and third acid treatments, then decreased to the level of the control.

The liming plus acid treatment, as compared with liming alone, induced a loss in activities.

Ironworks

Enzymological research in Sweden

Rühling and Tyler (1979) have carried out enzymological studies in the areas of two ironworks (at Avesta and Trollhätten), both emitting chromium as quantitatively most important pollutant element. The Avesta emissions also contain large amounts of Mo and Ni and small amounts of V, Cu, Co, Pb, Mn, Cd, and Zn. At Trollhätten, Cr and also Ca are major, whereas other metals (Mo, Ni, V, Cu, Mn) are minor components of the emissions.

Both areas are dominated by spruce forests. Needle mor was sampled at 60 sites in the Avesta area in October 1977 and at 88 sites in the Trollhätten area in April 1978. The sampling sites were selected in such a way that all parts of the metal gradient were well represented.

The soil phosphatase activity and the respiration (CO_2 evolution) rate were determined in both areas and, in addition, the urease activity and the N mineralisation rate were also estimated in the Trollhätten area.

In the Avesta mor samples degree of pollution (expressed as Cr concentration) significantly and negatively correlated with both phosphatase activity and respiration rate. The correlations were strongest at higher pH levels.

In the Trollhätten mor samples degree of pollution significantly and negatively correlated with phosphatase activity only at pH levels >4.2 and with urease activity only at pH levels >3.5. At the same time, no relationship was discernible between the degree of pollution and respiration or N mineralisation rate. But these two microbial indices showed close correlation with pH, the correlation being positive with the respiration rate and negative with the N mineralisation rate.

At comparable degrees of pollution, the enzymatic and microbial processes were affected to a larger extent in the Avesta than in the Trollhätten mors.

The stepwise regression analysis indicated that the Mo component of the emissions from both metallurgical plants had a greater negative effect on the enzymatic and microbial processes than the Cr component.

Enzymological research in Poland

The studies of Zwoliński *et al.* (1988) on the soil enzymological and microbiological effects of ironworks dust have already been mentioned on pages 79-81.

Ore enrichment works

Enzymological research in Sweden

Rühling (1981) has performed soil chemical, microbiological, and enzymological studies in the area of the lead mine in Laisvall. The underground mining of lead ore (PbS, galena) began here in 1943. The direct contribution of mining operations to pollution of air with fine particles of PbS and also of other heavy metal (Zn, Cu, Cr, Ni, Ag) minerals was small. The more important source of air pollution was the ore enrichment works as the emissions resulted mostly from the drying process.

During the 36th week of 1979, 80 mor samples were taken in the surrounding coniferous forests at different (up to 10-km) distances from Laisvall. Pb content in the samples varied between 11 and 10,900 ppm (this very high value was registered in a sample taken from the close vicinity of the mine). Around the mine, on an about 2-km^2 territory, the Pb content was higher than 1000 ppm. At 3 and 5 km from the mine, the Pb content was still higher than 500 and 100 ppm, respectively. The contents of other heavy metals were also highest in the vicinity of mine.

The Pb and other heavy metal contents very significantly ($P < 0.001$) correlated with each other. The acid phosphatase and urease activities, and respiration rate also correlated ($P < 0.001$) with each other, but not with the Pb content. The lack of correlation suggests that the lead despite of its high content in the polluted mor is, biologically, nearly completely unavailable which should be evenly valid for the Pb in form of PbS and the Pb strongly bound to organic matter.

The correlation of Cr content was negative and significant ($P < 0.01$) with phosphatase activity and respiration rate, and negative, but insignificant with urease activity. At the same time, each of these three indices gave significant ($P < 0.001$) positive correlation with soil pH and Ca content.

Enzymological research in the Russian Federation

The ore enrichment works at Kostomuksha (northern Karelia) emits to the atmosphere heavy metals and sulphur causing pollution of the surrounding boreal forests. The most polluted area around the works (impact zone) has a radius of 6-8 km. The heavy metal and sulphur pollutions affect, to a lesser extent, broader areas up to 10 and 50-70 km from the works, respectively (Zagural'skaya, 1992; Zagural'skaya and Zyabchenko, 1994). These investigators have described soil microbiological and enzymological studies carried out during the first 9 years since the works has been operating. At three forest sites on podzolic soils located at 0.5 and 5 km from the works (impact zone) and at 23 km (considered as background zone), samples were collected from the litter layer (medium thickness: 4-5 cm) and from the upper, 5-12-cm thick layer of the mineral soil.

The results of these long-term investigations have shown that only the litter microorganisms and enzymes were evidently affected by the emissions and these effects appeared rather stimulatory even in the 9th year of the works. Thus, the mean values and variation ranges of the numbers of ammonifying, mineral nitrogen-assimilating, oligonitrophilic, oligotrophic, and cellulolytic bacteria, as well as those of the total numbers of actinomycetes and microfungi were higher in the litter of the impact zone than in the background zone. The intensity of cellulose decomposition, ammonification, and production of free amino acids were also more pronounced in the impact than in the background zone. The catalase activity behaved similarly, but the protease and urease activities remained at the same level in the two zones.

The conclusion drawn from these findings was that the heavy metals from the emissions, during the first 9 years of the works, played the role of microelements (trace elements) in the nutrition of litter microorganisms. How long will this stimulatory effect of emissions on the life of litter microorganisms last? For elucidating this problem further investigations are needed.

Coking plants

The emissions from coking plants contain a large variety of organic and inorganic substances (e.g. phenol, thiocyanates, SO_2, H_2S, NH_3, pyridine) bound or unbound to soot (Dolgova, 1973, 1975, 1978a, b; Dolgova and Kuchma, 1983; Killham and Wainwright, 1981; Wainwright and Killham, 1982; Kuchma and Grishko, 1989).

Enzymological research in Ukraine

The enzymological study of soils affected by emissions from coking plants was initiated by Dolgova (1973). First, she studied phenol oxidase and peroxidase activities in such soils; later, other soil enzyme activities were also dealt with. The enzyme activity measurements were accompanied by microbiological analyses. For the remediation of such soils, biotechnologies were tested and evaluated; for evaluation, enzyme methods were also applied (Dolgova, 1973, 1975, 1978a, b; Dolgova and Kuchma, 1983; Kuchma and Grishko, 1989). All these studies will be summarised below.

A polluted tree plantation growing at 100-150 m from a coking plant was compared with a similar plantation on an unpolluted area. The dominant tree species were small-leaf lime (*Tilia cordata*), horse chestnut (*Aesculus hippocastanum*), and common lilac (*Syringa vulgaris*). In both areas, the soil is of the same type (chernozem). During the vegetation period (April-November), samples were taken from the rhizospheric and nonrhizospheric soil of each tree species for determination of phenol concentration, phenol oxidase and peroxidase activities, and number of phenol-oxidising microorganisms (capable of growing on mineral media to which phenol was added as sole carbon and energy source). All determinations gave higher values in the polluted than in the unpolluted soil under each tree species. The pollutant phenol enhanced the growth of phenol-oxidising microorganisms and induced the microbial synthesis of phenol oxidase. As expected, most enzyme and microbial values were higher in rhizospheric than nonrhizospheric soil.

Concerning the other soil enzyme activities studied, it was found that ascorbate oxidase activity increased, but other, both oxidoreductase (dehydrogenase, catalase, nitrate reductase, nitrite reductase) and hydrolase (invertase, urease, protease) activities decreased in the polluted soil as compared with the unpolluted one. For exemplification, Table 9 presents soil dehydrogenase activity values measured in the polluted and unpolluted soils during the vegetation period.

Table 9
Dynamics of soil dehydrogenase activity in polluted and unpolluted plantations, during the vegetation period

Tree species	Sampling area	Dehydrogenase activity (mg TPF 100 g^{-1} soil)			
		April	June	August	November
Small-leaf lime	Polluted	0.182	0.060	0.230	0.310
(*Tilia cordata*)	Unpolluted	0.220	0.155	0.783	0.760
Horse chestnut	Polluted	0.243	0.087	0.035	0.105
(*Aesculus hippocastanum*)	Unpolluted	0.428	0.108	0.760	0.240
Common lilac	Polluted	0.120	0.137	0.093	0.150
(*Syringa vulgaris*)	Unpolluted	0.253	0.255	0.835	0.800

Reproduced from: L.G. Dolgova, Pochvovedenie, 1975, No. 4, p. 115.

Significant negative correlations were registered between the soil dehydrogenase activity and the soil concentrations of both phenols and thiocyanates.

For the remediation of the polluted soil, two biopreparations (cultures of the bacteria *Pseudobacterium lacticum* strain 392 and *Pseudomonas liquefaciens* strain 399) were tested. Experimental plots were set up in the vicinity of the coking plant. Farmyard manure (7 t ha^{-1}) and mineral fertilisers (90 kg N, 60 kg P, and 90 kg K ha^{-1}) were added to the plots, then they were sown with bluegrass (*Poa pratensis*) seeds previously treated with strain 392 or 399 (at a rate of 2 g biopreparation for 20 g seeds). Seeds not submitted to the bacterial treatment served for comparison. The phenol oxidase activity, number of phenol-oxidising microorganisms, and phenol-degrading capacity of soil were determined during the vegetation period. Each of these three indices increased in the bacterial treatments as compared with the untreated control. In addition, the bacterial preparations exhibited an increasing effect on soil catalase and invertase activities, too. Strain 399 was always more efficient than strain 392.

In another remediation experiment, the plots were fertilised with complex NPK (500 kg ha^{-1}), sown with bluegrass or orchard grass (*Dactylis glomerata*) and wetted with water or Na humate solution (at a rate of 0.01% Na humate on NPK weight basis). The control plots received no NPK and no humate. During the vegetation period, phenol oxidase and peroxidase activities in the soil of plots showed the following order: NPK + humate > NPK > control. Number of phenol-oxidising microorganisms and phenol-degrading capacity were also highest in the humate-treated plots.

In both remediation experiments, a more vigorous growth of plants in the bacterial and humate treatments could also be observed; the improvement in growth was of 20-30%.

Enzymological research in the United Kingdom

Killham and Wainwright (1981) performed investigations in the area of a coking plant at Chapeltown, South Yorks, England. Soil (brown earth podzol) was taken from beneath the canopies of sycamore (*Acer pseudoplatanus*) trees growing 500 m downwind of the coking plant. Polluted sycamore leaves are covered with a thin layer of black, atmospheric pollution deposit (APD) largely composed of soot. Such leaves were randomly picked from trees; APD was collected from their surface by dry brushing, then dry-sieved to remove leaf debris and aphids, and used in a soil incubation experiment. Unsterilised (native) and sterilised (autoclaved) 200-g samples of the soil taken from beneath the canopies of sycamore trees were amended with sterilised APD (1% weight/weight) and the sterilised samples were, in addition, inoculated with *Fusarium solani* spores. Unsterilised samples left unamended were the controls. All samples (containing 30% moisture) were incubated under aerobic conditions, at 25°C for 30 days, after which they were submitted to enzyme and other analyses.

Table 10 shows that in the unsterilised, APD-amended soil, as compared with the control soil, arylsulphatase activity markedly increased during incubation. The increase was lower in the sterilised, APD-amended and inoculated soil. The activity increase was accompanied by increases in the concentrations of LiCl-extractable sulphur oxyanions: thiosulphate ($S_2O_3^{2-}$), tetrathionate ($S_4O_6^{2-}$) and, particularly, sulphate (SO_4^{2-}). Based on these findings, it has been assumed that APD contains both elemental and reduced forms of sulphur and organic sulphur (ester-sulphate) and, during the incubation of soil, microbially and enzymatically mediated oxidation and mineralisation processes were taking place. Due to these processes, the APD reaching the topsoil below sycamore canopies can give rise to S-containing ions more available to microorganisms and plants. In other words, the oxidation and mineralisation processes may result in decontamination of APD-polluted soils.

Table 10
Arylsulphatase activity in unamended and APD-amended soil samples after their aerobic incubation at 25°C for 30 days

Soil treatment	Arylsulphatase activity (μg *p*-nitrophenol g^{-1} soil h^{-1})
Unsterilized, unamended	83.53
Unsterilized, amended with APD	142.51
Sterilized, amended with APD and inoculated with *F. solani* spores	94.60
Sterilized, amended with APD	0

Reproduced from: K. Killham and M. Wainwright, Journal of Applied Ecology, 1981, Vol. 18, p. 894.

Enzymological aspects of these investigations were referred to in another paper, too (Wainwright and Killham, 1982).

Refractory brickworks

Enzymological research in the United Kingdom

The emissions from refractory brickworks, like those from coking plants, contain reduced forms of sulphur. Wainwright (1979, 1982) described studies on the effect of brickworks emissions on the activity of one of those soil enzymes that participate in the biological cycle of S. The studied enzyme, rhodanese catalyses the conversion of $S_2O_3^{2-}$ + CN^- to SO_3^{2-} + SCN^-, an early reaction in the microbial oxidation of $S_2O_3^{2-}$ to SO_4^{2-} in soils.

Polluted and unpolluted soils were sampled at Loxley, Sheffield, England. The polluted samples were collected from a wooded area situated 500 m downwind of a refractory brickworks. Leaves of the predominant tree species (sycamore) growing in this area showed signs of SO_2-induced damage and were generally covered with a thin layer of soot. The unpolluted samples were obtained from a similar wooded area 2500 m upwind of the brickworks. Here no signs of SO_2-induced leaf injury were seen. The vegetation and soil type on both polluted and unpolluted areas were similar. Soil was sampled below and at a distance of 5 m from the canopy of sycamore.

Sampling depths and the numerical results (and standard deviations) obtained in determination of rhodanese activity are presented in Table 11. This table shows that the polluted soil samples consistently had the highest rhodanese activities. In the top 6 cm, the greatest activities were found in soil samples taken from below the canopy of polluted sycamore. Below 6 cm, however, the highest activities were recorded in soil samples collected away from the polluted tree. In the unpolluted area, the highest activities were always found in soil samples taken away from the canopy. The pollution-induced high rhodanese activity suggests that such soils become capable of rapid oxidation of $S_2O_3^{2-}$, an important step in the decontamination process.

Table 11
Rhodanese activity in soil profiles in an area polluted by emissions from a refractory brickworks as compared with an unpolluted area

Soil depth (cm)	Rhodanese activity (nmoles SCN^- g^{-1} soil h^{-1})			
	Polluted area		Unpolluted area	
	Below canopy	Away canopy	Below canopy	Away canopy
0-2	135.3±13.5	133.7±2.6	137.0±0	178.7±10
2-4	308.7±5.3	250.2±39	171.0±33	193.0±12
4-6	336.7±8.4	271.2±0	135.0±28	174.2±7.7
6-8	189.1±12	289.7±3.7	135.0±19	165.0±10
8-10	167.5±18	403.0±26	130.0±10	154.0±4.5

Reproduced from: M. Wainwright, Soil Biology & Biochemistry, 1979, Vol. 11, p. 97, with kind permission from Elsevier Science.

Wainwright (1980) has also performed an experiment in which columns (height: 10 cm; diameter: 6.5 cm) were removed from the top of soil beneath the canopy of sycamore at an unpolluted site, then transported to a polluted site and buried beneath the canopy of a similar sized sycamore. The original litter was replaced and the introduced soil columns were left exposed to the effects of brickworks emissions for one year. Thereafter, the activity of a series of soil enzymes participating in the C, N, P, and S cycles were determined, columns prepared in the same way but left at the unpolluted site serving as controls (Table 12).

Table 12
Effects of exposure to refractory brickworks emissions on soil enzyme activities

Enzyme activities*	Exposed soil columns	Control soil columns
Dehydrogenase	4.43±1.7	4.46±1.1
Carboxymethylcellulase	631±95	539±46
Urease	7.12±1.3	7.14±2.3
Phosphatase	114.9±0.3	113.9±2.1
Arylsulphatase	5.68±0.3	5.62±0.5
Rhodanese	6.37±13.7	601±15.9

Reproduced from: M. Wainwright, Plant and Soil, 1980, Vol. 55, p. 202, with kind permission from Kluwer Academic Publishers.

* Expression of enzyme activities: dehydrogenase in μM of H g^{-1} soil 24 h^{-1}; carboxymethylcellulase in μg of glucose g^{-1} soil 24 h^{-1}; urease in μg of NH_4^+-N g^{-1} soil h^{-1}; phosphatase and arylsulphatase in μg of *p*-nitrophenol g^{-1} soil h^{-1}; and rhodanese in nmoles of SCN^- g^{-1} soil h^{-1}.

It is evident from this table that the one-year (i.e. a relatively short-term) exposure did not affect soil enzyme activities. Neither occurred significant changes in the microbial numbers, the rates of respiration, nitrification and solubilisation of insoluble phosphate. The rate of ammonification was, however, higher in the exposed than in the control soil columns.

Pulp and paper mills

Enzymological research in the Russian Federation

The soil enzymological and microbiological effects of emissions, containing methyl mercaptan (CH_3SH), carbon disulphide (CS_2), hydrogen sulphide (H_2S), phenol, and other pollutants, from a pulp and paper mill operating near the southern shore of the Baikal Lake (southern Siberia) were studied by Barykova (1992). Experimental plots were installed in forests on brown earths, along the lake shore at 0.1, 0.3, 1, 5, and 12 km from the mill. Plots at 25 km were the controls.

The soil gelatinolytic and cellulolytic activities were the highest in the plots at 0.1 or 0.3 km from the mill, but the ureolytic (urease) activity did not exhibit marked changes depending on the distance from the mill.

Synthetic fibre factories

Enzymological research in Belorussia

Soil enzymological effects of the emissions from the synthetic fibre factory "Khimvolokno" at Mogilev were the subject of several studies (Prokopenko *et al.*, 1983; Ivleva *et al.*, 1985, 1986; Efremov, 1989; Lovchii *et al.*, 1990). This factory produces the polyester fibre poly(ethylene terephthalate). The emissions contain methanol, *p*-xylene, ethylene glycol, dimethyl terephthalate, etc., among which methanol has the highest concentration.

During the vegetation period of 1981, Prokopenko *et al.* (1983) have determined dehydrogenase activity and some microbiological parameters in soil samples taken at 0.3, 1, 3, 5, 10, and 15 km from the factory. Along this transect, the dehydrogenase activity in the 0-5-cm soil layer was lowest at 1 km and tended to increase with the increasing distance from the factory; phase I nitrification capacity was lowest at the 3-km and highest at the 15-km distance; total number of bacteria increased, the numbers of sporogenic bacteria and actinomycetes did not change considerably and that of microfungi decreased with increasing distance from the factory.

Activities of other soil enzymes were studied in both croplands (Ivleva *et al.*, 1986) and spruce forests (Ivleva *et al.*, 1985; Efremov, 1989; Lovchii *et al.*, 1990) on podzolic soils around the factory. Along a transect, the sampling sites were established at 1, 5, 10, 15, and 35 km from the factory. The site at 35 km was the unpolluted control. The sampling depths were 0-5 and 5-20 cm in croplands and 0-30 cm (litter + mineral soil) in forests.

It has been found that the pollutants (methanol, etc.) affected more consistently the 0-5-cm part of the arable layer and the forest litter than the deeper soil layers. In the croplands, the soil protease activity was significantly lower, whereas the urease activity was higher at 1-5 km than at 35 km. Due to pollution, the forest litters exhibited lower protease, urease, catalase, polyphenol oxidase, and higher invertase and phosphatase activities.

In a microplot experiment, 25-cm thick, 1-m^2 surface soil layers were taken from the unpolluted site (at 35 km from the factory) and introduced to polluted cropland sites in place of the original polluted layers. The introduced soil layers were kept uncropped to be exposed to the effects of emissions for one year. During this (short) period, evident changes in enzyme activities did not occur (Ivleva *et al.*, 1986).

Oil production plants

Enzymological research in Sweden

The Kvarntorp area, studied by Nohrstedt (1985a), is situated in the central part of southern Sweden. From 1942 to 1966 this area was heavily exploited for production of oil fuels from the local alun shale containing 7% S. During these years, huge amounts (somewhere between $6 \cdot 10^5$ and $1.2 \cdot 10^6$ t) of S, mainly in the form of SO_2, were released into the atmosphere, damaging the surrounding forests.

Nohrstedt's studies started in 1982, the year before in the Kvarntorp area a central plant for destruction of hazardous chemical residual products from the industry went into operation; the studies were planned to be repeated at regular intervals in the future. The 1982 studies aimed at determining biological activities in the soil of 10 plots located in the surrounding forests. The plots were sized 300 m^2 and were covered by closed deciduous stands. The forest floor (0-3 cm) was sampled in June and August. Four plots, closest (within 2 km) to the previous oil production plant (nearby plots) were compared with the other plots located at greater distances (distant plots). In the period of oil production, the air pollution-induced reduction of tree growth was 30% or more in the nearby plots and about 10% or less in the distant plots. In the zone of nearby plots, the damage was so severe that it resulted in the total harvesting of coniferous trees, and the birch became the dominant tree species.

The analyses of the forest floor samples have shown that the organic C content, the C-to-N ratio and the total heavy metal content (sum of Cd+Cr+Cu+Pb+Zn) were significantly higher in the nearby than in the distant plots, whereas the pH and the dehydrogenase activity (nmol TTC g^{-1} C h^{-1}) gave significantly lower values in the nearby plots (3.59 and 87.7, respectively) than in the distant plots (4.35 and 546.1, respectively). The dehydrogenase activity significantly ($P < 0.01$) decreased with the decreasing distance from the previous oil production plant. At the same time, the two groups of plots did not differ significantly with regard to the respiration (CO_2 evolution), the N mineralisation (net production of inorganic N), and the phosphatase activity.

The finding that, in the forest floor samples from the Kvarntorp area, dehydrogenase activity strongly and positively correlated with the pH has been confirmed in another study (Nohrstedt, 1985b), in which the 0-3-cm deep floor samples taken from 20 coniferous and deciduous forest stands in central Sweden were analysed. As dehydrogenase activity also correlated with stemwood productivity, it was suggested that this enzymatic activity may be used as a biological index for soil fertility and, consequently, the decreased activity may serve as an indicator of soil acidification.

Other chemical factories

Enzymological research in the Russian Federation

Filippova *et al.* (1980) have determined soil dehydrogenase, catalase, invertase, amylase, urease, acid, neutral, and alkaline phosphatase activities in close vicinity of a factory manufacturing urea, formaldehyde, and cyclohexane and at 4 and 30 km from the factory. Soil was sampled from the rhizosphere of herbaceous plants in May, July, and September 1972 and 1973. The results obtained have clearly shown that the seasonal variations in enzyme activities were, in many cases, much higher than the differences between the activity values recorded in the unpolluted soils and in the polluted ones. Therefore, an accurate evaluation of the soil enzymological effects of pollutants was not possible.

Around a factory manufacturing phosphate fertilisers from Karatau phosphorite and emitting mainly phosphorus anhydride, fluoride, and heavy metals to the atmosphere, soil studies were carried out by Kuz'mina *et al.* (1988). Along four transects (northern, north-western, southern, and south-western), soil (serozem) was sampled from the 0-5-cm depth at 0.5, 3, 5, 7, and 10 km from the factory. The organic matter content showed a nearly two-fold decrease along the northern and north-western transects and also an approximately two-fold increase along the other two transects. The dehydrogenase and phosphatase activities expressed on soil weight basis behaved similarly; the dehydrogenase activity when reported to the organic matter content remained the highest and the phosphatase activity when reported to mobile P content became the lowest at 0.5-3 km from the factory. It has been concluded that the pollutants create anaerobic conditions stimulating dehydrogenase activity and enrich the soil in mobile P inhibiting phosphatase activity.

Aseeva *et al.* (1989) have compared enzyme activities in cropped podzolic soils at 0.4, 10, and 40 km from a factory producing mineral fertilisers. All activities studied were the lowest in the vicinity of the factory during the whole vegetation period (May-August). The dehydrogenase and urease activities were more sensitive to pollution than were catalase and invertase activities. The sensitivity also depended on the nature of crop plants, being much greater in soils under barley and oats than under timothy. Microbial biomass, N_2-fixing, and denitrifying capacities also decreased in the polluted soils.

Kovalenko *et al.* (1996) have studied the effect of the emissions from the cryolite (sodium hexafluoroaluminate, $Na_3[AlF_6]$) factory located near the town of Polevskoi on the catalase activity in litter (01 and 02 horizons) and soil (A_1 horizon) of pine plantations growing around the factory. These emissions pollute the atmosphere with fluoride and S-containing compounds. Experimental plots were set up at 1, 1.5, 4, 7, and 26 km from the factory.

Sum of the catalase activities measured in the three horizons (01 + 02 + A_1) decreased with decreasing distance from the factory. The decrease was more pronounced in the A_1 horizon than in the litter horizons. Thus, the percent ratio of catalase activity in A_1 to the sum of catalase activities in the three horizons analysed (this ratio being termed catalase coefficient by Kovalenko *et al.*) had the following mean values: 6.6 (in plots at 1 and 1.5 km), 18.8 (in plots at 4 and 7 km), and 20.2 (in the plot at 26 km from the factory). Kovalenko *et al.* conclude that the two bioindicators (sum of catalase activities and catalase coefficient) may be used for "express diagnosis" of the status of forest soils.

Enzymological research in Austria

A chemical factory at Mürztal, Upper Styria had polluted the surrounding spruce forest with SO_2 during 30 years when Härtel and Cerny's (1981) investigations began in this area. Based on SO_2 content in air and signs of spruce injury, unpolluted, slightly, and heavily polluted sites were selected for soil sampling. The sites had the same aspect and soil type and similar vegetation.

Figure 9 shows that each of the four enzyme activities measured was negatively affected by the emissions of the factory. This effect was particularly evident on dehydrogenase and catalase activities in the higher horizons and on invertase and urease activities in the lower horizons. The soil respiration (CO_2 evolution) and cellulolytic capacity were also negatively affected. Another observation should also be emphasised: when the spruce forest broke down due to the emissions, a deciduous bushwood grew up spontaneously, enabling the soil to acquire activity levels approaching those of the unpolluted site, even under the continuous influence of the emissions.

Enzymological research in Poland

As Żurawska (1984) specifies, for revegetation of the areas denuded by emissions (SO_3, SO_2, HF, HCl)* from the inorganic chemical factory "Polchem" at Torun, experimental plots (each of 5×20m) on the poor soils (loose and slightly loamy sands) of these areas were installed in 1976, at 500-m distance from the factory, The experiment comprised many variants: control (not treated); minerally, NPK-fertilised (150 kg N, 150 kg P_2O_5, and 150 kg K_2O ha^{-1}); NPK + sewage sludge (100 and 200 t ha^{-1}) + ash (10 t ha^{-1}). The sludge originated from a municipal-industrial and an industrial effluent treatment plant and the ash from a brown coal power plant. The fertilisers, sludge, and ash were mixed into the top 30-cm soil layer. Half of each plot (i.e. 50 m^2) was sown with a mixture of grasses and legumes, and the other half was planted with trees and bushes.

In 1979, Żurawska has analysed enzymologically and microbiologically the soils from plots sown with the grass-legume mixture. The samples were taken from the 5-10-cm soil

---

* They pollute the environment along a distance of 5.5, 3.5, 1.2, and 1 km from the factory, respectively.

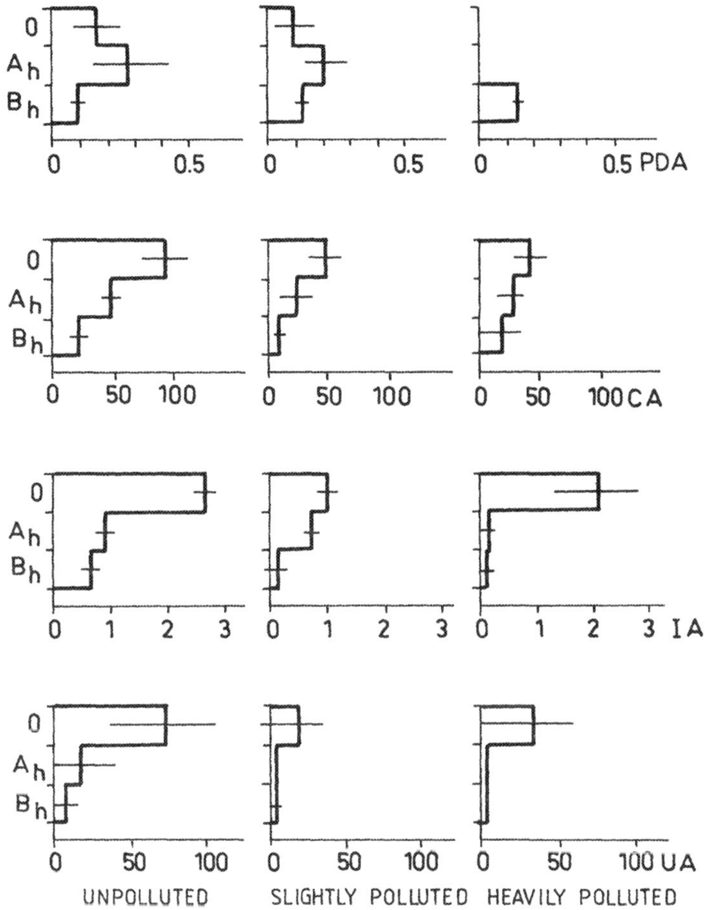

Figure 9. Enzyme activities in soil profiles at unpolluted, slightly, and heavily polluted sites in the Mürztal area.

Redrawn from O. Härtel and M. Cerny, Mitteilungen der Forstlichen Bundesversuchsanstalt Wien, 1981, No. 137, p. 239.

PDA – Potential dehydrogenase activity. CA – Catalase activity.

IA – Invertase activity. UA – Urease activity.

The horizontal lines indicate standard deviations.

depth on day 15 of each month in the May-November period. Each treatment led to increased dehydrogenase activity as compared with the untreated control. The highest increase was recorded in a complexly treated plot, namely in that treated with NPK + 100 t ha^{-1} sludge from industrial effluents + 10 t ha^{-1} ash. The activity values significantly correlated with the numbers of heterotrophic microorganisms. These results fully coincided with those concerning development of herbaceous vegetation in the plots.

Enzymological research in Ukraine

The nitrogen oxides, ammonia and other N-containing pollutants emitted to the atmosphere from a nitrogen fertiliser factory strongly affected soil enzyme activity, too, as established in long-term investigations conducted by Dolgova and Pavlyukova (1985, 1989), Pavlyukova (1989) and Pavlyukova and Dolgova (1993). Soil N contents (total, easily hydrolysable, nitrate, nitrite, and ammonium N) were higher and much higher in the vicinity of the factory than at slightly polluted and unpolluted (control) sites, respectively. The reverse was true for soil dehydrogenase, catalase, protease, and invertase activities. But urease, nitrate reductase, and nitrite reductase activities were higher in the heavily than in the slightly polluted and unpolluted soils. These increased activities should play a role in the decontamination of soils.

Stefurak (1988, 1990) studied – from both enzymological and microbiological viewpoints – the soddy-podzolic soils in the areas of two of the enterprises belonging to the "Khlorvinil" Industrial Complex, located in the Carpathian region of the country. (Centre of the Complex is in the town of Kalush). One of the two selected enterprises is a chemical factory, the other is a magnesium plant.

The emissions from both works contain SO_3, nitrogen oxides, and chlorinated hydrocarbons. The chemical factory also emits to the atmosphere CO and a series of heavy metals (Pb, Zn, Cd, Cu, Al, Mn, Hg). The emissions from the magnesium plant also contain heavy metals (Al, Pb, Cu) as well as formaldehyde, ethylenediamine and other pollutants.

The area of the chemical factory was studied during the vegetation periods of 1984-1986. Soil samples were collected from the 0-10-, 10-20-, and 20-30-cm depths of experimental plots installed many years previously on the territory of the factory as well as at 2 km downwind and at the 25 km upwind of the factory. The plot at 25 km was the control.

In the strongly polluted soil on the territory of the factory, the catalase activity could not be detected, the polyphenol oxidase and peroxidase activities were markedly lower at each soil depth in comparison with the activities measured in the soil of the control plot.

The total numbers of bacteria and actinomycetes, numbers of cellulose-decomposing microorganisms, ammonifying bacteria, *Azotobacter* cells, and N_2-fixing clostridia, and intensity of cellulose decomposition and N_2 fixation (acetylene reduction) increased, whereas numbers of oligonitrophilic and mineral N-assimilating bacteria, and the total number of microfungi decreased with increasing distance from the factory.

The area of the magnesium plant was studied during 1984-1988. The experimental plots were installed, many years previously, at distances of 0, 0.5, 2, and 5 km from the plant. Sampling depths were 0-4 and 4-12 cm.

Pollution had decreasing effect on each of the five enzyme activities determined, the cellulase, protease and peroxidase being more sensitive to pollution than the polyphenol oxidase and catalase.

All microbial indices examined (total number and biomass of bacteria, total mycelial length and biomass of microfungi, number of cellulose-decomposing microorganisms, as well as the intensity of respiration – CO_2 evolution –, decomposition of cellulose and plant residues, ammonification, nitrification, and N_2 fixation) were negatively influenced by the pollution.

Stefurak has drawn the conclusion that the long-term industrial pollution in the studied areas exerted a detrimental effect on the biological activity of soil, and some indices of the soil biological activity may be used for evaluating the degree of technogenic disturbance of the environment.

Biotechnological plants

Enzymological research in the Russian Federation

Gradova *et al.* (1994) have initiated studies to evaluate the soil biological effects of dusts emitted into the atmosphere during the yeast-drying (-inactivating) process in bio-technological plants producing fodder yeasts. These dusts contain up to 50% proteins and have allergenic properties.

Besides these dusts, inactivated yeast biomasses (commercial products) were also studied. Both dusts and yeast biomasses were taken from many biotechnological plants located in different regions.

The soil samples were collected at 5-7-km distance from the plants. Samples of a soddy-podzolic soil and of a chernozem not affected by emissions were also used. The soil samples were treated with dust or yeast biomass at rates ranging from 0.005 to 10% (on dry soil weight basis), then moistened to 60% of water-holding capacity, incubated at 20°C for 22 days and analysed periodically during the incubation.

The treatment resulted in increased soil protease activity. For example, in the samples of soddy-podzolic soil treated with yeast biomass at a rate of 0.2%, the maximum activity was registered on the 5th day of incubation and exceeded nearly 8 times the initial activity; when the yeast biomass was added at 1% rate, the maximum activity appeared on day 7 and exceeded 6 times the initial activity. Even after 22 days of incubation, the protease activity was higher (5 and 3 times, respectively) than the initial activity.

The treatment of soil samples with dust or yeast biomass also resulted in the increased numbers of heterotrophic bacteria and in the stimulated proliferation of saprophytic microfungi. The multiplication of the sanitary indicator bacterium *Escherichia coli* inoculated into soil samples was also stimulated by the treatment. At the same time, the germination of seeds of oats, wheat, and cucumber in the treated soil samples was inhibited. The inhibition which increased with the rate of yeast biomass addition (0.02-10%) was attributed to the toxic substances produced by the saprophytic microfungi, the proliferation of which – as already mentioned – was stimulated by the treatment.

Oil- and coal-fired power plants

Enzymological research in Sweden

Tyler's (1977) analyses have shown that the emission dust collected from the oil-fired power plant at Karlshamn in the autumn of 1976 contained on average 16% metals, namely 3.8% V, 2.6% Ni, 2.4% Fe, 4.8% Na, 1.3% Ca, and several other metals in smaller amounts. One-g samples (on air-dry basis) of a spruce needle mor with low metal contents were treated with 1-500 mg emission dust and immediately submitted to the determination of acid phosphatase activity in reaction mixtures not buffered (initial pH 5.75) or buffered (with acetate) to pHs 5.0 and 3.6, and incubated at 22.8°C for 3 h. Other 1-g mor samples were treated with 1-100 mg dust and preincubated at 3°C for 6 days before the phosphatase activity determination in unbuffered reaction mixtures.

The emission dust strongly inhibited phosphatase activity in mor samples. Thus, at the lowest rate (1 mg dust g^{-1} mor), the inhibition was ~20% in the unbuffered reaction mixtures (not preincubated and preincubated) and ~5% in the reaction mixtures buffered to pH 5.0; at this dust rate, the phosphatase activity was not affected in the reaction mixtures buffered to pH 3.6. At the dust rate of 100 mg g^{-1} mor, the inhibition was ~70% in reaction mixtures unbuffered or buffered to pH 5.0 and ~60% in those buffered to pH 3.6. At the highest rate (500 mg dust g^{-1} mor), the inhibition reached ~90% in all reaction mixtures.

Enzymological research in the Russian Federation

Nikitina *et al.* (1988) and Naprasnikova (1993) have described soil enzymological and microbiological investigations in the area of the network of power plants constructed in 1967 in the Nazarovo Basin which belongs to the Kansk-Achinsk Fuel-Energetic Complex (Siberia). For electricity generation, these power plants use the brown coal stripmined in this basin. The emissions, consisting mainly of calcium, silicon, iron, and aluminium oxides, affect a large area (>100 km^2), increase the Ca carbonate and hydrocarbonate contents of soils even up to 9% and pH up to 8.2. Tha sampling places were selected on grey forest soils at 5, 40, and 100 km from the power plants.

In soils sampled in the affected area in 1977, the urease activity exhibited, on average, a 50% increase. In contrast, the neutral and acid phosphatase activities decreased in the affected soils, the averaged decrease being of 31 and 78%, respectively. The changes in enzyme activities were attributed to pH changes. Cellulolytic and gelatinolytic capacities of soils were not significantly affected by the emissions.

Enzymological research in Poland

Zwoliński and co-workers' (1988) studies related to the soil enzymological and microbiological effects of power plant dust have already been referred to on pages 79-81.

Atomic power plants

Enzymological research in the Russian Federation

The effects of the Chernobyl accident (1986) on soil enzyme activities were studied by Egorova (1995). In July 1993, she took 23 samples from soddy-podzolic soils in the Zlynka and Vygonichi districts (Bryansk region) for determination of ^{137}Cs, Zn, and Cd concentrations and invertase, dehydrogenase, and catalase activities. The Zlynka soils showed radioactive pollution, their γ-activity varying between 10^{-7} and 10^{-6} Ci kg^{-1} soil, whereas the unpolluted, Vygonichi soils exhibited a much lower γ-activity (10^{-8} Ci kg^{-1} soil). A significant positive correlation was found between the concentrations of ^{137}Cs and Zn. The soil samples were grouped along the ^{137}Cs and Zn gradient.

The activity of each of the three enzymes studied increased in soil samples with increasing ^{137}Cs and Zn concentrations up to a mximum value, then two of the activities (invertase and dehydrogenase) decreased in soil samples with higher ^{137}Cs and Zn concentrations. Thus, each activity was the highest in the samples containing around 60 mg Zn kg^{-1} soil, but in the samples, in which the Zn concentration was around 110 mg kg^{-1} soil, invertase and dehydrogenase activities represented only 23 and 30% of their maximum values, respectively, whereas the catalase activity maintained its maximum value.

In six soil samples, in which the Cd concentration ranged from 0.7 to 3.8 mg kg^{-1} soil, the relationship between enzyme activities and Cd concentration was also studied. Maximum invertase and dehydrogenase activities were measured at 1.7-2.4 and at 2.4 mg Cd kg^{-1} soil, respectively, while the catalase activity continuously decreased with the increasing Cd concentration. Each activity was lowest at 3.8 mg Cd kg^{-1} soil.

ADDENDA

Military waste disposal operations

Enzymological research in the United States of America

Kuperman and Carreiro (1997) have conducted soil chemical, enzymological, and microbiological investigations on an area in the U.S. Army's Proving Ground at Aberdeen, Maryland. This area is the Toxic Burning Pits (TBP) area, an open field of approximately 3.6 ha. Between the late '40s and '80s, the pits were used to dispose of chemical agents, bulk chemical wastes, high explosives, nerve, incapacitating and blister agents, and chlorinated solvents. The methods applied for disposal included open burning and open detonation. The pits were maintained by pushing burned soil and ash toward an adjacent area referred to as the "pushout" area (PA).

In October-November 1994, soils were sampled from the 0-10-cm depth on the PA, near the TBP and in the local background site (LBS), approximately 30 m from the TBP. A reference site (RS), with similar soil characteristics, was selected in Gunpowder Falls State Park, 7 km west of TBP.

The three sites are grasslands which showed significant differences among the mean above-ground vegetation biomasses: 22.6, 155.9, and 676.1 g m^{-2} in PA, LBS, and RS, respectively.

Table 13
Enzyme activities in soils of three grassland sites near or in Aberdeen Proving Ground, Maryland, USA

Enzyme activities	Reference site	Local background site	Pushout area
Endocellulase*	88.7 a	16.3 b	9.2 b
	1912 a	769 ab	361 b
β-Glucosidase**	0.629 a	0.089 b	0.012 c
	13.567 a	4.191 b	0.445 c
N-Acetylglucosaminidase**	0.268 a	0.065 b	0.016 c
	5.785 a	3.044 b	0.623 c
Acid phosphatase**	1.782 a	0.402 b	0.081 c
	38.814 a	19.098 b	3.011 c
Alkaline phosphatase**	0.296 a	0.118 b	0.145 b
	6.439 a	5.914 a	5.901 a

Reproduced from: R.G. Kuperman and M.M. Carreiro, Soil Biology & Biochemistry, 1997, Vol. 29, p. 183, with kind permission from Elsevier Science.

* Enzyme activity is expressed as viscosimetric units g^{-1} dry weight soil h^{-1} (first row) and g^{-1} ash-free dry weight soil h^{-1} (second row).

** Enzyme activity is expressed as mmoles of converted substrate g^{-1} dry weight soil h^{-1} (first row) and g^{-1} ash-free dry weight soil h^{-1} (second row).

Numbers followed by the same letter in a row are not significantly different at $P = 0.05$.

The concentration of each of the seven metals analysed (As, Cd, Cr, Cu, Ni, Pb, Zn) and expressed in either mg or mmol kg^{-1} dry weight soil was significantly higher in PA than in LBS and, except for As, significantly higher in LBS than in RS. The total mean metal concentrations (mmol kg^{-1} dry weight soil) were 39.50 in PA, 7.69 in LBS, and 1.75 in RS, which means a 22.5-fold significant increase in PA and a 4.4-fold significant increase in LBS as compared with RS. The order of abundance of the metals (mmol kg^{-1} dry weight soil) was Zn > Cu > Pb > Cr > Ni > As > Cd in both PA and LBS and Zn > Cr > Ni > Cu > Pb > As > Cd in RS.

As shown in Table 13, the soil enzyme activities in the three sites decreased in the order: RS » LBS > PA, the differences, excepting alkaline phosphatase activity, being significant among the sites. Significant negative correlations were found between enzyme activities (excepting again alkaline phosphatase activity) and total metal concentrations.

The microbial indices such as fluorescein diacetate (FDA)-stained (active) bacterial biomass, fungal biomass, and fungal length, total fungal biomass and length, substrate (glucose)-induced respiration had the significantly lowest values in the soil of PA. The enzyme activities (excepting endocellulase and alkaline phosphatase activities) significantly correlated with FDA-stained bacterial biomass, and all enzyme activities significantly correlated with total fungal length and substrate-induced respiration.

In conclusion, the results obtained indicate that the military waste disposal operations had detrimental effects on soil enzyme activities and microorganisms, for which the high heavy metal concentrations appear primarily responsible.

Rocket destruction operations

Enzymological research in the Russian Federation

Gaponyuk and Klyueva (1990) have carried out a soil enzymological study in the Volgo-Akhtubinsk floodplain, some areas of which are used for destruction of rockets by open detonation. It was calculated that perchlorate (ClO_4^-), the main component of burning rockets, is deposited on the soil of detonation site in an amount of 3 mg kg^{-1} soil after each detonation. Will the detonation-generated perchlorate and, generally, will rocket destruction by detonation affect dehydrogenase activity considered as a global indicator of microbial life in soils? To answer this question, soil samples were taken for determining dehydrogenase activity before and after detonation of rockets on six sites.

Site-dependent variation of dehydrogenase activity (µl H_2 g^{-1} soil 24 h^{-1}) ranged from 15.0 to 80.1 before detonation and from 22.3 to 83.6 after detonation. Detonation resulted in 3-35% reductions in dehydrogenase activity on three sites and in 3-49% activity increases on the other three. The conclusion could be drawn that the detonation-generated perchlorate and, generally, destruction of rockets by detonation did not exhibit significant effects (at least instantaneous or short-term effects) on soil dehydrogenase activity. Neither were the soil chemical characteristics affected.

Chapter 6. STUDIES OF THE SOIL ENZYMOLOGICAL EFFECTS OF INDUSTRIAL EMISSIONS ORIGINATING FROM MULTIPLE SOURCES (MANY INDUSTRIAL PLANTS MANUFACTURING DIFFERENT PRODUCTS, BUT SITUATED IN THE SAME, INDUSTRIAL AREA)

These studies represent the third direction of the enzymology of soils affected by industrial emissions. Studies performed in nine countries are reviewed. Additionally, enzymology of urban soils and that of roadside soils are dealt with.

Enzymological research in Sweden

Rühling and Tyler (1973) were the first to perform soil enzymological studies in an area affected by industrial emissions from multiple, more precisely from two pollution sources. In the town of Finspång (central Sweden) there are two large industries, an alloy factory with a copper smelter and, in the close vicinity, a turbine industry. The surrounding woodland is dominated by spruce (*Picea abies*) growing on soils of podzolic types. For sampling (on October 7-9, 1971) 49 spruce sites were selected along a heavy metal concentration gradient up to a distance of 13 km from the main pollution source. At each site, three fractions of predominantly spruce needle litter were collected: fraction *a* consists of the uppermost part of the litter layer, fractions *b* and *c* correspond to the F_1 and F_2 layers of the mor, respectively.

The Cu concentrations of needle litter and needle mor from sites situated within 0.5 km from the pollution source were, on average, about 100 times higher than from sites situated more than 10 km away. The corresponding factors for other heavy metals were about 25 (for Zn), 30 (Cd), 10 (Ni and V), 4 (Pb), 2.5 (Fe and Co), and 2 (Cr). Opposite gradients with concentration increases with the distance from the pollution source were measured for Ca, Mg, K, and Mn.

The dehydrogenase activity significantly decreased with heavy metal concentration (increased with the distance from the pollution source) in fractions *b* and *c*. For example, as an average in fraction *b*, about 45 μg of TTC were reduced in samples with >10 μmol

Zn+Cu+Cd+Ni per g dry matter and about 85 μg of TTC in samples with <10 μmol Zn+Cu+Cd+Ni g^{-1} dry matter. The corresponding values in fraction *c* were about 40 and 70 μg of TTC, respectively. Respiration (CO_2 evolution) behaved like dehydrogenase activity. The lack of correlations in fraction *a* was attributed to low water content in most samples of this fraction.

Enzymological research in Poland

Olszowski (1976a, b, 1980, 1981) described 7-year fertilisation experiments in an about 40-year-old pine (*Pinus sylvestris*) forest stand strongly damaged by emissions from the Upper Silesian Industrial District. The soils of the forest belong to the medium-podzolised soil group made up by loose sands. The experiments comprised two stages. In the first stage (1971-1976), experimental plots were installed in eight variants:

1. control (not treated);
2. fertilised with NPK (100 kg N, 120 kg P_2O_5, and 100 kg K_2O ha^{-1});
3. treated with bentonite (10 t ha^{-1}) to improve the loose sandy texture;
4. treated with bentonite (30 t ha^{-1});
5. limed with caustic CaO (2 t ha^{-1});
6. NPK plus bentonite (10 t ha^{-1});
7. NPK plus bentonite (30 t ha^{-1}); and
8. NPK plus lime (2 t CaO ha^{-1}).

Bentonite and lime were applied in the autumn of 1971 and NPK in the spring of 1972.

During 1972-1974, - besides botanical and plant morphological studies, plant and soil chemical analyses-, soil enzymological and microbiological analyses were also carried out. Dehydrogenase, catalase, invertase, β-glucosidase, urease, and asparaginase activities as well as total numbers of bacteria, actinomycetes, and microfungi, and numbers of ammonifying, nitrifying, denitrifying, N_2-fixing, and cellulolytic microorganisms were determined in soils sampled from the 0-5-cm depth (A_1 horizon) 8-9 times in each year.

At the beginning of the second stage (1976-1978), namely in the spring of 1976, NPK fertilisation was repeated (N, P_2O_5, and K_2O each at a rate of 150 kg ha^{-1}), then the same studies and analyses were carried out as in the first stage. In addition, soil phosphatase activity was also determined.

The results have shown that during the first year (1972), soil enzyme activities and microbial numbers gave similar values in all variants, but increases occurred in 1973 and further increases in the next years in the treated variants. The increases were always highest in variant 8 (NPK-fertilised and CaO-limed). The enhanced soil enzymatic and microbial potential was accompanied by a 14.7% yearly forest stand volume increment and by improvement in the morphological properties of pines.

Enzymological research in the Russian Federation

Marusina and Vazhenin (1981) and Krasnova (1982) have studied the soddy-podzolic sols strongly affected by emissions from large metallurgical, chemical, and power plants in the area of Cherepovets (Vologda region). The soil samples were collected from experimental plots installed at 0.5, 2, 5, 7, and 10 km from the metallurgical plant in the direction of prevailing winds. The plot at 10 km served as control. Sampling depths were: 0-1, 1-5, 5-10, and 10-20 cm. The 0-1- and 1-5-cm soil layers contained 1.5-4 times more Fe, Mn, Zn, Cr, Pb in the plots closest to the plant as compared with the control plot. Considerable changes in enzyme activities occurred only in these two layers. According to Marusina and Vazhenin (1981), invertase and urease activities were significantly lower ($P = 0.05$) in the polluted plots than in the control one. But the numerical data published by Krasnova (1982) show that invertase, urease, and dehydrogenase activities were highest at 5 km, and catalase and sulphoxide oxidase activities at 2 km from the plant. The level of enzyme activities in plot at 7 km and control plot was similar.

The area of Meshchera, in which Belitsyna and Dronova (1981) and Belitsyna *et al.* (1989) performed soil studies, is located in the north of the Moscow region. The soils in this area (humic gley and soddy-podzolic soils) are affected by the emissions from factories manufacturing chemicals and pharmaceutical products, or preparing ceramics. The main polluting elements are Zn, Pb, and Cd. The soil dehydrogenase and invertase activities were found to decrease with the increasing level of heavy metal pollution. For example, in a humic gley soil sample containing 45 mg Zn, 43 mg Pb, and 3.4 mg Cd kg^{-1} soil, dehydrogenase and invertase activities were relatively high (15.5 mg TPF 10 g^{-1} soil 24 h^{-1} and 22.3 mg glucose g^{-1} soil 24 h^{-1}, respectively). The activities were only 1.1 and 5.6, respectively, in another humic gley soil sample in which the Zn, Pb, and Cd concentrations were 308, 76, and 3 mg kg^{-1} soil, respectively. Both activities disappeared in a third sample containing 407 mg Zn, 71 mg Pb, and 16.7 mg Cd kg^{-1} soil, but the dehydrogenase activity, as compared with the invertase activity, was, in general, more sensitive to heavy metals.

Enzymological research in Germany

Hüttermann *et al.* (1983) have determined β-glucosidase, aminopeptidase, acid phosphatase, and phosphodiesterase activities and nitrification rate in brown earth sampled from different depths within soil profiles, at Haard, Nordrhein-Westfalen (North Rhine-Westphalia). Haard is a small hilly forest area located north of the Ruhr region; it was heavily affected by industrial emissions from the Ruhr region for several decades. Beech trees fail here to regenerate naturally; they die off within a few years. Only red oak and pine forests survive.

It was found that the enzyme activities and nitrification rate within soil profiles in both red oak and pine forests gave measurable values in the top 15-cm layer and were practically not detectable in the mineral soil below 15 cm of depth. This finding was interpreted as evidence proving that "Life has retired here almost completely from the mineral soil". This interpretation was supported by the observation that the depth-dependent decline of enzyme activities and nitrification rate within the 10-30-cm layer of similar soil profiles in an unpolluted forest (Göttinger Wald) was not so steep.

Enzymological research in the Czech Republic

SO_2 pollution of forest stands in the Ore Mountains (north-western Bohemia) led to withering of the spruce, which – according to Lettl's (1984, 1986a, b, 1987) investigations performed during 1982-1985 – was not accompanied by disappearance of microbial life in the soil. Development of grassy vegetation and natural replacement of spruce with stands of more resistant, deciduous trees, namely mountain ash or birch have a clearly positive effect on the soil microflora and enzyme activities.

For the investigations four areas were selected in the vicinity of Nová Ves. The areas located close to each other are covered by a residual original spruce stand (I), a withered spruce stand (II), a sparse mountain ash stand (III), and a birch stand (IV). Area I has no ground vegetation, whereas on the other areas an abundant ground vegetation dominated by *Calamagrostis villosa* has developed.

The samples were taken from the fermentation, humus, and mineral A horizons for chemical, microbiological, and enzymological analyses.

The numbers of aerobic and ammonifying bacteria in the fermentation and humus horizons, but not in the A horizon, were evidently lower in the residual spruce stand without ground vegetation than in the other stands. The number of microfungi, rates of respiration, ammonification, nitrification, and elemental sulphur oxidation gave similar values in the corresponding horizons of all stands. Of the enzyme activities determined in the fermentation and humus horizons, only rhodanese activity in the fermentation horizon was markedly lower in the two spruce stands than in the other two, while invertase, α-amylase, urease, and asparaginase activities in both horizons showed only minor differences among the stands.

Rejšek (1991) has determined ectomycorrhizal acid phosphatase activity in mature spruce (*Picea abies*) pure stands affected by repeated short periods of heavy air pollution, in four localities at the south-eastern outskirts of the Ostravsko-Karvinský Coalfield (Moravian-Silesian Beskids). The pollution degree was different at the four localities selected: Řečice is the least polluted, Salajka is medium-polluted, and Smrčiny and Albinovo námešti are heavily polluted. For enzyme analysis, ectomycorrhizal roots were sampled from each locality in June 1989. The analytical results showed that acid phosphatase activity of ectomycorrhizal spruce roots significantly decreased with increasing degree of soil pollution.

Enzymological research in the United Kingdom

Press *et al.* (1985), who measured arylsulphatase activity in blanket peats from 29 sites in northern England and Wales, point out that blanket peats, on these territories, differ markedly in their proximity to major sources of atmospheric pollution. The southern Pennines have extensive blanket peatlands, some of which are close to large industrial towns, and which have been subjected to high rates of acidic deposition for two centuries (for example, the peatland at Holme Moss). Other blanket bogs are more remote from urban and industrial areas and, thus, are relatively unpolluted (for example, the peatland in the Berwyn Mountains and that in Migneint, both in North Wales).

At each of the 29 sites studied, surface (0-2 cm) samples were collected and at two sites (Holme Moss and Berwyn Mountains) deeper samples from the 0-30-cm layer were also taken, between February and May 1983.

The arylsulphatase activity in the 0-2-cm layer was low in the sites close to industrial towns, the lowest values being recorded in the Holme Moss peat. The sites remote from industrial areas had activities an order of magnitude higher.

At Holme Moss, arylsulphatase activity was evenly low at each depth within the 0-30-cm layer. But the activity, high in the 0-2-cm layer, decreased rapidly with depth in the Berwyn Mountain site, and at 30 cm the activity was less than 20% of that measured at the surface.

The effect of present-day pollution on the arylsulphatase activity was also studied. Surface peat was collected from the relatively unpolluted Migneint site in February 1983. Seed trays were filled with this peat to a depth of 2 cm, then placed ("transplanted") on the bog surface in the Berwyn Mountain and Holme Moss sites. The transplanted peat had only minimal contact with the underlying peat.

The arylsulphatase activity, determined in the transplanted peat over 12 weeks (Figure 10), showed little changes in the relatively unpolluted Berwyn Mountain site, but in the Holme Moss site, exposed to pollution, the activity declined rapidly and in 12 weeks the transplanted peat lost 50% of its initial activity. In other words, arylsulphatase activity was sensitive indicator even of a short-term (12-week) present-day pollution.

Enzymological research in the United States of America

Arylsulphatase activity in peat exposed to pollution (acid precipitation) was studied in the USA, too, by Jarvis *et al.* (1987). Their study site was Big Run Bog, a 15-ha *Sphagnum*-dominated wetland in the Appalachian Mountains of West Virginia. In May 1984, 30-cm deep cores of *Sphagnum*-derived peat were collected, then cut into 6 sections (0-5 cm, 5-10 cm, etc.). The arylsulphatase activity and total S and ester-sulphate S concentrations were determined in each section.

The mean arylsulphatase activity was lowest in surface peat (0-5 cm) and generally increased with depth to a maximum value at 25-30 cm. Total S and ester-sulphate S concentrations were lower at 0-5 and 25-30 cm than at intermediate depths. But, near peat surface, the SO_4^{2-} availability was higher due to the high loading from both incident acid precipitation and minerotrophic runoff from the surrounding forested watershed.

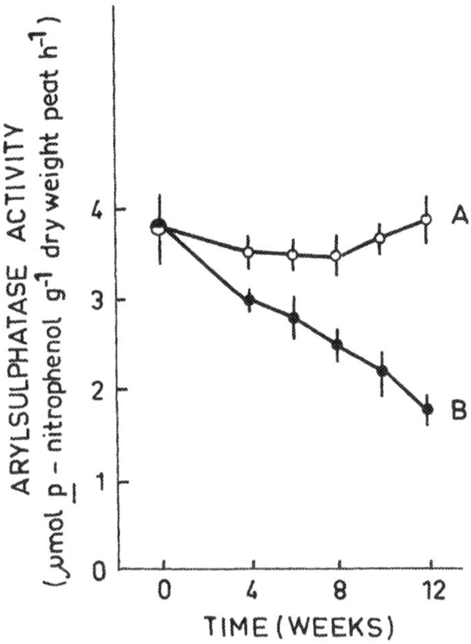

Figure 10. Arylsulphatase activity in an initially unpolluted peat from Migneint after its transplantation to a relatively unpolluted peatland site (Berwyn Mountains) (A) and to the Holme Moss site, exposed to pollution (B).

Redrawn from: M.C. Press, J. Henderson and J.A. Lee, Soil Biology & Biochemistry, 1985, Vol. 17, p. 102, with kind permission from Elsevier Science.

Vertical bars represent the standard errors.

The low arylsulphatase activity in the surface peat may be the result of the high SO_4^{2-} availability: the free sulphate ions inhibited, by feedback mechanism, the microbial synthesis of arylsulphatase. In deeper peat layers, such an inhibition does not occur because of the lower SO_4^{2-} availability which would explain the depth-dependent increase of arylsulphatase activity.

Enzymological research in Austria

Kinzel and co-workers (Kinzel *et al.*, 1989; Kinzel, 1991; Baumgarten and Kinzel, 1991; Zechmeister-Boltenstern *et al.*, 1991) have conducted investigations in the Wienerwald (Vienna Woods), namely in 11 beech forest sites, in the west and north-west of Vienna, to assess the effects of pollutants (acid precipitation, heavy metals, especially Pb and Cd, soot) on soil properties.

For soil sampling, the stemflow (Stammabfluß) method has been applied: the soil samples are taken from a zone between the stems or far from the stems and from a zone at the stem (trunk) base which is more polluted owing to the rainwater flowing down on the surface of stems and containing a great part of pollutants collected by the canopy. After the removal of litter, the uppermost 0-5 cm of the Ah horizon is sampled downhill and uphill, right and left of the stem (trunk) at different distances from it.

Besides a series of chemical and microbiological parameters, many enzyme activities were also determined in the soil samples collected in the 1986-1988 period.

Figure 11 presents a star plot of the mean values of eight enzyme activities in the soil of the strongly polluted beech forest site at Exelberg. One can see from the star plot that dehydrogenase and alkaline phosphatase activities were measurable in the less polluted zone between the stems, but were lacking in the intensely polluted zone near the stems. At the same time, the phosphatase activity measured at the natural pH of soil (i.e. without addition of buffer solution to the reaction mixtures) was less affected by the pollutants in stemflow. The enzymes participating in the N cycle (protease, urease) were more sensitive to pollutants than were those taking part in the C cycle (invertase, xylanase, pectinase).

The microbial biomass was more abundant, the respiration and ammonification were slightly stimulated, whereas the N_2 fixation and nitrification were markedly inhibited in the soil of the stemflow zone as compared with the soil of the intermediate (between-stem) zone.

Another finding was that the differences in the enzyme activities between sites were greater than the seasonal differences at the same site.

At the Schöneben beech forest site in the Upper Austrian Böhmerwald remote from industrial emission sources, the stemflow did not bring about any evident changes in soil properties, including phosphatase (at soil pH), dehydrogenase, urease, protease, and β-glucosidase activities as well as microbial biomass and respiration.

Using the stemflow method, Zehner *et al.* (1994) have studied chemically and enzymologically a beech forest site in the Wienerwald. Around six beech trees, soil was sampled along a 1-m line in the direction of stemflow and also from the opposite direction. The sampling depth was 0-5 cm.

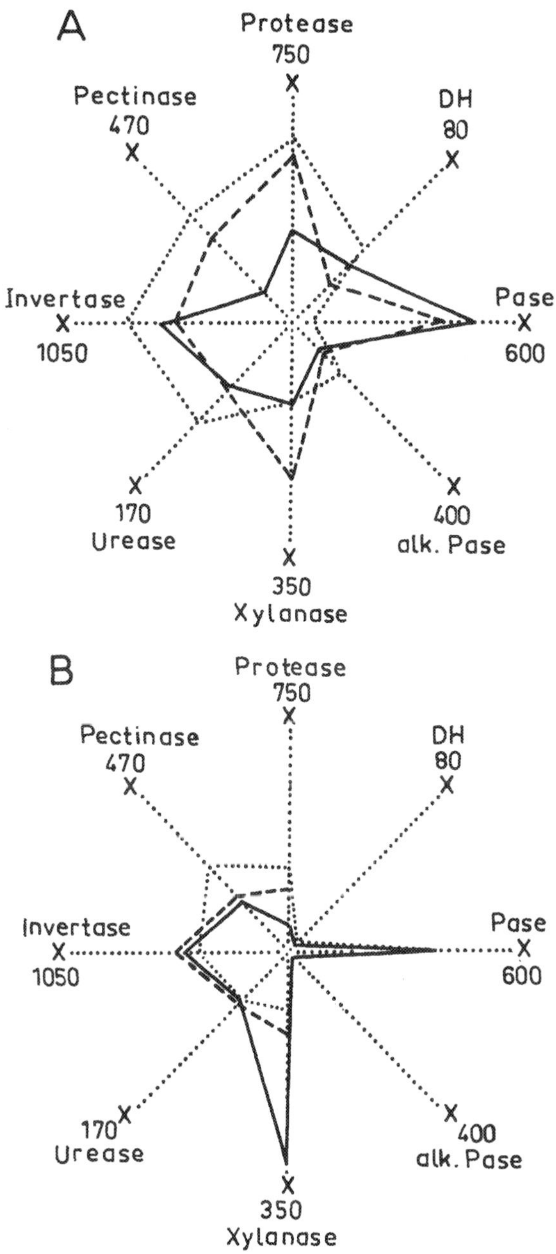

Figure 11. Star plot of eight enzyme activities in soil from the beech forest at Exelberg (north-west of Vienna).

Redrawn from: H. Kinzel, A. Baumgarten, K. Spadinger and S. Zechmeister-Boltenstern, Ecological Impact of Acidification, Magyar Tudományos Akadémia, Budapest, 1989, p. 72.

A - Soil between stems.
B - Soil at the stem base.
Full line - April 1987.
Dotted line - July 1987.
Dashed line - September 1987.
DH - Dehydrogenase
Pase - Phosphatase at soil pH.
alk.Pase - Alkaline phosphatase.

Enzyme activities are expressed in μg reaction product: TPF (dehydrogenase), phenol (both phosphatases), glucose (xylanase), NH_4-N (urease), glucose (invertase), galacturonic acid (pectinase), tyrosine (protease) g^{-1} dry weight soil h^{-1}.

The dimethylsulphoxide reductase, dehydrogenase, and arylsulphatase activities in the soil of stemflow zone of each tree were strongly depressed in comparison with the activities measured in the opposite zone. The depressions were of up to 55, 80, and 87%, respectively, and significantly correlated with soil pH decrease.

For estimating the effects of airborne pollutants on soil microbial and enzymatic activities in the Federal Region of Salzburg, Tscherko *et al.* (1996) have selected four long-term observation areas, namely a "pure air" area, an area far from and two areas near to emission sources. In soil samples taken from the 0-5- and 5-10-cm depths in May and September 1995, besides a lot of chemical parameters, the following seven microbial and enzymatic activities were determined: substrate-induced respiration, N mineralisation, potential denitrification, xylanase, urease, phosphatase, and arylsulphatase activities.

The numerical values of activities were grouped into three categories comprising low, medium, and high activities, respectively. Then, the percent distribution of the three categories for each area was calculated (Table 14).

Table 14 shows that the microbial and enzymatic activities in the soil of the four areas decreased in the order: I < II < III < IV, i.e. in concordance with their degree of pollution.

Table 14
Distribution of low, medium, and high soil microbial and enzymatic activities at four long-term observation areas in the Federal Region of Salzburg, Austria

Areas	Low	Medium	High
		activities (%)	
I. St. Koloman, "pure air" area, grassland	11.9	40.5	47.6
II. Saalfelden-Ramseiden, area far from emission sources, grassland	19.1	52.4	28.5
III. Hallein-Gamp, area near emission sources, grassland	40.5	38.1	21.4
IV. Salzburg Freisaal, area near emission sources, arable land	78.6	7.1	14.2

Reproduced from: D. Tscherko, R. Öhlinger, E. Hackl and E. Kandeler, Mitteilungen der Deutschen Bodenkundlichen Gesellschaft, 1996, Vol. 81, p. 358.

The differences in activities among the three grassland areas were marked in the 0-5-cm soil layer, whereas the activities and differences among areas decreased in the 5-10-cm layer.

Based on the well-documented effect of acid precipitation to decrease soil pH and mobilise toxic Al ions, Illmer and Schinner (1996) have selected forest sites in the Federal Region of Tirol for studying the relationship between Al concentration and some microbial and enzymatic parameters. The acid soils in the selected sites contained different amounts of Al (between 5 and 85 μmol g^{-1} dry matter), but their chemical, pedological, phytosociological, climatic, and topographic characteristics were similar.

Significant negative correlations ($P < 0.001$) were found between the Al concentration and microbial biomass, N mineralisation, and protease activity. These three microbial and enzymatic parameters significantly correlated also with each other.

Enzymological research in Finland

In 1987, Ohtonen and co-workers (Ohtonen and Markkola, 1989, 1991; Ohtonen *et al.*, 1990, 1992, 1994) have initiated complex soil biological investigations on 20 sites (of about 100 m^2 each), all representing mature pine (*Pinus sylvestris*) forest stands belonging to the middle boreal vegetation zone, around the industrialised city of Oulu located near the coast of the Bothnia Bay in northern Finland. A pulp mill and a chemical factory as well as local heating plant and motor vehicles emit to the atmosphere large amounts of SO_2, NO_x, CO, H_2S, dust and smaller amounts of heavy metals, polluting the area for the last 30-40 years. For example, the annual emissions of SO_2, NO_x, H_2S, and Pb are 6275, 7010, 507, and 20 t, respectively.

The 20 sites selected lie on sulphur and nitrogen concentration gradients from 1.3 to 2.8 g S kg^{-1} and from 7.9 to 16.3 g N kg^{-1} in mor humus. Ten of the sites are considered to belong to the less polluted zone (5-40 km from the emission sources) and 10 sites to the more polluted zone (1-5 km from the emission sources).

The dehydrogenase and cellulase activities were determined from the total mor humus layer, the thickness of which was from 2.3 to 8.9 cm and increased towards the emission sources. Both activities were significantly higher at the less polluted than at the more polluted sites. For example, the dehydrogenase activity determined in mor humus sampled from the 10 + 10 sites many times during 1987 and 1988 gave the following mean values and standard errors (expressed in μmol TPF g^{-1} organic matter 24 h^{-1}): 2.08±0.06 (less polluted sites) and 1.68±0.05 (more polluted sites).

Other soil biological variables, namely the basal respiration rate, the microbial biomass, the amount of mycorrhizal pine roots, the diversity and the sporophore production of mycorrhizal fungi, the numbers of enchytraeids and nematodes have also been found to decrease towards the polluted end of the S and N concentration gradients. But the length of the FDA-stained (living) fungal hyphae was not significantly affected by the pollution degree.

The dehydrogenase activity significantly and positively correlated with pH ($CaCl_2$) and number of nematodes in mor humus. The cellulase activity significantly correlated with the basal respiration and decreased simultaneously with the increase in the thickness of the humus layer which indicates that the decomposition of organic matter has been retarded as a result of pollution.

ADDENDA

Urban soils

In urban areas, constant pressure from human activity is a feature of the soils. The emissions from industrial and heating plants, exhausts from motor vehicles, presence of manufactured materials, housing constructions, trampling by humans, etc. all are disturbing factors for urban soils, too (Harris, 1991).

Enzymological research in Poland

Żukowska-Wieszczek (1979, 1980), Zimny and Żukowska-Wieszczek (1983a, b, c) and Zimny *et al.* (1992a, b) have described investigations, in which enzyme activities, microbial, and chemical parameters were analysed in soils of Warsaw during 1974-1990.

Of the enzyme activities, the dehydrogenase was always determined. In many studies, the urease was also assayed. The invertase activity was also determined in the first studies, but this activity – contrarily to dehydrogenase and urease activities – proved to be insensitive to assess differences among urban soils. Therefore, in further studies the invertase activity was not determined. The 5-10-cm soil layer was analysed in all studies, and the 10-20- and 20-40-cm layers were submitted to enzyme analysis only in some studies.

Different urban soils were compared enzymologically.

a) Streetside lawns and park lawns. The streetside lawns are lawns in open areas, unsheltered by trees and/or bushes, near blocks of flats, whereas park lawns are far from highways, sheltered by trees and/or bushes, far from buildings.

Table 15 comprises the average and range of enzyme activities measured monthly in the April-November 1976 period in soils of three streetside lawns and five park lawns. It is evident from this table that both the dehydrogenase and urease activities were much lower in the unsheltered streetside lawn soils than in the sheltered park lawn soils at each sampling depth. It should be added that the differences were persistent not only in the average values but in the monthly values, too.

Table 15
Average and range of dehydrogenase and urease activities in soils of three streetside lawns and five park lawns at three depths

Soils	Depth (cm)					
	5-10		10-20		20-40	
	Average	Range	Average	Range	Average	Range
	Dehydrogenase activity (μl H_2 10 g^{-1} dry weight soil)					
Streetside lawn soils	6.8	0.2-29	3.4	0.2-16	1.8	0.0-13
Park lawn soils	49.6	7.5-93	24.7	7.5-52	12.4	1.0-31
	Urease activity (mg NH_3-N 10 g^{-1} dry weight soil)					
Streetside lawn soils	0.54	0.001-1.69	0.69	0.01-1.72	0.49	0.01-1.68
Park lawn soils	2.93	2.13-3.43	2.72	2.20-3.36	2.67	1.99-3.46

Reproduced from: Żukowska-Wieszczek, Ekologia Polska, 1980, Vol. 28, pp. 275-276.

b) Trees along streets and park trees. Nine sites were selected for this study. In April, July, and November 1978-1979, soils were sampled from the 5-10-cm depth at 20-30-cm distance from the trunk of trees (*Acer platanoides, A. pseudoplatanus,* and *A. negundo*) growing along streets and in parks. The average dehydrogenase activity was the lowest in soils near trees growing inside pavements, along busy streets, with a free non-grassy surface of about 1 m^2. A little higher activity was registered in soils near trees growing inside pavements, along busy streets, but with a free grassy surface area of about 1 m^2. As expected, the activity was highest in soil near trees growing in parks.

c) Allotment gardens. Nine gardens, located in three housing estates, were studied. The cultivated plants were tomatoes, roses, and apple trees. The 5-10-cm soil layer was sampled in April, July, and November 1978-1979. The garden soils presented the following increasing order of the average dehydrogenase activity: apple trees < roses < tomatoes.

The comparison of dehydrogenase activities in all urban soils studied made it possible to establish the following order: streetside soils < park soils < garden soils, and within each soil group the soils under trees were less dehydrogenase-active* than those under other plants, including weeds.

The total number of microorganisms, cellulose decomposition and respiration (CO_2 evolution) rates were also higher in less polluted than in more polluted urban soils.

Inorganic fertilisation of streetside and park lawn soils at a rate of 100 kg N, 60 kg P, and 80 kg K ha^{-1} $year^{-1}$ during 1976-1979 led to increased dehydrogenase and urease activities, and cellulose decomposition rate in the first year, but later the activity-increasing effect of NPK remained significant only in the more polluted streetside lawn soils.

Significant negative correlations were found between the soil dehydrogenase and urease activities and heavy metal, particularly Zn, Cu, and Pb contents.

Enzymological research in Germany

In studying the urban soils in Trier and Bonn-Bad Godesberg, Weritz and Schröder (1988, 1989, 1990) and Weritz (1989) selected, for chemical, enzymological, and microbiological analyses, 70 places in the autumn of 1987 and 66 places in the spring of 1988. The places included park lawns, streetside green stripes and other green areas as well as laylands, grounds covered with rubbles or used for depositing scrap iron. The soils were sampled from the 0-15-cm depth.

Catalase, xylanase, and urease activities, microbial biomass, and substrate (glucose)-induced respiration (SIR) were higher in soils of the green areas than in those of the other grounds. Variation of these indices, like that of the organic C content, in urban soils of the same use was characterised by a broad amplitude, which was most marked in soils covered with rubbles. When the results obtained in all soils were evaluated together, significant correlations were found between the three enzyme activities and the microbial biomass, between catalase and urease activities and the SIR as well as between catalase and urease activities and between xylanase and urease activities. At the same time, these indices did not significantly correlate with the organic C and heavy metal (total and 0.1 M $CaCl_2$-extractable Pb, Cu, Cd, and Zn) contents of soils. But in the soils of the green areas taken separately from the other soils, the organic C content significantly correlated with the SIR.

---

* The soils of other tree species growing along streets (*Betula verrucosa, Fraxinus excelsior, Populus nigra, Salix alba, Tilia cordata, Quercus robur*) were also found to be less dehydrogenase- and urease-active.

Using the data of Weritz (1989) on the catalase, xylanase, and urease activities and the microbial biomass in soils of 12 places and determining the abundance of earthworms, enchytraeids, and snails in soils of 42 places in Bonn-Bad Godesberg in 1987, Fründ *et al.* (1988) found that the enzyme activities and microbial biomass did not significantly correlate with the abundance of any of the three animal groups. It should be added that the correlations between abundances of the animal groups were also insignificant.

Brief mentions were made of these soil enzymological investigations in several sections of the book by Schulte *et al.* (1989).

Studying soils in the city of Dorsten, located at the northern rim of the Ruhr industrial area, Keplin and Broll (1997) have found that soil dehydrogenase activity and biomass of earthworms decreased from ancient garden land to grassland and arable land.

Enzymological research in Finland

The studies dealing with enzyme activities in soils (mor humus) of 20 sites around the city of Oulu were cited on pages 127-128. It should be mentioned that the sites closest to the city are located on urban soils.

Enzymological research in the Russian Federation

Semenov *et al.* (1996) have determined cellulase activity in two degraded soddy-podzolic soils in Moscow. One soil is under forest (birch-pine) in a suburb, the other in a park (birch-mountain ash). The park soil was found to be a little more cellulase-active which might be related to periodical manuring of the park soil with a peat-soil mixture.

Roadside soils

Enzymological research in the United States of America

Neal and Herbein (1983) studied the effect of vehicle disturbance on sulphatase activity in arctic tundra soils at a wet and a drier site located in upland tussock tundra east of Slope Mountain, 175 km south of Prudhoe Bay, Alaska. Tracked vehicles had partially removed the vegetative mat and compacted the tundra surface approximately 9 years prior to the study. The soils are permanently underlain with permafrost. At both sites, samples were collected to a depth of 10 cm from under tussocks in vehicle track and adjacent undisturbed tundra in the autumns of 1979 and 1980. At the wet site, the sulphatase activity in both years was significantly less in disturbed (vehicle track) than in the adjacent undisturbed soil. At the drier site, the sulphatase activity of the soil sampled in 1979 was not significantly different between disturbed and undisturbed tundra, but in 1980 (in which the soil moisture levels were higher throughout the growing season) the disturbed soil was significantly less sulphatase-active than the undisturbed soil.

In another study, in which the disturbed (vehicle track) and adjacent undisturbed arctic tundra soils sampled during the second week of August were compared, Herbein and Neal (1990) found that at the wet site, approaching water saturation, vehicle disturbance caused a significant decrease in acid phosphomonoesterase and phosphodiesterase activities. In contrast, the activities were not significantly affected by vehicle disturbance at the drier, well drained site.

In Alaska, the effect of road dust on soil enzyme activities was also studied, more precisely Moorhead *et al.* (1996) have studied the effect of dust raised from the Dalton highway on soil endocellulase, exocellulase, and acid phosphatase activities. The Dalton highway is a gravel road stretching 577 km from Fairbanks to Prudhoe Bay. Dust raised from

Table 16
Soil characteristics along a dust depositional gradient in a tussock tundra, Alaska

Distance from road (m)	Organic C content (% dry weight)	Dust loading (g m^{-2} day^{-1})	Enzyme activities*		
			Endocellulase	Exocellulase	Acid phosphatase
5	14.01	0.750	30	90	2900
7.5	15.12	0.660	40	120	3100
10	15.78	0.580	120	205	4100
50	34.55	0.075	175	275	4250
500	39.11	0.010	250	350	5250

Reproduced from: D.L. Moorhead, A.E. Linkins and K.R. Everett, Arctic and Alpine Research, 1996, Vol. 28, p. 348, by permission of the Regents of the University of Colorado.

* Expressed as units of viscosity changes (endocellulase), μg of glucose (exocellulase), and μmoles of *p*-nitrophenol (acid phosphatase), respectively, g^{-1} ash-free dry weight soil h^{-1}.

this road is deposited on a total area of about 1154 km^2. A 500-m transect was established perpendicular to the highway in tussock tundra near the Toolik Lake Long-Term Ecological Research site, on which the vegetation is dominated by vascular plants (55%) and mosses (35%) and among the vascular plants *Eriophorum vaginatum* accounts for the greatest cover.

Soil samples were collected between *E. vaginatum* tussocks (0-3-cm depth) at 5-, 7.5-, 10-, 50-, and 500-m distances from the road, along the perpendicular transect.

It results from the analytical data summarised in Table 16 that the enzyme activities increased with distance from the road, increasing with organic C content and decreasing with dust loading. The sensitivity of activities to dust showed the order: endocellulase > exocellulase > acid phosphatase. Thus, the endocellulase, exocellulase, and acid phosphatase activities within 5 m of the road were reduced by 88, 74, and 45%, respectively, of the activity levels at 500 m. It was also found that the relationships among activity levels, soil organic C content, and dust loading were remarkably similar among enzymes.

Fenn *et al.* (1992) have devoted a study to urease activity in two arid, silty clay loam soils (from Harkey and Saneli), each represented by a cultivated, cotton (*Gossypium hirsutum*) field and an adjacent roadbed site (40 years without cultivation). Urease activity was found to be much lower and more variable in the roadbed soils than in the cultivated ones.

Enzymological research in the Russian Federation

Samoilova (1988) and Samoilova and Gorbatyuk (1988), studying the soil biological effects of motor vehicle exhausts along a highway with a very high traffic density (50,000 vehicles 24 h^{-1}), took soil samples in an adjacent field cultivated with oats, at 15 and 250 m from the road. The amylase activity was 1.6 times, acid and alkaline phosphatase activities were 1.4 and 2 times higher at 250 m than at 15 m from the road. The invertase activity tended to increase with increasing distance, whereas the urease activities at 15 and 250 m did not differ significantly. The total numbers of bacteria and actinomycetes increased, but that of the microfungi did not change significantly with the increasing distance from the road.

Enzymological research in India

Joshi *et al.* (1993) have determined enzyme activities in extracts of leaf litters of alder (*Alnus nepalensis*) and pine (*Pinus kesiya*) growing in a subtropical forest stand polluted due to its location at 15-m distance from the Shillong-Jowai highway (with a traffic density of 8,000-9,000 vehicles 24 h^{-1}) in the East Khasi Hills district of Meghalaya, North-East India. Another, relatively unpolluted forest stand about 500 m away from the road served for comparison. The two forest stands were similar with regard to the soil, vegetational composition, tree height, extent of canopy closure, depth of forest floor, and microenvironmental conditions. Alder and pine litters were sampled monthly during 1990 and 1991 and submitted to chemical, enzymological, and microbiological analyses.

Cellulase, amylase, and invertase activities in extracts of both litters from both forests exhibited monthly variations, and each activity was higher in the alder than in the pine litter. The activities, like numbers of bacteria and microfungi, were lower in the polluted than in the unpolluted forest stand. The correlations between cellulase and amylase activities and bacterial and fungal numbers were, in general, more significant in polluted than unpolluted litter. At the same time, the invertase activity never correlated significantly with either bacterial or fungal numbers, but always gave significant correlations with total N and total sugar contents of litters.

Enzymological research in Bulgaria

Deribeeva (1995) determined enzyme activities in the 0-20-, 20-40-, and 40-60-cm depths of a leached smolnitsa soil located on the side of the E-80 highway, in the Slivnitsa district near Sofia. The sampling places had distances of 10, 50, 100, and 150 m from the road, in an uncultivated and a cultivated land. The invertase, urease, and protease activities in the uncultivated smolnitsa increased with the increase in distance from the road. The same was true for the invertase, protease, and catalase activities in the cultivated smolnitsa. These trends were most evident in the 0-20-cm soil layer. Surprisingly, there was no correlation between decrease in enzyme activities and amounts of Pb, Zn, and Cd in the studied smolnitsa soil.

In another study, Deribeeva and Belichki (1995) have established a transect originating at the Sofia-Kulata highway. Along the transect on a leached cinnamonic forest soil under different crop plants, samples were taken at 10, 50, 100, and 150 m from the road. Sampling depth was 0-20 cm. Of the enzyme activities assayed, invertase was found to be a sensitive indicator of pollution. This activity increased, the urease activity decreased, the protease and catalase activities did not change considerably with the increasing distance from the road. The numbers of bacteria, oligonitrophilic microorganisms, and *Azotobacter* cells, like invertase activity, increased, while the numbers of actinomycetes, microfungi, and cellulolytic microorganisms remained at a similar level along the transect.

CONCLUSIONS

The studies performed in many countries (in alphabetical order: Austria, Belorussia, Bulgaria, Canada, Czech Republic, Finland, Germany, India, Kazakhstan, Poland, Romania, Russian Federation, Sweden, Switzerland, Ukraine, United Kingdom, United States of America) have proved that enzyme activities are, in most situations, sensitive indicators of *a*) soil pollution caused by industrial emissions (and motor vehicle exhausts) and *b*) efficiency of the decontamination technologies applied. In addition, there are accumulated soil enzymes capable of participating even directly in the decontamination of some polluted soils. But soil enzymes, like soil microorganisms, are not infallible. This is why the prevention of pollution should remain the best way for environmental protection.

REFERENCES

Aristovskaya, T.V. and Chugunova, M.V. (1989) Ekspress-metod opredeleniya biologicheskoi aktivnosti pochvy (Express method for determination of biological activity in soil). *Pochvovedenie*, No. 11, 142-147.

Aseeva, I.V., Lavrent'eva, V.A. and Konovalova, O.E. (1989) Vliyanie promyshlennykh vybrosov na biokhimicheskuyu aktivnost' dernovo-podzolistoi pochvy v agroekosistemakh (Effect of industrial emissions on the biochemical activity of soddy-podzolic soil in agro-ecosystems). *Tezisy Dokladov 8. Vsesoyuznogo S'ezda Pochvovedov* (Novosibirsk, 1989), **2**, 273.

Bååth, E. (1989) Effects of heavy metals in soil on microbial processes and populations (a review). *Water, Air, and Soil Pollution*, **47**, 335-379.

Badura, L., Galimska-Stypa, R., Górska, B. and Smylla, A. (1984) Wplyw emisji huty cynku na mikroorganizmy glebowe (Effect of zinc smelter emission on soil microorganisms). *Acta Biologica* (Katowice), **15**, 112-127.

Barykova, I.N. (1992) Vliyanie tekhnogennykh vybrosov Baikal'skogo tsellyulozno-bumazhnogo kombinata na pochvennye mikroorganizmy (Effect of technogenic emissions from the Baikal pulp and paper mill on soil microorganisms). *Geografiya i Prirodnye Resursy*, No. 2, 80-84.

Baumgarten, A. and Kinzel, H. (1991) Mikrozonen im Stammfußbereich von Buchen: Untersuchungen der bodenbiologischen Aktivität (Microzones in the stem base area of beeches: Investigations of the soil biological activity). In *Zustandserhebung Wienerwald. Pflanzenphysiologische und bodenökologische Untersuchungen zur Bioindikation* (R. Albert, K. Burian and H. Kinzel, Eds.), pp. 247-288. Verlag der Österreichischen Akademie der Wiessenschaften, Vienna.

Belitsyna, G.D. and Dronova, N.Ya. (1981) Vliyanie tekhnogennoi nagruzki na svoistva pochv dernovo-podzolistoi zony (Effect of technogenic loading on properties of soils in the soddy-podzolic zone). *Tezisy Dokladov 6. Delegatskogo S'ezda Vsesoyuznogo Obshchestva Pochvovedov* (Tbilisi, 1981), **2**, 124.

Belitsyna, G.D., Dronova, N.Ya., Skvortsova, I.N. and Tomilina, L.N. (1989) Izmenenie nekotorykh pokazatelei biologicheskoi aktivnosti pochv pod vliyaniem antropogennoi nagruzki (Changes in some indices of the biological activity of soils under the influence of anthropogenic loading). *Pochvovedenie*, No. 1, 140-144.

Deribeeva, D. (1995) Enzimna aktivnost na izluzhena smolnitsa podlozhena na zamursyavane (Enzyme activity of leached smolnitsa soil subjected to pollution). *Selskostopanska Nauka i Proizvodstvo* (Sofia), No. 4-5, 49-51.

Deribeeva, D. and Belichki, I. (1995) Biologichna aktivnost na izluzhena kanelena gorska pochva, razlozhena krai avtomagistrala Sofiya-Kulata (Biological activity of cinnamonic forest soil located along the Sofia-Kulata highway). *Pochvoznanie, Agrokhimiya i Ekologiya* (Sofia), **30**, 185-186.

Doelman, P. (1978) Lead and terrestrial microbiota. In *The Biogeochemistry of Lead in the Environment* (J.O. Nriagu, Ed.), pp. 343-353. Elsevier, Amsterdam.

Dolgova, L.G. (1973) O fenoloksidaznoi aktivnosti pochvy v usloviyakh promyshlennogo zagryazneniya (Phenol oxidase activity of soil under conditions of industrial pollution). *Pochvovedenie*, No. 9, 64-69.

Dolgova, L.G. (1975) Biokhimicheskaya aktivnost pochvy pri zagryaznenii (Biochemical activity of soil affected by pollution). *Pochvovedenie*, No. 4, 113-118.

Dolgova, L.G. (1978a) Aktivnost' nekotorykh oksidoreduktaz kak diagnosticheskii pokazatel', kharakterizuyushchii pochvy, zagryaznennye promyshlennymi vybrosami (Activity of some oxidoreductases as a diagnostic indicator characterising the soils polluted by industrial emissions). *Pochvovedenie*, No. 5, 93-98.

Dolgova, L.G. (1978b) Aktivnost' oksidoreduktaz kak diagnosticheskii pokazatel' zagryazneniya pochvy promyshlennymi otkhodami (Activity of oxidoreductases as an indicator of soil pollution by industrial wastes) . *Pochvovedenie*, No. 7, 152-157.

Dolgova, L.G. and Kuchma, V.N. (1983) Vliyanie gumata natriya na biologicheskuyu aktivnost' pochvy v usloviyakh tekhnogennykh territorii (Effect of sodium humate on the biological activity of soil under conditions of technogenic territories). In *Teoriya Deistviya Fiziologicheski Aktivnykh Veshchestv* (A.A. Kolbasin, Ed.), pp. 125-126. Sel'skohozyaistvennyi Institut, Dnepropetrovsk.

Dolgova, L.G. and Pavlyukova, N.F. (1985) Biokhimicheskaya aktivnost' pochv tekhnogennykh territorii (Biochemical activity of soils of technogenic territories). *Introduktsiya i Eksperimental'naya Ekologiya Rastenii* (Dnepropetrovsk), pp. 12-16.

Dolgova, L.G. and Pavlyukova, N.F. (1989) Biokhimicheskaya aktivnost' pochv, zagryaznennykh soedineniyami azota (Biochemical activity of soils polluted with nitrogen compounds). In *Okhrana Truda i Okruzhayushchei Sredy v Tekhnologicheskikh Protsessakh Energetiki i Chernoi Metallurgii* (V.M. Zadara, Ed.), pp. 81-86. Gosudarstvennyi Universitet, Dnepropetrovsk.

Dumontet, S., Dinel, H. and Lévesque, P.E.N. (1992) The distribution of pollutant heavy metals and their effect on soil respiration and acid phosphatase activity in mineral soils of the Rouyn-Noranda region, Québec. *The Science of the Total Environment*, **121**, 231-245.

Ebregt, A. and Boldewijn, J.M.A.M (1977) Influence of heavy metals in spruce forest soil on amylase activity, CO_2 evolution from starch and soil respiration. *Plant and Soil*, **47**, 137-148.

Efremov, A.L. (1989) Vliyanie otkhodov khimicheskogo proizvodstva na biologicheskuyu aktivnost' lesnykh pochv (Effect of the wastes of chemical industry on the biological activity of forest soils). In *Okhrana Truda i Okruzhayushchei Sredy v Tekhnologicheskikh Protsessakh Energetiki i Chernoi Metallurgii* (V.M. Zadara, Ed.), pp. 132-136. Gosudarstvennyi Universitet, Dnepropetrovsk.

Egorova, E.I. (1995) Fermentativnaya aktivnost' pochv Bryanskoi oblasti, postradavshei v rezul'tate avarii na Chernobyl'skoi AES (Enzymatic activity of soils in the Bryansk region, suffering as a result of the accident at the Chernobyl Atomic Energy Station). *Izvestiya Vuzov. Yadernaya Energetika*, No. 3, 72-77.

Eliade, G., Ionescu, A., Ghinea, L. and Ştefanic, G. (1977) Poluarea cu fluor şi repercursiunile sale asupra solului şi plantelor (The fluorine pollution and its repercussions on the soil and plants). *Analele Institutului de Cercetări pentru Cereale şi Plante Tehnice* (Fundulea), **42**, 447-454.

Evdokimova, G.A. (1982) Mikrobiologicheskaya aktivnost' pochv pri zagryaznenii tyazhelymi metallami (Microbiological activity of soils polluted by heavy metals). *Pochvovedenie*, No. 6, 125-132.

Evdokimova, G.A. (1989) Vosstanovlenie mikrobiologicheskikh i biokhimicheskikh svoistv pochvy posle khimicheskogo zagryazneniya (Restoration of the microbiological and biochemical properties of soil after chemical pollution). In *Plodorodie Pochv Murmanskoi Oblasti* (V.N. Pereverzev, Ed.), pp. 74-80. Kol'skii Nauchnyi Tsentr, Apatity.

Evdokimova, G.A. (1992) Ekologicheski dopustimye promyshlennye vozdeistviya na pochvennuyu biotu i vozmozhnost' vosstanovleniya biologicheskikh svoistv pochvy (Ecologically admissible industrial influences on the soil biota and possibility of restoration of the biological properties of soil). In *Ekologo-Geograficheskie Problemy Kol'skogo Severa*, pp. 47-57. Kol'skii Nauchnyi Tsentr, Apatity.

Evdokimova, G.A. (1995) *Ekologo-Mikrobiologicheskie Osnovy Okhrany Pochv Krainego Severa* (Ecological-Microbiological Bases of Soil Protection in the Far North). Kol'skii Nauchnyi Tsentr, Apatity.

Evdokimova, G.A., Kislykh, E.E. and Mozgova, N.P. (1984) *Biologicheskaya Aktivnost' Pochv v Usloviyakh Aerotekhnogennogo Zagryazneniya na Krainem Severe* (Biological Activity of Soils under the Conditions of Technogenic Air Pollution in the Far North). Izdatel'stvo Nauka, Sankt-Petersburg.

Fenn, L.B., Tipton, J.L. and Tatum, G. (1992) Urease activity in two cultivated and non-cultivated arid soils. *Biology and Fertility of Soils*, **13**, 152-154.

Filippova, K.F., Varankina, G.I. and Gorbunova, R.V. (1980) Vliyanie proizvodstvennykh gazov na mikrobiologicheskie protsessy v pochve (Effect of industrial gases on the microbiological processes in soil). In *Gazoustoichivost' Rastenii* (V.S. Nikolaevskii, Ed.), pp. 131-152. Izdatel'stvo Nauka, Sibirskoe Otdelenie, Novosibirsk.

Finkernagel, R. and Schinner, F. (1986) Einfluss von Kalkung und schwefelsaurer Beregnung auf bodenbiologische Aktivitäten eines belasteten Waldstandortes (Influence of liming and simulated sulphuric acid rain on the soil biological activities of a polluted forest site). *Veröffentlichungen der Landwirtschaftlich-chemischen Bundesanstalt Linz/Donau*, **18**, 193-207.

Francis, A.J. (1986) Acid rain effects on soil and aquatic microbial processes. *Experientia*, **42**, 455-465.

Francis, A.J. (1989) Effects of acidic precipitation on soil microorganisms. In *Acidic Precipitation. Vol. 2. Biological and Ecological Effects* (D.C. Adriano and A.H. Johnson, Eds.), pp. 305-326. Springer, New York.

Freedman, B. and Hutchinson, T.C. (1980a) Pollutant inputs from the atmosphere and accumulations in soils and vegetation near a nickel-copper smelter at Sudbury, Ontario, Canada. *Canadian Journal of Botany*, **58**, 108-132.

Freedman, B. and Hutchinson, T.C. (1980b) Effects of smelter pollutants on forest leaf litter decomposition near a nickel-copper smelter at Sudbury, Ontario. *Canadian Journal of Botany*, **58**, 1722-1736.

Fründ, H.-C., Ruszkowski, B., Söntgen, M. and Graefe, U. (1988) Besiedlung städtischer Böden durch Regenwürmer, Enchytraeiden und bodenlebende Gehäuseschnecken (Colonisation of urban soils by earthworms, enchytraeids and terrestrial snails). *Mitteilungen der Deutschen Bodenkundlichen Gesellschaft*, **56**, 351-356.

Galiulin, R.V. (1989) Indikatsiya zagryazneniya pochv tyazhelymi metallami putem opredeleniya aktivnosti pochvennykh fermentov (Indication of soil pollution by heavy metals through determination of the activity of soil enzymes). *Agrokhimiya*, No. 11, 133-142.

Gaponyuk, E.I. and Klyueva, O.O. (1990) Svoistva pochv kak faktor ustoichivosti perkhloratov (Soil properties as a factor of the resistance of perchlorates). *Trudy Instituta Eksperimental'noi Meteorologii*, No. 17, 60-64.

Gianfreda, L. and Bollag, J.-M. (1996) Influence of natural and anthropogenic factors on enzyme activity in soil. In *Soil Biochemistry, Vol. 9* (G. Stotzky and J.-M. Bollag, Eds.), pp. 123-193. Dekker, New York.

Gradova, N.B., Belov, A.P., Kozhevin, P.A. and Myakshina, I.V. (1994) Vybor indikatornykh biosistem pri otsenke vozdeistviya biotekhnologicheskikh proizvodstv na pochvu (Selection of indicator biosystems for evaluation of the influence of biotechnological plants on the soil). *Biotekhnologiya*, No. 9-10, 34-38.

Greszta, J. (1988) The effect of dusts from electro-filters of different industrial works on forest ecosystems. *Zeszyty Naukowe Akademii Rolniczej v Krakowie, Leśnictwo*, No. 18, 3-21.

Greszta, J., Braniewski, S., Marczyńska-Galkowska, K. and Nosek, A. (1979) The effect of dusts emitted by non-ferrous metal smelters on the soil, soil microflora and selected tree species. *Ekologia Polska*, **27**, 397-426.

Greszta, J., Braniewski, S., Chrzanowska, E., Nosek, A., Chlodny, J., Olszowski, J. and Zwoliński, J. (1988) The influence of dusts from chosen industrial plants on particular links of forest ecosystem of the Niepolomice Forest. *Ekologia Polska*, **35**, 291-326.

Harris, J.A. (1991) The biology of soils in urban areas. In *Soils in the Urban Environment* (P. Bullock and P.J. Gregory, Eds.), pp. 139-152. Blackwell, Oxford.

Härtel, O. and Cerny, M. (1981) Veränderungen in Fichtenwaldböden durch Langzeiteinwirkung von SO_2 (Changes in spruce forest soils by long-term influence of SO_2). *Mitteilungen der Forstlichen Bundesversuchsanstalt Wien*, No. 137, 233-240.

Herbein, S.A. and Neal, J.L. (1990) Phosphatase activity in arctic tundra soils disturbed by vehicles. *Soil Biology & Biochemistry*, **22**, 853-858.

Hüttermann, A., Fedderau-Himme, B. and Rosenplänter, K. (1983) Biochemical reactivity in forest soils as indicators for environmental pollution. In *Effects of Accumulation of Air Pollutants in Forest Ecosystems* (B. Ulrich and J. Pankrath, Eds.), pp. 257-270. Reidel, Dordrecht.

Illmer, P. and Schinner, F. (1996) Einfluß von Aluminium auf bodenmikrobiologische Kenngrößen in sauren Waldböden (Influence of aluminium on soil biological parameters in acid forest soils). *Mitteilungen der Deutschen Bodenkundlichen Gesellschaft*, **81**, 271-274.

Ionescu, A., Moga, I. and Eliade, G. (1977) Interférences agriculture – pollution – zones urbaines (Interferences agriculture – pollution – urban zones). In *Probleme ale Agriculturii Contemporane* (I. Ceauşescu and A. Ionescu, Eds.), pp. 111-122. Editura Ceres, Bucharest.

Ivleva, S.N., Lovchii, N.F. and Efremov, A.L. (1985) Vliyanie levsano-efirnykh zagryaznenii na proteoliticheskuyu i ureaznuyu aktivnost' pochv el'nikov (Effect of lavsan ester pollution on the proteolytic and urease activity of soils in spruce forests). *Lesovedenie*, No. 6, 79-81.

Ivleva, S.N., Lovchii, N.F. and Efremov, A.L. (1986) Proteoliticheskaya i ureaznaya aktivnost' okul'turennykh pochv v zone vybrosov predpriyatiya poliefirnogo proizvodstva (Proteolytic and urease activity of cultivated soils in zone of emissions from a polyester-manufacturing enterprise). *Pochvovedenie*, No. 12, 133-136.

Jarvis, B.W., Lang, G.E. and Wieder, R.K. (1987) Arylsulphatase activity in peat exposed to acid precipitation. *Soil Biology & Biochemistry*, **19**, 107-109.

Joshi, S.R., Sharma, G.D. and Mishra, R.R. (1993) Microbial enzyme activities related to litter decomposition near a highway in a sub-tropical forest of North East India. *Soil Biology & Biochemistry*, **25**, 1763-1770.

Kandeler, E., Lüftenegger, G. and Schwarz, S. (1992) Bodenmikrobiologische Prozesse und Testaceen (Protozoa) als Indikatoren für Schwermetallbelastung (Soil microbiological processes and Testacea (Protozoa) as indicators of heavy metal pollution). *Zeitschrift für Pflanzenernährung und Bodenkunde*, **155**, 319-322.

Keplin, B. and Broll, G. (1997) Earthworms and dehydrogenase activity in urban biotopes. *Soil Biology & Biochemistry*, **29**, 533-536.

Killham, K. and Wainwright, M. (1981) Microbial release of sulphur ions from atmospheric pollution deposits. *Journal of Applied Ecology*, **18**, 889-896.

Kinzel, H. (1991) Bodenbiologische Parameter als Hilfsmittel der Bioindikation (Soil biological parameters as auxiliary measurements of the bioindication). *VDI Berichte*, No. 901, 1003-1016.

Kinzel, H., Baumgarten, A., Spadinger, K. and Zechmeister-Boltenstern, S. (1989) Influence of environmental pollution on nitrogen metabolism and other metabolic activities in forest soils. In *Ecological Impact of Acidification* (I. Szabolcs, Ed.), pp. 69-77. Magyar Tudományos Akadémia, Budapest.

Kiss, S. (1958) Talajenzimek (Soil enzymes). In *Talajtan* (M.J. Csapó, Ed.), pp. 491-622. Mezőgazdasági és Erdészeti Könyvkiadó, Bucharest.

Kiss, S. (1997) Enzymology of soils affected by industrial emissions. *Studia Universitatis Babeş-Bolyai, Biologia*, **42** (1-2), 3-62.

Kobus, J. (1975) Wplyw uprzemyslowienia na równowage biologiczna gleby (Influence of industrialisation on the biological equilibrium of soil). In *Procesy Mikrobiologiczne w Glebie* (E. Matusiewicz, Ed.), pp. 102-113. Akademia Rolnicza, Poznan.

Kobzev, V.A. (1980) Vzaimodeistvie zagryaznyayushchikh pochvu tyazhelykh metallov i pochvennykh mikroorganizmov (obzor) (Interaction of the soil-polluting heavy metals and soil microorganisms (a review)). *Trudy Instituta Eksperimental'noi Meteorologii*, No. 10, 51-66.

Kovalenko, L.A. (1996) Izmeneniya v biologicheskoi aktivnosti lesnoi podstilki i pochvy sosnovykh nasazhdenii v zone zagryazneniya serusoderzhashchimi soedineniyami i tyazhelymi metallami (Changes in the biological activity of forest litter and soil of pine plantations in the zone of pollution by sulphur-containing compounds and heavy metals). *Tezisy Dokladov Mezhdunarodnogo Soveshchaniya "Biologicheskaya Rekul'tivatsiya Narushennykh Zemel'"* (Ekaterinburg, 1996), pp. 67-68.

Kovalenko, L.A., Babushkina, L.G. and Shebalova, N.M. (1996) Summarnaya aktivnost' katalazy i koeffitsient katalazy pochvy kak bioindikatory stepeni zagryazneniya i podavleniya biologicheskoi aktivnosti pochvy razlichnymi po khimicheskoi prirode pollyutantami (Summed catalase activity and catalase coefficient of soil as bioindicators of the degree of pollution and depression of soil biological activity by pollutants of different chemical nature). *Tezisy Dokladov Mezhdunarodnogo Soveshchaniya "Biologicheskaya Rekul'tivatsiya Narushennykh Zemel'"* (Ekaterinburg, 1996), pp. 69-71.

Krasnova, N.M. (1982) Aktivnost' pochvennykh fermentov v usloviyakh tekhnogennogo zagryazneniya (Activity of soil enzymes under the conditions of technogenic pollution). *Khimiya v Sel'skom Khozyaistve*, No. 3, 28-30.

Kuchma, V.N. and Grishko, V.N. (1989) Degidrogenaznaya aktivnost' – pokazatel' zagryazneniya pochv soedineniyami ftora i rodanidami (Dehydrogenase activity – an indicator of the pollution of soils with fluorine compounds and thiocyanates). In *Ekologicheskie Aspekty Okhrany i Ratsional'nogo Ispol'zovaniya Biologicheskikh Resursov*, pp. 69-76. Gosudarstvennyi Universitet, Dnepropetrovsk.

Kulhavý, J. (1987) The effect of sulphur pollutants on microorganisms and biochemical processes in forest soils. In *Proceedings, 9th International Symposium on Soil Biology and Conservation of the Biosphere* (Sopron, 1985) (J. Szegi, Ed.), pp. 635-642. Akadémiai Kiadó, Budapest.

Kuperman, R.G. and Carreiro, M.M. (1997) Soil heavy metal concentrations, microbial biomass and enzyme activities in a contaminated grassland ecosystem. *Soil Biology & Biochemistry*, **29**, 179-190.

Kuz'mina, T.I., Reut, G.M., Gaponyuk, E.I. and Morshina, T.N. (1988) Vliyanie tekhnogennykh vybrosov fosfornogo proizvodstva na sostav i svoistva pochvy (Effect of technogenic emissions from a phosphate fertiliser factory on soil composition and properties). *Trudy Instituta Eksperimental'noi Meteorologii*, No. 16, 95-102.

Lettl, A. (1984) Vliv rhizosféry travin na biochemické aktivity heterotrofních baktérii z lesní půdy (Effect of grass rhizosphere on the biochemical activity of heterotrophic bacteria from forest soil). *Lesnictví* (Prague), **30**, 351-362.

Lettl, A. (1986a) Vývoj mikrobního osídlení lesních půd zatížených imisemi SO_2 (Development of microbial population in forest soils exposed to SO_2 emissions). *Lesnictví* (Prague), **32**, 593-610.

Lettl, A. (1986b) Biochemical activities of soil microflora in SO_2 polluted forest stands. *Folia Microbiologica* (Prague), **31**, 220-227.

Lettl, A. (1987) Thiosulphate sulfurtransferase (rhodanese) in forest soils. *Folia Microbiologica* (Prague), **32**, 334-338.

Lovchii, N.F., Efremov, A.L., Shimko, N.A. and Ivleva, S.N. (1990) Fermentativnaya aktivnost' pochv, zagryaznennykh vybrosami predpriyatiya po proizvodstvu lavsana (Enzymatic activity of soils polluted by emissions from an enterprise manufacturing lavsan). *Pochvovedenie*, No. 4, 120-124.

Marusina, N.M. and Vazhenin, I.G. (1981) Vliyanie tekhnogennykh vybrosov cherez atmosferu na fermentativnuyu aktivnost' dernovo-podzolistykh pochv (Effect of the technogenic emissions into the atmosphere on the enzymatic activity of soddy-podzolic soils). *Povyshenie Plodorodiya Pochv i Proizvoditel'noi Sposobnosti Zemel' v Intensivnykh Sistemakh Zemledeliya* (Minsk), p. 46.

Moorhead, D.L., Linkins, A.E. and Everett, K.R. (1996) Road dust alters extracellular enzyme activities in tussock tundra soils, Alaska, U.S.A. *Arctic and Alpine Research*, **28**, 346-351.

Nannipieri, P., Badalucco, L., Landi, L. and Pietramellara, G. (1997) Measurement in assessing risk of chemicals to the soil ecosystem. In *Ecotoxicology: Responses, Biomarkers and Risk Assessment, an OECD Workshop* (J.T. Zelikoff, Ed.), Chapter 34. SOS Publications, Fair Haven, New Jersey.

Naprasnikova, E.V. (1993) Biokhimicheskie i mikrobiologicheskie pokazateli ekologicheskikh funktsii pochv (Biochemical and microbiological indices of the ecological functions of soils). In *Pochvy i Povyshenie Ikh Proizvoditel'noi Sposobnosti*, pp. 57-60. Krasnoyarskii Nauchno-Issledovatel'skii Institut Sel'skogo Khozyaistva, Novosibirsk.

Neal, J.L. and Herbein, S.A. (1983) Abiontic enzymes in arctic soils: changes in sulphatase activity following vehicle disturbance. *Plant and Soil*, **70**, 423-427.

Nikitina, Z.I., Naprasnikova, E.V. and Kislitsyna, V.P. (1988) Sostoyanie mikrobioty pochv tekhnogennykh landshaftov (State of microbiota in soils of technogenic landscapes). In *Geografiya Pochv i Geokhimiya Landshaftov Sibiri* (E.G. Nechaeva and V.A. Snytko, Eds.), pp. 81-94. Institut Geografii, Irkutsk.

Nohrstedt, H.-Ö. (1985a) Studies of forest floor biological activities in an area previously damaged by sulphur dioxide emissions. *Water, Air, and Soil Pollution*, **25**, 301-311.

Nohrstedt, H.-Ö. (1985b) Biological activity in soil from forest stands in central Sweden as related to site properties. *Microbial Ecology*, **11**, 259-266.

Nordgren, A., Bååth, E. and Söderström, B. (1983) Microfungi and microbial activity along a heavy metal gradient. *Applied and Environmental Microbiology*, **45**, 1829-1837.

Nordgren, A., Kauri, T., Bååth, E. and Söderström, B. (1986) Soil microbial activity, mycelial lengths and physiological groups of bacteria in a heavy metal polluted area. *Environmental Pollution, Series A*, **41**, 89-100.

Ohtonen, R. and Markkola, A.M. (1989) Effect of local air pollution on the sporophore production of mycorrhizal fungi, mycorrhizae and microbial activity in Scots pine forests. *Meddelelser fra Norsk Institut for Skogforskning*, **42**, 121-132.

Ohtonen, R. and Markkola, A.M. (1991) Biological activity and amount of FDA mycelium in mor humus of Scots pine stands (*Pinus sylvestris* L.) in relation to soil properties and degree of pollution. *Biogeochemistry*, **13**, 1-26.

Ohtonen, R., Markkola, A.M., Heinonen-Tanski, H. and Fritze, H. (1990) Soil biological parameters as indicators of changes in Scots pine forests (*Pinus sylvestris* L.) caused by air pollution. In *Acidification in Finland* (P. Kauppi, K. Kenttämies and P. Anttila, Eds.), pp. 373-393. Springer, Berlin.

Ohtonen, R., Ohtonen, A., Luotonen, H. and Markkola, A.M. (1992) Enchytraeid and nematode numbers in urban, polluted Scots pine (*Pinus sylvestris*) stands in relation to other soil biological parameters. *Biology and Fertility of Soils*, **13**, 50-54.

Ohtonen, R., Lähdesmäki, P. and Markkola, A.M. (1994) Cellulase activity in forest humus along an industrial pollution gradient in Oulu, northern Finland. *Soil Biology & Biochemistry*, **26**, 97-101.

Olszowski, J. (1976a) The effect of fertilization on a pine forest ecosystem in an industrial region. VI. Biological activity of the soils. *Ekologia Polska*, **24**, 345-358.

Olszowski, J. (1976b) Nawożenie boru sosnowego w rejonie przemyslowych zanieczyszczeń powietrza. II. Wplyw na siedlisko (Fertilization of a pine forest in an area of industrial air pollution. II. Effect on the site). *Sylwan* (Warsaw), No. 11, 9-18.

Olszowski, J. (1980) Wplyw nawożenia na aktywność biochemiczna gleb leśnych w rejonie przemyslowym (Effect of fertilization on the biochemical activity of forest soils in an industrial area). *Archivum Ochrony Środowiska*, No. 3-4, 107-112.

Olszowski, J. (1981) Auswirkung der Düngung eines Kiefernbestandes in einem Rauchschadensgebiet (Effect of fertilisation of a pine stand in a smoke-damaged region). *Mitteilungen der Forstlichen Bundesversuchsanstalt Wien*, No. 137, 247-251.

Pancholy, S.K., Rice, E.L. and Turner, J.A. (1975) Soil factors preventing revegetation of a denuded area near an abandoned zinc smelter in Oklahoma. *Journal of Applied Ecology*, **12**, 337-342.

Pavlyukova, N.F. (1989) Aktivnost' proteazy v pochve, zagryaznennoi soedineniyami azota (Protease activity in soil polluted by nitrogen compounds). In *Ekologicheskie Aspekty Okhrany i Ratsional'nogo Ispol'zovaniya Biologicheskikh Resursov*, pp. 64-69. Gosudarstvennyi Universitet, Dnepropetrovsk.

Pavlyukova, N.F. and Dolgova, L.G. (1993) Indikatsiya edafotopov, zagryaznennykh tekhnogennymi veshchestvami, po aktivnosti fermentov (Enzyme activity as an indicator for edaphotopes polluted by technogenic substances). *Pochvovedenie*, No. 1, 45-47.

Pohla, H., Palzenberger, M., Krassnigg, F., Kandeler, E., Schwarz, S. and Kasperowski, E. (1991) Bodenbiologische, -chemische und –physikalische Parameter entlang eines Schadstoffgradienten auf Grünlandstandorten in der Umgebung von Brixlegg (Tirol) – Vorstellung eines Pilotprojekts (Soil biological, chemical, and physical parameters along a pollutant gradient on grassland sites in the surroundings of Brixlegg (Tirol) – Presentation of a pilot project). *VDI Berichte*, No. 901, 1083-1094.

Polomski, J. (1985) Fluorbedingte Veränderungen von chemischen und biologischen Gleichgewichten im Boden (Changes in chemical and biological equlibria in fluoride-contaminated soil). *Landwirtschaftliche Forschung*, **38**, 139-146.

Press, M.C., Henderson, J. and Lee, J.A. (1985) Arylsulphatase activity in peat in relation to acidic deposition. *Soil Biology & Biochemistry*, **17**, 99-103.

Prokopenko, N.P., Prokazov, G.F. and Zimenko, T.G. (1983) Deistvie pylegazovybrosov MPO "Khimvolokno" na mikrobiologicheskoe sostoyanie pochvy (Effect of dust-gas emissions from the "Khimvolokno" synthetic fibre factory on the microbiological state of soil). *Okhrana Okruzhayushchei Sredy* (Minsk), No. 2, 105-108.

Răuţă, C., Ianculescu, M., Mihăilescu, A., Cârstea, S., Toti, M., Buceag, E., Gamenţ, E., Mihalache, G., Dancău, H. and Tisescu, A. (1988) Contribuţii la cunoaşterea poluării industriale a solului şi vegetaţiei forestiere în zona Copşa Mică (Contributions to the knowledge of industrial pollution of soil and forest vegetation in the Copşa Mică area). *Ziridava* (Arad), **17**, 114-116.

Rejšek, K. (1991) Acid phosphomonoesterase activity of ectomycorrhizal roots in Norway spruce pure stands exposed to pollution. *Soil Biology & Biochemistry*, **23**, 667-671.

Rühling, Å. (1981) Effekter av bly på nedbrytningsprocesser i skogsmark (Effects of lead on decomposition in forest soils). *Institutet för Vatten- och Luftvårdsforskning* (Stockholm), *Publ. B*, **595**, 1-15.

Rühling, Å. and Tyler, G. (1973) Heavy metal pollution and decomposition of spruce needle litter. *Oikos*, **24**, 402-416.

Rühling, Å. and Tyler, G. (1979) Effekter av tungmetallförorening på nedbrytningsprocesser i skogsmark. IV. Kromemitterande industri (Effects of heavy metal pollution on decomposition in forest soils. IV. Chromium-emitting industry). *Statens Naturvårdsverk, PM*, **1150**, 1-32.

Samoilova, T.S. (1988) Vliyanie vybrosov avtotransporta na mikrofloru pochvy (Effect of emissions from motor vehicle transport on the soil microflora). *Tezisy Dokladov "Mikrobiologicheskie Metody Zashchity Okruzhayushchei Sredy"* (Pushchino, 1988), pp. 162-163.

Samoilova, T.S. and Gorbatyuk, M.S. (1988) Vliyanie antropogennogo faktora na fermentativnuyu aktivnost' pochvy (Effect of anthropogenic factor on the enzymatic activity of soil). *Tezisy Dokladov 3. Vsesoyuznogo Simpoziuma "Biodinamika Pochv"* (Harku-Tallinn, 1988), p. 142.

Schulte, W., Fründ, H.-C., Söntgen, M., Graefe, U., Ruszkowski, B., Voggenreiter, V. and Weritz, N. (1989) *Zur Biologie städtischer Böden. Beispielraum: Bonn – Bad Godesberg* (Biology of Urban Soils. An Example: the Bonn – Bad Godesberg Area). Kilda-Verlag, Greven.

Semenov, A.M., Batomunkueva, B.P., Nizovtseva, D.V. and Panikov, N.S. (1996) Method of determination of cellulase activity in soils and microbial cultures, and its calibration. *Journal of Microbiological Methods*, **24**, 259-267.

Skujinš, J. (1978) History of abiontic soil enzyme research. In *Soil Enzymes* (R.G. Burns, Ed.), pp. 1-49, Academic Press, London.

Skvortsova, I.N., Li, S.K. and Voroteikina, I.P. (1980) Zavisimost' nekotorykh pokazatelei biologicheskoi aktivnosti pochv ot urovnya kontsentratsii tyazhelykh metallov (Dependence of some indices of the biological activity of soils on the level of the concentration of heavy metals). In *Tyazhelye Metally v Okruzhayushchei Srede* (V.V. Dobrovol'skii, Ed.), pp. 121-125. Izdatel'stvo Moskovskogo Universiteta, Moscow.

Soreanu, I. (1982) The influence of pollution on soil enzyme activity in forest stands. In *Fourth Symposium on Soil Biology* (Cluj-Napoca, 1977) (M.P. Nemeş, S. Kiss, P. Papacostea, G. Ştefanic and D. Nemeş, Eds.), pp. 193-198. Editura Ceres, Bucharest.

Starzecka, A. (1989) Effect of industrial dusts on the development and activity of microorganisms in soils of the Niepolomice Forest (southern Poland). *Acta Hydrobiologica* (Cracow), **31**, 167-206.

Stefurak, V.P. (1988) Biologicheskaya aktivnost' dernovo-podzolistykh pochv v usloviyakh aerotekhnicheskogo zagryazneniya v Prikarpat'e (Biological activity of soddy-podzolic soils under the conditions of industrial air pollution in the Carpathian region). *Mikrobiologicheskii Zhurnal* (Kiev), No. 1, 7-11.

Stefurak, V.P. (1990) Vliyanie promyshlennykh vybrosov magnievogo zavoda na biologicheskuyu aktivnost' pochvy (Effect of the industrial emissions of a magnesium plant on the biological activity of the soil). *Mikrobiologicheskii Zhurnal* (Kiev), No. 4, 93-98.

Tabatabai, M.A. (1985) Effect of acid rain on soils. *CRC Critical Reviews in Environmental Control*, **15**, 65-110.

Tiller, K.G. (1989) Heavy metals in soils and their environmental significance. *Advances in Soil Science*, **9**, 113-142.

Tscherko, D., Öhlinger, R., Hackl, E. and Kandeler, E. (1996) Bodenmikrobiologisches Monitoring im Rahmen der Salzburger Dauerbeobachtungsflächen (Soil microbiological monitoring within the long-term observation areas in Salzburg). *Mitteilungen der Deutschen Bodenkundlichen Gesellschaft*, **81**, 357-360.

Tyler, G. (1973) Heavy metal pollution and soil enzymatic activity. *Meddelanden från Avdelningen för Ekologisk Botanik, Lunds Universitet*, **1** (9), 1-12.

Tyler, G. (1974) Heavy metal pollution and soil enzymatic activity. *Plant and Soil*, **41**, 303-311.

Tyler, G. (1975a) Heavy metal pollution and mineralisation of nitrogen in forest soils. *Nature*, **225**, 701-702.

Tyler, G. (1975b) Effects of heavy metal pollution on decomposition in forest soils. II. Decomposition rate, mineralization of nitrogen and phosphorus, soil enzymatic activity. *Statens Naturvårdswerk, PM*, **542E**, 1-47.

Tyler, G. (1975c) Effect of heavy metal pollution on decomposition and mineralization rates in forest soils. In *International Conference on Heavy Metals in the Environment* (Toronto, 1975) (T.C. Hutchinson, A.L. Page and J.C. van Loon, Eds.), pp. 217-226.

Tyler, G. (1976) Heavy metal pollution, phosphatase activity, and mineralization of organic phosphorus in forest soils. *Soil Biology & Biochemistry*, **8**, 327-332.

Tyler, G. (1977) Effekter av tungmetallförorening på nedbrytningsprocesser i skogsmark. III. Utvidgade undersökningar och sammanfattande utvärdering (Effects of heavy metal pollution on decomposition in forest soils. III. Extended studies and concluding evaluation). *Statens Naturvårdsverk, PM*, **861**, 1-105.

Tyler, G. (1981) Heavy metals in soil biology and biochemistry. In *Soil Biochemistry, Vol. 5* (E.A. Paul and J.N. Ladd, Eds.), pp. 371-414. Dekker, New York.

Tyler, G. (1984) The impact of heavy metal pollution on forests: a case study of Gusum, Sweden. *Ambio*, **13**, 18-24.

Tyler, G. and Westman, L. (1979) Effekter av tungmetallförorening på nedbrytningprocesser i skogsmark. VI. Metaller och svavelsyra (Effects of heavy metal pollution on decomposition in forest soils. VI. Metals and sulphuric acid). *Statens Naturvårdsverk, PM*, **1203**, 1-33.

Wainwright, M. (1979) Microbial S-oxidation in soils exposed to heavy atmospheric pollution. *Soil Biology & Biochemistry*, **11**, 95-98.

Wainwright, M. (1980) Effect of exposure to atmospheric pollution on microbial activity in soil. *Plant and Soil*, **55**, 199-204.

Wainwright, M. (1982) Microbial oxidation of sulphur in soils subject to atmospheric sulphur deposition. In *Proceedings, International Sulphur Conference* (London, 1982) (A.I. More, Ed.), pp. 427-437. British Sulphur Co., London.

Wainwright, M. and Killham, K. (1982) Microbial transformations of some particulate pollution deposits in soils – A source of plant-available nitrogen and sulphur. *Plant and Soil*, **65**, 297-301.

Weritz, N. (1989) Untersuchungen zur Biologie städtischer Böden. 6. Mikroflora (Investigations on the biology of urban soils. 6. Microflora). *Verhandlungen der Gesellschaft für Ökologie* (Essen, 1988), **18**, 211-213.

Weritz, N. and Schröder, D. (1988) Mikrobielle Aktivität in Stadtböden unterschiedlicher Nutzung (Microbial activity in urban soils under different use). *Mitteilungen der Deutschen Bodenkundlichen Gesellschaft*, **56**, 399-404.

Weritz, N. and Schröder, D. (1989) Mikrobielle Aktivitäten in Stadtböden und ihre Bewertung unter besonderer Berücksichtigung von Schwermetallbelastungen (Microbial activities in urban soils and their evaluation with special regard to heavy metal contaminations). *Mitteilungen der Deutschen Bodenkundlichen Gesellschaft*, **59**, 1015-1020.

Weritz, N. and Schröder, D. (1990) Die Bewertung mikrobieller Aktivitäten in Stadtböden als Beitrag zum städtischen Bodenschutz (Evaluation of the microbial activities in urban soils as a contribution to urban soil protection). *Mitteilungen der Deutschen Bodenkundlichen Gesellschaft*, **61**, 149-152.

Zagural'skaya, L.M. (1992) Ispol'zovanie biokhimicheskikh pokazatelei dlya otsenki stepeni tekhnogennoi degradatsii pochv (Use of biochemical indices for evaluating the degree of the technogenic degradation of soils). In *Pochvennye Resursy Karelii, Ikh Ratsional'noe Ispol'zovanie i Okhrana*, pp. 132-142. Institut Lesa, Petrozavodsk.

Zagural'skaya, L.M. and Zyabchenko, S.S. (1994) Vozdeistvie promyshlennykh zagryaznenii na mikrobiologicheskie protsessy v pochvakh boreal'nykh lesov raiona Kostomukshi (Impact of industrial pollution on the microbiological processes in soils of boreal forests of the Kostomuksha district). *Pochvovedenie*, No. 5, 105-110.

Zechmeister-Boltenstern, S., Spadinger, K. and Kinzel, H. (1991) Bodenenzymatische Aktivitäten an verschieden stark belasteten Buchenwaldstandorten (Soil enzymatic activities on beech forest sites affected by differently heavy pollution). In *Zustandserhebung Wienerwald. Pflanzenphysiologische und bodenökologische Untersuchungen zur Bioindikation* (R. Albert, K. Burian and H. Kinzel, Eds.), pp. 209-246. Verlag der Österreichischen Akademie der Wissenschaften, Vienna.

Zehner, R., Mentler, A., Pfeffer, M. and Blum, W.E.H. (1994) Bodenbiologische Aktivitätsmessungen im Stammablaufbereich eines immissionsbelasteten Buchenbestandes im Wienerwald (Soil biological activity measurements in the stemflow zone of an emission-contaminated beech stand in the Vienna Woods). *Mitteilungen der Österreichischen Bodenkundlichen Gesellschaft*, No. 48/49, 435.

Zimny, H. and Żukowska-Wieszczek, D. (1983a) Dehydrogenase activity depending on the kind of urban greens. *Polish Ecological Studies*, **9**, 113-122.

Zimny, H. and Żukowska-Wieszczek, D. (1983b) Enzymatic activity in soils of urban lawns depending on sources of degradation. *Polish Ecological Studies*, **9**, 123-130.

Zimny, H. and Żukowska-Wieszczek, D. (1983c) The effect of inorganic fertilization on the enzymatic activity of soils of urban lawns. *Polish Ecological Studies*, **9**, 131-142.

Zimny, H., Korzeniewska, E. and Stawicki, M. (1992a) Enzymatic activity of the Warsaw soils in dependence on the degree of anthropopressure and plant cover. In *Proceedings, 15th International Meeting of Specialists in Air Pollution Effects on Forest Ecosystems "Air Pollution and Interactions between Organisms in Forest Ecosystems"* (Tharandt/Dresden, 1992), pp. 259-267.

Zimny, H., Żukowska-Wieszczek, D. and Korzeniewska, E. (1992b) Biological activity of soils as index of environment degradation under urban conditions. *Annals of Warsaw Agricultural University, Horticulture*, No. 16, 53-57.

Żukowska-Wieszczek, D. (1979) Aktywność biologiczna gleb w aglomeracji warszawskiej (Biological activity of soils in Warsaw agglomeration). *Materialy z Konf. "Trawniki, Nawierzchnie Darniowe i Runo Parkowe"* (Gdansk, 1979), pp. 20-33.

Żukowska-Wieszczek, D. (1980) Bioindication of soil pollution on urban area. *Ekologia Polska*, **28**, 267-284.

Żurawska, J. (1984) Wplyw substancji użyżniajacej na aktywność mikrobiologiczna gleby w strefie bezroślinnej zakladów przemyslu nieorganicznego "Polchem" w Toruniu (The influence of fertilising substances on the microbiological activity of soil in the plantless zone of the "Polchem" inorganic industry works in Torun). *Czlowiek i Środowisko*, **8**, 85-107.

Zwoliński, J., Olszowski, J., Olszowska, G. and Zwolińska, B. (1988) The effect of industrial dusts from different emission sources on the biological activity of soils. *Zeszyty Naukowe Akademii Rolniczej v Krakowie, Leśnictwo*, No. 18, 105-123.

PART III. ENZYMOLOGY OF TECHNOGENIC SOILS

INTRODUCTION

Technogenic soils are soils that form during the technical and biological recultivation of overburdens, tailings, and other spoils and wastes resulting from strip (opencast, surface) and shaft (underground, deep) mining and other industrial activities. At the same time, all these wastes constitute a dangerous source of environmental pollution. The evolution of technogenic soils is, by definition, the process of transforming all these wastes into agricultural and forest soils or into soils used for other purposes (parks, sports fields, etc.). Simultaneously, this process is accompanied by reduction or elimination of the polluting effects of wastes on the environment. The practical importance of this process is growing because the development of mining and other industries leads to increasing amounts of wastes and, therefore, the recultivation of wastelands becomes more and more a major economic necessity. It is estimated that up to 1980 about 1600×10^9 m^3 of mine spoils accumulated on the earth's surface and this amount has increased yearly by about 40×10^9 m^3. For comparison it should be mentioned that water erosion affects a smaller amount of soil (about 13×10^9 m^3 per year). There are also estimates showing that the surfaces degraded during extraction of each million ton of coal in one of the basins of our country comprise up to 40-50 ha of agricultural land. The evolution of technogenic soils presents questions of theoretical importance, too, which is related to a better understanding of the evolution of landscape as a whole (Greszta, 1973; Nastea *et al.*, 1973; Khazanov, 1975; Bradshaw and Chadwick, 1980; Szegi, 1983; Sutton and Dick, 1987; Răuţă, 1988; Panas, 1989; Harris, 1991).

The evolution of technogenic soils, which reflects the efficiency of recultivation, is studied using many physical, chemical, and biological methods. Enzymological methods have also been applied, and it has been found that the level of enzymatic activity is a good indicator of the degree of evolution of technogenic soils (Kiss *et al.*, 1985, 1989a, 1991, 1993, 1996).

In the present review work, data from the literature concerning the enzymology of technogenic soils will be grouped according to the nature of the raw material from the mining and processing of which the wastes resulted (coal; peat; oil shale; iron, manganese, lead and zinc, and gold ores; apatite and phosphorite; sulphur; lime and dolomite; refractory clay; bentonitic clay; pumice; sand; and gravel). Then, the enzymology of the recultivation of overburdens remaining after pipeline construction and of soils remaining after "mining" (removal) of their topsoil for use in landscape improvement, and the enzymology of the revegetation of soils used for covering rubble dumps and of landfill soils will be dealt with. Finally, enzymological aspects of the transformation of harbour muds into agricultural soils and of the recultivation of soils disturbed by different industrial activities will be described.

Chapter 7. TECHNOGENIC SOILS FROM COAL MINE SPOILS

Enzymological research in Ukraine

Keleberda (1973) determined the H_2O_2-splitting capacity in the Aleksandrii spoil heap at the Baidakov brown coal strip mine (located in the northern steppe zone of Ukraine, Kirovograd, formerly Elizavetgrad, region) and in the spoil heaps of the abandoned Yurkov brown coal strip mine (located in the forest-steppe zone of Ukraine, Cherkassy region). The spoils consisted predominantly of loess. At both mines, a part of the spoils were recultivated with Scotch pine (*Pinus sylvestris*) for 8 and 3 years, respectively. In some spoil plots the plants grew well, whereas in others their growth was inhibited. The H_2O_2-splitting capacity, and especially its heat-labile component (catalase activity), was always higher in the plots with well-developed plants than in those on which the growth of plants was inhibited. Keleberda has drawn the conclusion that the forest-growing properties of recultivated spoils can be evaluated by catalase activity measurements.

Verbin and Keleberda (1974) compared urease and catalase activities in the Yurkov spoil plots recultivated only with black alder (*Alnus glutinosa*), with black alder plus Scotch pine, or only with Scotch pine. In pure stands, the growth of black alder was good and that of Scotch pine bad. In mixed stands, the black alder stimulated the growth of the Scotch pine. After 3 years of recultivation, enzyme activities, and nitrogen (N) content (total, NH_4^+, NO_3^-) in the 0-20- and 20-40-cm layers of the spoil plots showed the following order: black alder plots > black alder plus Scotch pine plots > Scotch pine plots. The values were highest in the rhizosphere and especially in the zone of root nodules of black alder.

Keleberda *et al.* (1974) and Keleberda and Dan'ko (1975) reported on a new enzymological study of the recultivated spoils at the Yurkov mine. Spoils (sandy loams or loamy sands) of the Strizhev coal strip mine (located in the central forest zone of Ukraine, Zhitomir region) were also studied. Invertase, urease, and catalase activities were determined in soils of spoil plots cultivated with a green manure crop, perennial lupine (at both mines), or with black alder (at Yurkov). Uncultivated spoil plots served as controls. It was found that the enzyme activities decreased with the sampling depth in each plot. Recultivation led to evident, sometimes manifold, increases of enzyme activities as compared to the uncultivated controls. The effect of perennial lupine was stronger than that of the black alder (Table 17). Owing to the lupine, the humus and total N contents also increased, 2.54-2.87 and 2.94-3.30 times, respectively. In the alder plots, the increase of humus and N contents of the soils was 1.95- and 1.81-fold, respectively. The results indicate that recultivation with black alder or, even better, with the green manure plant perennial lupine makes it possible to increase, in a relatively short period of time, the fertility level of technogenic soils.

Table 17
Enzyme activities in technogenic soils resulting from the recultivation of spoil plots at the Yurkov and Strizhev coal strip mines

Plant stand and its age	Depth (cm)	Invertase*	Urease*	Catalase*
Spoil plots at the Yurkov mine (mean values of analytical data over 3 years)				
Perennial lupine	0-2	39.48	136.9	209.8
(5 years)	2-10	17.57	97.4	135.5
	10-20	6.45	61.5	84.7
Black alder	0-2	30.82	125.6	168.2
(5 years)	2-10	16.76	74.5	106.6
	10-20	3.68	42.7	74.3
Control	0-2	6.56	70.5	124.0
(uncultivated)	2-10	2.65	45.2	90.1
	10-20	1.56	15.2	59.4
Spoil plots at the Strizhev mine (mean values of analytical data from three experiments)				
Perennial lupine	0-2	29.60	126.0	179.3
(3 years)	2-10	11.58	70.0	132.5
	10-20	5.68	56.0	98.3
Control	0-2	3.84	46.6	83.3
(uncultivated)	2-10	3.39	42.0	76.3
	10-20	2.76	28.0	68.3

Reproduced from: T.N. Keleberda and V.N. Dan'ko, Rekul'tivatsiya Zemel' Narushennykh pri Dobyche Poleznykh Iskopaemykh, Tartu, 1975, p. 246.

* Expression of enzyme activities: invertase in mg of inverted sugar g^{-1} soil 40 h^{-1}; urease in mg of NH_4^+-N 100 g^{-1} soil 40 h^{-1}; and catalase in ml of O_2 100 g^{-1} soil 5 min^{-1}.

In another study, Keleberda (1976) found that the soils of the Yurkov spoil plots cultivated with Scotch pine for 9 years or with black locust (*Robinia pseudacacia*) for 4 years also manifested higher invertase activity than did the uncultivated spoil. The activity increase in the Scotch pine soil was close to that in the lupine soil, but in the locust soil the activity increase was lower as compared with that in the lupine soil. The humus and total N contents increased in the Scotch pine and locust soils, also. It was also found (Keleberda and Dan'ko, 1975; Keleberda, 1976) that jack pine (*Pinus banksiana*), Scotch pine, weeping birch (*Betula verrucosa*), and pedunculated oak (*Quercus pedunculata*) grew much better on the green (perennial lupine)-manured than on the nonmanured Strizhev spoil plots.

These investigations were reviewed in the book of Zaitsev *et al.* (1977).

Keleberda (1977) has pointed out that in addition to invertase, urease, and catalase activities, proteinase and amylase activities also reflect the evolution of spoil heaps toward technogenic soils.

The Yurkov spoil heaps recultivated with different tree and shrub species became primitive soils in a relatively short time (9-14 years), acquiring a humus layer and an increased N content and enzymatic potential (Keleberda, 1978) (Table 18).

The spoils, consisting predominantly of loess, at the Baidakov brown coal strip mine located in the northern steppe zone of Ukraine, were also studied multilaterally by Mikhnovskaya (1981) and Eterevskaya *et al.* (1985). They found that invertase, urease, and proteinase activities and respiration (CO_2 evolution) were very low in the loess forming the walls of the quarry and in the loess heap not covered by vegetation and increased considerably with the age of the spontaneous vegetation covering the loess heaps, but even 23 years after the growth of spontaneous vegetation the activity values were lower than those measured in the undisturbed zonal soil (Table 19).

In the same mining zone, Mikhnovskaya and Lapta (1989) studied the transformation of nitrogen in loess spoil plots covered by spontaneous vegetation (for 1, 2, 4, and 20-25 years) and in those cultivated with alfalfa (for 1, 2, and 4 years), wheatgrass, bromegrass, or sainfoin (for 4 years). It was found that during 4 years, urease and proteinase activities increased to a larger extent in the plots cultivated with alfalfa than in those under spontaneous vegetation. After 4 years, values of the two enzymatic activities in the plots presented, depending on the nature of plants, the following order: spontaneous vegetation < bromegrass < wheatgrass < sainfoin < alfalfa. The amounts of total and easily hydrolysable N increased in a similar order, but the nonsymbiotic N_2 fixation capacity became more pronounced only in the plots cultivated with alfalfa or sainfoin.

Table 18
Humus and N contents and enzyme activities in primitive soils under forest vegetation developed on spoil plots at the Yurkov coal strip mine

Plant stand and its age	Depth (cm)	Humus (%)	N (%)	N in humus (%)	Invert-ase*	Ure-ase*	Protein-ase*
Scotch pine	0-2	4.74	0.24	6.0	29.70	1.96	0.56
(14 years)	2-5	2.33	0.12	5.1	10.00	0.56	0.30
	5-10	1.20	0.07	5.8	6.62	0.45	0.20
	10-20	0.41	0.02	5.0	4.42	0.34	0.16
	20-30	0.36	0.02	5.5	0.60	0.26	0.00
Black locust	0-2	2.08	0.18	8.6	22.30	1.64	0.48
(9 years)	2-5	1.02	0.09	9.0	8.02	0.62	0.28
	5-10	0.73	0.07	9.6	4.90	0.33	0.28
	10-20	0.16	0.01	6.2	1.12	0.28	0.16
	20-30	0.11	0.01	9.9	0.60	0.15	0.00
Sea buckthorn	0-2	1.65	0.19	11.5	44.50	3.68	0.96
(9 years)	2-5	0.75	0.06	8.0	11.20	0.72	0.34
	5-10	0.43	0.04	9.5	2.25	0.37	0.26
	10-20	0.26	0.02	8.0	1.71	0.16	0.15
	20-30	0.18	0.02	11.1	1.02	0.11	0.00
Black alder	0-2	1.30	0.13	10.0	28.50	2.86	0.76
(9 years)	2-5	1.03	0.12	11.6	13.50	0.84	0.36
	5-10	0.84	0.08	9.5	7.50	0.42	0.26
	10-20	0.17	0.03	17.6	4.70	0.32	0.12
	20-30	0.19	0.03	15.8	0.80	0.25	0.00
Control	0-2	0.72	0.05	6.9	10.80	1.08	0.34
(uncultivated)	2-5	0.43	0.02	4.6	5.20	0.71	0.15
	5-10	0.24	0.01	4.1	2.35	0.32	0.08
	10-20	0.23	0.01	4.3	0.10	0.20	0.00
	20-30	0.21	0.01	5.0	0.00	0.00	0.00

Reproduced from: T.N. Keleberda, Pochvovedenie, 1978, No. 9, p. 111.

* Expression of enzyme activities: invertase in mg of inverted sugar g^{-1} soil 40 h^{-1}; urease in mg of NH_4^+-N g^{-1} soil 40 h^{-1}; and proteinase in mg of NH_2-N g^{-1} soil 72 h^{-1}.

Table 19
Enzyme activities and respiration in spoils (loess) at the Baidakov coal strip mine

Analysed material	Invertase*	Urease*	Proteinase*	Respiration**
Loess from the walls of quarry	0.3	1.1	2.3	4.4
Loess (spoil) heap without vegetation	0.3	1.4	2.4	4.6
Loess (spoil) heaps spontaneously vegetated; age of vegetation:				
1 year	1.2	1.5	2.7	5.1
2 years	2.2	2.0	3.8	N.D.
4 years	4.1	2.3	5.1	28.7
23 years	7.0	4.5	9.1	35.3
Zonal soil (common chernozem)	20.5	15.6	28.7	217.0

Reproduced from: L.V. Eterevskaya, L.V. Lekhtsier, A.D. Mikhnovskaya and E.I. Lapta, Tekhnogennye Ekosistemy: Organizatsiya i Funktsionirovanie, Novosibirsk, 1985, p. 125.
* Expression of enzyme activities: invertase in mg of inverted sugar g^{-1} material 24 h^{-1}; urease in mg of NH_3 g^{-1} material 24 h^{-1}; and proteinase in mg of tyrosine 100 g^{-1} material 24 h^{-1}.
** Expressed in mg of CO_2 kg^{-1} material h^{-1}.
N.D. – Not determined.

Some enzymological aspects of the recultivation of coal shaft mine spoils situated in the valley of the Samara River (western Donets Basin, Donbass) have been dealt with by Gel'tser and co-workers (Gel'tser and Tsvetkova, 1982; Tsvetkova *et al.*, 1982; Ras'kova *et al.*, 1984; Gel'tser *et al.*, 1986, 1989; Gel'tser, 1990). The spoils were silts and clay shales with sands and coal inclusions; their pH ranged from 3.2 to 7.7. The control plot was not covered by any material. Other plots were covered, having the following structure (from top to bottom): 0.5 m of loess plus 0.5 m of sand plus mine spoils, and 0.5 m of chernozem plus 0.5 m of sand plus 0.5 or 1 m of loess plus mine spoils, respectively. The thickness of the mine spoil layer was 7-10 m. The plots had all been planted to black locust in 1976. In the control plot the growth of locust was unsatisfactory. In the other plots the locust grew well. For comparison, a zonal soil (common chernozem) under natural steppe vegetation or under a black locust plantation installed in 1966 was used. Five enzyme activities (invertase, urease, proteinase, dehydrogenase, and catalase) were measured in the 0.5-m layer of the plots. After 5-8 years of recultivation, the activities were present only in traces in the control plot. The highest activities were found in the litter of the covered plots. Urease, proteinase, and dehydrogenase activities were no lower in the 0-2- and 2-10-cm layers of these plots than at the same depths of the humus-enriched horizon of the zonal soil. The opposite was true for the invertase and catalase activities. In the 10-50-cm layer of the plot covered with loess on the surface, the activities decreased sharply, whereas in the same layer of the plot with chernozem on the surface, they remained, after a slight initial decrease, practically constant. The enzymological measurements made possible a more accurate differentiation of the layers and places of soil-forming processes within the soil profile than did its morphological description.

In another study, Gel'tser *et al.* (1985) found that proteinase and invertase activities were nearly twice as high in the litter of the black locust plots on loess and chernozem overlying mine spoils as in the litter of the black locust plantation on zonal soil.

Enzymological research in the United States of America

Reviewing the role of microorganisms in the revegetation of strip-mined land in the western United States, Cundell (1977) emphasised the observations of Pancholy *et al.* (1975), according to which the low urease and dehydrogenase activities in a completely denuded area near an abandoned zinc smelter in Oklahoma were an excellent indicator of the inability of this soil to revegetate. Therefore, Cundell advocated enzyme activity determinations for studies of mine soil recovery.

Miller (1978) described complex recultivation investigations in which enzymological methods were also applied. Four sites were studied. The first three were strip coal mine sites and the fourth, a coal refuse site, resulted from deep coal mining. The sites are located in different parts of the United States.

1. At the Indian Head mine (Zap, North Dakota), proteolytic and dehydrogenase activities were determined in topsoil stockpiled prior to replacement on the spoil. In the 0-2.5-cm layer of a single topsoil storage pile at 10 and 22 months of age, the proteolytic activity was only ~30 and 25%, respectively, in comparison to that measured in the same surface layer of the adjacent undisturbed soil. At 30-300-cm depths, this activity was also lower in the topsoil pile at 22 months of age than at 10 months of age. Dehydrogenase activity in the 0-2.5-cm layer was 20% greater in the 29-month-old topsoil pile examined than in the undisturbed soil, but in the 30-400-cm depths of the pile, dehydrogenase activity represented

only ~1-15% of that found in the 0-2.5-cm layer of the same pile. In other words, during storage the enzymatic potential diminishes to a large extent in most parts of the topsoil pile.

2. At the Jim Bridger mine near Rock Springs, Wyoming, three plots were studied enzymologically. In the first plot the spoil was covered with stored topsoil. In the second plot a native soil was immediately reapplied to the spoil without prior storage. Adjacent to these plots was the third (control) plot, an undisturbed native area composed of an *Atriplex* community and an *Artemisia* community. All plots had a west-facing aspect and a 30% slope. Plots 1 and 2 were established during the spring and autumn of 1976, with the reapplied stored topsoil plot having the extra growing season. They were seeded with wheatgrasses (*Agropyron* spp.) and fourwing saltbush (*Atriplex canescens*). No fertilisers were applied. On both plots, the dominant vegetation was *Halogeton glomeratus* followed by *Salsola kali*. Almost no wheatgrass survived, and no fourwing saltbush was evident. However, volunteer *Atriplex gardneri* and *Atriplex confertifolia* were encountered, with *A. gardneri* predominating. In July 1977, soils were sampled to determine their cellulase, urease, and dehydrogenase activities, which were found to decrease in the following order: undisturbed soil > stored topsoil applied on spoil > native topsoil applied on spoil. It should be mentioned that in a sample from the native topsoil plot, cellulase activity was not detectable at all.

3. Proteolytic and dehydrogenase activities were assayed in the surface layer of orphaned (nonrecultivated) spoil heaps at the Big Horn mine and at the abandoned Hidden Water Creek mine near Sheridan, Wyoming. The spoils were composed of semiconsolidated shale and sandstone. Surface soil samples collected from undisturbed grazed rangeland sites near the Big Horn mine served for comparison. The mean value of the proteolytic activity was ~7.5 times lower in the spoil than in the soil samples, and dehydrogenase activity was lacking in the spoil samples.

4. A coal refuse pile generated in the coal-cleaning process at a deep coal mine near the city of Staunton in Macoupin County, Illinois, was submitted, many decades after the abandonment of the mine, to technical and biological recultivation. The enzymological analyses performed at the end of the growing season in the first year of recultivation (1977) indicated the presence of two activities (urease and dehydrogenase) and the absence of proteolytic and cellulase activities in the coal refuse. These four activities gave high values in the adjacent field soil.

During the summers of 1975 and 1977, Hersman and Temple (1979) collected samples from six reclamation plots of the coal strip mine spoils in the Colstrip area (eastern Montana) and determined their ATP content, phosphatase, and pectinolyase activities, and rate of respiration (O_2 uptake). ATP content correlated significantly with respiration rate and pectinolyase activity for the 1975 samples, and with respiration rate and phosphatase activity for the 1977 samples. Of the two enzymes tested, phosphatase gave more positive values for correlation. Nevertheless, its positive correlation with both ATP content and respiration rate in one set of samples, but not in the other, suggests that phosphatase activity is more variable than either respiration rate or ATP content. Pectinolyase is more specifically related to plant decomposition and may have some implications for spoils that are in the early stage of pedogenesis. Pectinolyase activity showed only one significant correlation, suggesting that it may be primarily useful as an adjunct measurement, but not as a general indicator of microbial activity.

We think that this conclusion of Hersman and Temple's would not be valid if the evolution of a given reclamation spoil plot were considered, because the six plots studied by these investigators, as they emphasise, were distributed over a distance of many miles and had been submitted to a variety of treatments, but the differences between plots (and treatments) were not examined.

At the San Juan strip coal mine near Farmington, New Mexico, Fresquez and Lindemann (1982) determined dehydrogenase activity and several microbiological parameters (numbers of aerobic heterotrophic bacteria, streptomycetes, ammonium oxidisers, *Azotobacter*, and fungi, and distribution of fungal genera) in representative samples from four sites:

1. a spoil bank (~1 year in age; nonvegetated);

2. a topsoil stockpile (at least 1 year in age);

3. a reclaimed area (revegetated 3 years earlier by grading and levelling the spoil, spreading stockpile topsoil 18 cm deep, mulching with native hay, fertilising with 67 kg of N and 100 kg of P_2O_5 ha^{-1}, seeding with native grasses and shrubs, and irrigating for 2 years); and

4. an undisturbed surface soil (near the reclaimed area).

At each site the samples were collected to a depth of 18 cm.

The results of the determinations showed that dehydrogenase activity decreased in the following order: undisturbed soil >> reclaimed area > stockpiled topsoil > nonvegetated spoil. Microbial numbers and distribution of fungal genera were greater in the undisturbed soil and reclaimed area than in the stockpiled topsoil or nonvegetated spoil, the lowest values being registered in the nonvegetated spoil. *Azotobacter* was not found at any of the sites. One can draw the conclusion that stockpiling of topsoil leads to the diminution of enzymatic and microbial potential, whereas reclamation of spoil leads to an increase of this potential.

In agreement with this conclusion, Fresquez *et al.* (1985) found that, except for arylsulphatase, the other enzymes analysed (dehydrogenase, nitrogenase, urease, phosphatase, amylase, invertase, and protease) were less active in an older, 3- to 4-year topsoil stockpile than in the undisturbed soil.

In February 1979, Fresquez and Lindemann (1982) initiated a greenhouse experiment to study the influence of amendments on the enzymatic and microbial parameters of the nonvegetated spoil. The experimental variants, each in four repetitions, were as follows:

1. spoil;
2. spoil plus topsoil inoculant (224 t ha^{-1});
3. spoil plus alfalfa hay (22.4 t ha^{-1}) plus fertilisers (336 kg of urea and 336 kg of P_2O_5 as triple superphosphate ha^{-1});
4. spoil plus alfalfa hay plus fertilisers plus topsoil inoculant;
5. spoil plus sewage sludge (γ-irradiated sludge at a rate of 89.6 t ha^{-1}); and
6. spoil plus sewage sludge plus topsoil inoculant.

The spoil and the spoil-amendment mixtures placed in pots were seeded to blue grama grass (*Bouteloua gracilis*) and kept under favourable humidity and temperature conditions. In June 1979, nonrhizosphere spoil samples were taken for enzymatic and microbial analyses. In September 1979, the same pots were replanted to fourwing saltbush. In April 1980, rhizosphere samples were collected for microbial analyses.

The analytical data indicated that dehydrogenase activity remained at the same low level in pots with spoil and in those with topsoil-inoculated spoil but increased in the other pots, the highest increase being found in pots containing spoil plus sludge plus topsoil inoculant. In this experiment, *Azotobacter* could be detected in each pot, both in nonrhizosphere and rhizosphere samples. Comparison of pots 1 and 2 showed that the inoculation of spoil with stockpiled soil did not lead to greater numbers in nonrhizosphere and rhizosphere samples, except in the case of fungal number, which increased significantly in the nonrhizosphere sample. The inoculation slightly increased the distribution of fungal genera, also, in the nonrhizosphere sample. In the other pots, as compared to pots 1 and 2, the microbiological parameters gave higher values (except for *Azotobacter* number in nonrhizosphere samples from pots 5 and 6). Under the influence of the treatment with alfalfa hay plus fertilisers, the highest increase occurred in the numbers of ammonium oxidisers and *Azotobacter* in both nonrhizosphere and rhizosphere samples, whereas sewage sludge had the strongest effect on the increase in numbers of aerobic heterotrophic bacteria and streptomycetes in rhizosphere samples.

Thus, it is evident that amendment with organic matter (alfalfa hay or sewage sludge) was more effective for increasing the enzymatic and microbial potential of spoil than was topsoil inoculation alone.

At the San Juan mine, a reclamation field experiment was also carried out (Lindemann *et al.*, 1984). In May 1979, plots were established on graded and levelled spoil. To increase water movement into the spoil, sterile bottom ash from an electrical generating plant was spread to a depth of 10 cm over the entire area and incorporated to a depth of 20 cm.

Nine treatments, each in four repetitions, were applied to the spoil and bottom ash mixtures:

1. control: unamended spoil;
2. topsoil: stockpiled topsoil at least 1 year old was applied to a depth of 30 cm;
3. topsoil inoculum: topsoil collected from around plants in an undisturbed area and applied as a source of microorganisms, including spores of vesicular-arbuscular (VA) mycorrhizal fungi, primarily *Glomus fasciculatum*, at a rate of 14.5 t of topsoil ha^{-1};
4. hay: native hay (mostly grasses) at a rate of 2.2 t ha^{-1};
5. sludge: dried and γ-irradiated sewage sludge at a rate of 11.2 t ha^{-1};
6. *Glomus mosseae* root inoculum: sorghum roots containing *G. mosseae* mycelium, vesicles, and spores at a rate of 1.2 t ha^{-1};
7. *Glomus mosseae* soil inoculum: soil from sorghum pots containing *G. mosseae* spores at a rate of 9.1 t ha^{-1};
8. *Glomus fasciculatum* root inoculum: sorghum roots containing *G. fasciculatum* mycelium, vesicles, and spores at a rate of 0.7 t ha^{-1}; and
9. *Glomus fasciculatum* soil inoculum: soil from sorghum pots containing *G. fasciculatum* spores at a rate of 14.5 t ha^{-1}.

In addition to these amendments, 34 kg ha^{-1} each of N (NH_4NO_3) and P_2O_5 (triple superphosphate) was applied to each plot. The plots were planted with a mixture of native grasses and shrubs in May 1979 and replanted in July 1979.

In August 1979 and May 1980, nonrhizosphere spoil (or soil) samples were taken from the 0-20-cm depth of plots 1-5 for enzymatic and microbial analyses. In March and September 1980, samples were collected from roots of fourwing saltbush (*Atriplex canescens*) and alkali sacaton (*Sporobolus aëroides*) growing on plots 1-5 for enumeration of rhizosphere microorganisms. In March and September 1980, the mycorrhizal formation was also evaluated in all plots.

The analyses of nonrhizosphere samples indicated that dehydrogenase activity, numbers of streptomycetes and fungi, and distribution of fungal genera were higher in the hay- and sludge-amended and topsoiled plots than in the unamended or topsoil-inoculated plots. The number of ammonium oxidisers increased significantly only in the hay- and sludge-amended plots (August 1979) or in the sludge-amended plots (May 1980). The aerobic heterotrophic bacteria were least affected by treatments.

It is clear from the results of this field experiment that hay or sludge amendments were more effective in increasing dehydrogenase activity and microbial parameters than was topsoil inoculation, which is in good agreement with the results of the greenhouse experiment of Fresquez and Lindemann (1982).

The analyses of rhizosphere samples showed that microbial numbers and distribution of fungal genera in the rhizosphere of *A. canescens* and *S. aëroides* were not significantly affected by any of the treatments. *Azotobacter* was absent in the nonrhizosphere samples and in those from the *A. canescens* rhizosphere, but was present in great number in the *S. aëroides* rhizosphere; its growth was strongly stimulated in the organically amended and topsoiled plots.

Formation of VA mycorrhizae on the planted grass species was, in general, much weaker in plots 3 and 6-9 (spoil inoculated with topsoil or with sorghum soil and roots containg mycorrhizal fungi) than in plot 2 (spoil covered with stockpiled topsoil). Practically no mycorrhizal infection occurred on plants from plots 1, 4, and 5.

At the San Juan mine, Fresquez *et al.* (1987) also studied the effects of topsoiling and reclamation age on enzyme activities in three reclaimed non-topsoiled spoil areas and five reclaimed topsoiled areas where stockpiled topsoil material was placed 20-30 cm deep over the spoil (surface structure was clay, sandy loam, clay loam or sandy clay loam; pH was 6.67-7.73). A surrounding undisturbed soil (loamy sand; pH 7.83) served as reference. The non-topsoiled areas were reclaimed in 1974, 1975, and 1976, respectively, and the topsoiled areas in 1978, 1979, 1980, 1981, and 1982, respectively. Except for the differences among sites in soil treatments (non-topsoiled and topsoiled) and in the number of native plant species seeded (four in the older non-topsoiled sites and ten in the younger topsoiled sites), all eight areas had been mulched (4.5 t of grass hay ha^{-1} crimped 15 cm deep), fertilised (72 kg of N and 90 kg of P ha^{-1}) and irrigated (for the first two growing seasons) in the same manner.

On a single date in July 1982, samples were taken from the 13-cm depth of each of the eight reclaimed sites and of the undisturbed soil. Eight enzyme activities (dehydrogenase, nitrogenase, phosphatase, arylsulphatase, amylase, cellulase, invertase, and urease) were determined.

The results showed that 3 months after topsoiling and revegetation, activities of enzymes in the reclaimed areas, with the exception of dehydrogenase, were statistically equal to activities of the undisturbed soil. Most enzymes, including dehydrogenase, peaked in the next 1 or 2 years after reclamation with topsoiling and declined thereafter. The 4-year-old topsoiled site (revegetated in 1978) was statistically similar to the undisturbed soil. Amylase activity, however, was significantly lower after the fourth year compared to the undisturbed soil.

The non-topsoiled areas, even after 6, 7, and 8 years, appeared to have lower enzyme activities than the younger topsoiled areas or the undisturbed soil. This trend was supported by the finding that the 4-year-old topsoiled site was enzymatically more similar to the undisturbed soil than was the 8-year-old non-topsoiled site (revegetated in 1974).

The low enzyme activities registered in the non-topsoiled areas may be a result of their adverse chemical and physical properties as well as the low diversity of microorganisms.

Thus, these studies demonstrated the value of topsoil use for early establishment of soil processes in reclaimed areas.

Stroo and Jencks (1982) have studied 11 coal strip mine soils distributed within a distance of 3 km in Preston County, West Virginia. These mine soils were or were not excessively acid depending upon how the pyritic and nonpyritic materials were mixed when replaced. They tended to be high in sandstone fragments. The 11 mine soil sites studied were varied in age, type, and degree of plant cover and in type of postmining treatments. Two adjacent native soils were included for comparison. Amylase, urease, and phosphatase activities and respiration (O_2 uptake) rate were measured in samples taken in early April 1980. The top layer (0-10 cm) was collected after removing the loose unincorporated litter at the surface. Selected physical and chemical soil properties (pH, clay, oxidisable C, total N, C-to-N ratio, mineralisable N, acid-extractable P and K) were also analysed. The description of the study sites and the results of the enzyme activity and respiration measurements are presented in Tables 20 and 21.

One can deduce from these tables that the enzyme activities and respiration rate were generally lower in the mine soils than in the adjacent native soils. The activities and respiration recovered with time, which was attributed to organic matter and N accumulation, but these indices in a 20-year-old mine soil were still lower than in native soils. Vegetation was critical to the recovery of activities and respiration. As long as legumes were present and actively fixing N_2, there seemed to be little difference between grassland and locust vegetation. The only vegetated site with activity levels as low as the barren sites was a mine soil (L-2) with a high clay content that was heavily compacted, resulting in slow organic matter and N accumulation.

On the unamended locust sites, amylase and phosphatase activities and respiration rate were all significantly correlated with each other. These three indices were dependent on the levels of oxidisable C and total and mineralisable N. Significant correlations were found between amylase and phosphatase activities and mine soil age. Amylase activity was also correlated with clay content, whereas urease activity correlated only with respiration rate. There were no significant correlations between activities or respiration rate and pH or acid-extractable P or K.

When all sites were considered together, amylase activity correlated with urease activity, pH, and acid-extractable P, too. Correlations also appeared between urease activity and oxidisable C and total N, whereas respiration rate correlated only with amylase activity, pH, and P.

The results obtained suggest several potential problems in reclamation. Compaction, particularly of fine-textured overburdens, should be avoided because it slows recovery. Phosphatase activity was too low in the studied mine soils, which indicates possible difficulties in phosphorus mineralisation. The case of the B-2 barren mine soil suggests that

Table 20
Description of mine soil sites in Preston County, West Virginia

Site	Age (years)	Dominant vegetation	Surface texture	Treatments
			Barren mine soils	
B-1	17	Barren	Sandy clay loam	Unamended, unplanted
B-2	18	Barren	Sandy loam	Limed and fertilised after mining, planted to tall fescue and birdsfoot trefoil; the latter died out within the first 5 years after seeding
			Locust mine soils	
L-1	9	Black locust, goldenrod	Sandy loam	Unamended, planted to locust
L-2	11	Black locust, blackberry	Clay loam	Unamended, planted to locust; heavily compacted
L-3	11	Black locust, tall fescue	Silty clay	Unamended, planted to locust
L-4	17	Black locust, fire cherry	Clay loam	Unamended, planted to locust
L-5	18	Black locust, blackberry	Clay loam	Unamended, planted to locust
L-6	20	Black locust, joe-pye weed	Clay	Unamended, planted to locust
			Grass-legume mine soils	
G-1	17	Tall fescue, birdsfoot trefoil	Sandy clay loam	Unamended, planted to grass-legume mixture
G-2	3	Tall fescue, orchard grass, mixed clovers	Clay	Limed, fertilised, topsoil replaced, planted to grass-clover mixture; cut for mulch
G-3	5	Same as above	Clay	Same as above
			Native soils	
N-1	-	Red oak, black cherry	Silt loam	Native soil, undisturbed forest
N-2	-	Maize	Silt loam	Limed, fertilised, manure added annually; continuous corn production with annual ploughing

Reproduced from: H.F. Stroo and E.M. Jencks, Soil Science Society of America Journal, 1982, Vol. 46, p. 549.

the goal of recreating productive, self-sustaining ecosystems is not being achieved on such sites.

Table 21
Average enzyme activities and respiration of mine soils in Preston County, West Virginia

Site*	Amylase**	Urease**	Phosphatase**	Respiration***
B-1	0.6 a	13.6 bc	1.2 b	0.58 a
B-2	1.0 b	13.6 bc	0.8 a	1.65 b
L-1	1.8 c	20.4 de	2.3 d	1.88 b
L-2	1.1 b	6.8 a	0.6 a	0.84 a
L-3	2.3 d	23.6 fg	1.9 c	1.80 b
L-4	3.4 e	19.3 def	3.6 f	2.83 e
L-5	2.4 d	17.0 cd	2.6 de	1.59 b
L-6	3.5 e	11.2 b	4.8 g	1.96 b
G-1	3.6 e	18.6 de	2.8 e	2.07 bc
G-2	6.5 h	17.2 cd	2.9 e	5.17 g
G-3	4.4 f	19.1 def	2.8 c	3.53 f
N-1	4.6 f	17.0 cd	8.6 h	2.77 cd
N-2	5.5 g	25.1 g	5.1 g	2.74 cd

Reproduced from: H.F. Stroo and E.M. Jencks, Soil Science Society of America Journal, 1982, Vol. 46, p. 550.
* See Table 20.

** Expression of enzyme activities: amylase in μmoles of reducing sugars 100 g^{-1} soil h^{-1}; urease in μg of urea g^{-1} soil h^{-1}, and phosphatase in μmoles of *p*-nitrophenol g^{-1} soil h^{-1}.
*** Expressed in μl of O_2 g^{-1} soil h^{-1}.

Numbers followed by the same letter in a column are not significantly different at $P = 0.05$.

For a further study, Stroo and Jencks (1985) selected the B-2 barren mine soil (see Tables 20 and 21). The acidic, infertile soil was sparsely vegetated with tall fescue (*Festuca arundinacea*). Different plots of this mine soil were treated in four replications with lime, fertilisers, lime plus fertilisers, and lime plus fertilisers plus sewage sludge. Limestone sufficient to raise the pH from 4.6 to 6.2 was spread in late February. Ammonium nitrate, triple superphosphate, and KCl sufficient to raise N, P, and K to average levels of 40, 34, and 66 kg per ha, respectively, were spread in mid-March. Fifty or 100 kg of air-dried, aerobically digested sewage sludge were added per ha, 14 days after fertilisation, as a slurry. In August, soil was sampled from the top 10 cm of each plot for determining enzyme activities and respiration rate. The yield of tall fescue was also estimated.

The data of Table 22 show that the treatments, except for liming, led to significantly increased amylase activity as compared to that in the untreated control. Urease and phosphatase activities increased significantly only in the plots treated with lime plus fertilisers plus sludge applied at the higher rate. All treatments produced significant increases in respiration rate and tall fescue yield, this effect being strongest in the plots treated with lime, fertilisers, and sludge at the higher rate.

Table 22
Average enzyme activities, respiration, and tall fescue yield in an amended mine soil in Preston County, West Virginia

Treatment	Amylase*	Urease*	Phosphatase*	Respiration**	Yield***
Control	26 a	35.5 cd	2.2 bc	0.89 a	217 a
Lime	24 a	35.5 cd	0.6 a	1.31 b	337 b
Fertilisers	43 bc	20.9 b	1.5 b	2.17 c	669 c
Lime+fertilisers	40 b	9.0 a	1.9 bc	2.71 c	703 c
Lime+fertilisers+ 50 kg of sludge ha^{-1}	48 cd	45.0 d	2.5 c	4.42 d	838 d
Lime+fertilisers+ 100 kg of sludge ha^{-1}	55 d	66.6 e	5.3 d	6.38 e	1043 e

Reproduced from: H.F. Stroo and E.M. Jencks, Journal of Environmental Quality, 1985, Vol. 14, p. 303.

* Expression of enzyme activities: amylase in μmoles of reducing sugars kg^{-1} soil h^{-1}; urease in mg of urea kg^{-1} soil h^{-1}; and phosphatase in μmoles of *p*-nitrophenol g^{-1} soil h^{-1}.
** Expressed in ml of O_2 kg^{-1} soil h^{-1}.
*** Recorded as kg ha^{-1}.

Numbers followed by the same letter in a column are not significantly different at $P = 0.05$.

It is evident from this study that lime, mineral fertilisers, and sewage sludge might be used profitably to restore seeded vegetation diminished over time because of acidity and low fertility in mine soil. As amylase activity was in general strongly, whereas urease and phosphatase activities were poorly, related to respiration rate, it can be concluded that respiration rate and/or amylase activity might serve as a reliable indicator of microbial activity in amended mine soils, although urease and phosphatase activities appear to be of little value in evaluating mine soil microbial activity. This conclusion is not surprising, because only a part of the activity of these enzymes is due to the proliferating microorganisms that determine the respiration of soil; the other part is the result of the accumulated enzymes, which are independent of the momentary proliferation of soil microorganisms (Kiss *et al.*, 1975).

Persson and Funke (1986) studied, enzymologically and microbiologically, the topsoil pile R26T03, which is located on the Baukol-Noonan lignite strip mine near Center, North

Dakota. The pile was formed in July 1983 and vegetative cover was established on its surface. Samples were taken from depths of 0-7.6 and 114-122 cm in May 1984 and July 1985. Alkaline phosphatase and dehydrogenase activities were found to decrease at increasing depths and to show further declines during the storage of the topsoil pile. Numbers of bacteria and actinomycetes were similar in all samples, whereas the number of fungi decreased with depth but increased with storage time.

Persson and Funke (1988) performed similar studies on three topsoil piles located at the lignite strip mine Glenharold near Stanton, North Dakota. Topsoil pile T-114 was formed in October 1985; it was sampled twice: during its formation and in April 1986, i.e. after 6 months of storage. Topsoil pile T-4, formed in 1973, was sampled after 11 years of storage, in August 1984. Topsoil pile T-1 was also formed in 1973, but in February 1983 its upper portion was removed and the remaining part was revegetated in May 1983 to minimise erosion. This pile was sampled in August 1985, i.e. after 12 years of storage. An undisturbed prairie site at the mine served for comparison; it was sampled in November 1984. The samples were taken from the 0-47-cm depth during formation of pile T-114 and from the 0-8- and 110-120-cm depths of the stored piles and undisturbed soil. They were analysed enzymologically (alkaline phosphatase and dehydrogenase activities) and microbiologically (total numbers of bacteria, actinomycetes, and fungi; numbers of nitrifying and nonsymbiotic N_2-fixing bacteria and of vesicular-arbuscular mycorrhizal, VAM, fungi).

Results of the enzymological analyses made it possible to compare topsoil pile T-114 at the moment of its formation with the same pile after 6 months of storage; the 0-8-cm layer of the stored piles and undisturbed soil with their 110-120-cm layer; piles with the undisturbed soil; and alkaline phosphatase activity with the dehydrogenase activity.

In the 0-47-cm layer of pile T-114 at the moment of its formation, alkaline phosphatase activity was lower than in the 0-8- and 110-120-cm layers of the same pile after the 6-months of storage; dehydrogenase activity showed an increased value in the 0-8-cm layer, but a decreased one in the 110-120-cm layer after 6 months of storage. In other words, storage of the topsoil pile for 6 months did not bring about great changes in the enzymatic potential of soil.

Alkaline phosphatase and dehydrogenase activities were higher in the 0-8-cm layer of each pile and of the undisturbed soil than in their 110-120-cm layer, with the exception of a single case: alkaline phosphatase activity of pile T-1 was a little lower in the 0-8-cm layer than in the 110-120-cm one.

In the 0-8-cm layer, alkaline phosphatase activity presented the following order: undisturbed soil >> T-114 (age: 6 months) > T-4 (age: 11 years) > T-1 (age: 12 years), i.e. the activity was highest in the undisturbed soil and decreased with storage time. In the same layer, dehydrogenase activity showed the order: undisturbed soil ≈ T-114 > T-4 > T-1. This means that dehydrogenase activity decreased only in the soils of the 11- and 12-year-old piles.

In the 110-120-cm layer, the order of alkaline phosphatase activity (T-114 > T-1 > T-4 > undisturbed soil) and that of dehydrogenase activity (T-114 > T-4 > T-1 > undisturbed soil) indicate that the enzymatic activities tended to decrease with storage time in this layer, too. Nevertheless, the activities in the stored topsoil did not decrease to the level registered in the 110-120-cm layer of the undisturbed soil.

The decrease in alkaline phosphatase activity with the time of the topsoil storage was not so great as that in the dehydrogenase activity.

The microbiological analyses showed that total numbers of bacteria, actinomycetes, and fungi did not decrease markedly during the long-term storage of topsoil in piles and these numbers registered in the 110-120-cm layer of stored piles were usually higher than those recorded in the same layer of the undisturbed soil.

Nitrifying and nonsymbiotic N_2-fixing bacteria and VAM fungi were present in considerable numbers in both layers of the stored piles and in the 0-8-cm layer of the undisturbed soil, their smallest numbers being recorded in the 110-120-cm layer of the undisturbed soil. The nonsymbiotic N_2-fixing bacteria isolated were spore-formers (*Bacillus* and *Clostridium* spp.) and, suprisingly, they were most numerous in the 110-120-cm layer of the oldest piles (T-4 and T-1).

Based on the findings by Persson and Funke (1986, 1988), one can state that the enzymological analyses gave more concludent results in the evaluation of changes occurring in the biological potential of topsoil piles during their storage than did the microbiological analyses.

Shirey and Sexstone (1989) have determined denitrification enzyme activity (reduction of NO_3^- to N_2O) in technogenic soils at two abandoned strip coal mines in West Virginia. These soils, containing oxidisable pyrite, are acid; they were derived from acidic carbolithic shales; their texture is loamy. The Bald Knob mine (Mineral County) was actively mined from 1965 to 1970. In 1979, an experimental plot was amended with lime (10.2 t ha^{-1}) and fertilised with NH_4NO_3 (220 kg ha^{-1}) and P (43 kg ha^{-1}). A nonamended and nonfertilised area served for comparison. At the Osage mine (Monongalia County), three areas located on the refuse dump were studied, namely a nonamended area, an area amended with pH 12 fly ash and an area amended with both fly ash and lime. Fly ash (336 t ha^{-1}) was applied in 1971.

In 1985 and 1986 when samples were taken from the 0-20-cm layer of soils for analyses, the plant cover on nonamended area at both Bald Knob and Osage was less than 1%, consisting primarily of broomsedge (*Andropogon virginicus*). Plant cover on amended areas at both mines was estimated to be 80%, consisting primarily of a mixture of birdsfoot trefoil (*Lotus corniculatus*) and tall fescue (*Festuca arundinacea*). Only unvegetated places were sampled on nonamended areas, whereas on the amended area the samples were collected from vegetated places.

Liming (and fertilisation) of mine soil at Bald Knob resulted in an increase of pH from 3.0 to 6.8, whereas denitrification enzyme activity (expressed as μg of N_2O-N kg^{-1} soil h^{-1}) increased on average from 4.6 to 10.7. Amendments with fly ash and with fly ash plus lime improved the pH of mine soil at Osage: the initial pH (2.7) increased to 5.2 and 6.4, respectively. Denitrification enzyme activity assayed in the soil of nonamended and fly ash-amended areas gave average values of 11.1 and 67.6, respectively. This means that, under the influence of amendments and fertilisation, denitrification enzyme activity increased about twice at Bald Knob and 6 times at Osage. It should, however, be added that this activity in acid to neutral agricultural and forest soils is much higher (40-4000 μg of N_2O-N kg^{-1} soil h^{-1}).

Application of amendments (and fertilisers) also led to increases in total numbers of bacteria and fungi, in numbers of aerobic spore-forming bacteria and of bacteria reducing nitrate to nitrite ($NO_3^- \rightarrow NO_2^-$) as well as in nitrification capacity ($NH_4^+ \rightarrow NO_3^-$) in soils of both mines.

Enzymological research in Poland

Gilewska and Bender (1978, 1979) carried out detailed studies of the Patnów strip mine spoil heap located in the Konin Brown Coal Basin. They have determined the cellulolytic, invertase, urease, and catalase activities in the ploughed layer (0-25 cm) of spoil plots (medium-heavy loam; pH in H_2O 7.6) cultivated with spring barley (in the third year of their recultivation) and not fertilised or fertilised with NH_4NO_3, triple superphosphate, and potash salt in single (1 NPK: N 130, P_2O_5 240, and K_2O 140 kg ha^{-1}) or double (2 NPK) dosage. An adjacent arable soil that had not been affected by the strip mining was used for a control. Each activity was demonstrable, although at a low level, in the unfertilised plots and each gave higher values in the fertilised plots, particularly in those treated with the double NPK dosage. Cellulolytic activity in these plots exceeded, whereas invertase and urease activities approached those found in the arable soil. Catalase activity in fertilised plots increased to a lesser extent, reaching only 50% of that measured in the arable soil*. The barley grain yields were also higher in the fertilised spoil plots than in the unfertilised ones. The increased enzyme activities and barley yields were accompanied by the accumulation of humus and the beginning of cloddy structure formation in the ploughed layer of the spoil plots (Bender and Strzyszcz, 1978).

Mineral fertilisation, especially with the double NPK dosage, increased cellulolytic, invertase, urease, and catalase activities and led to humus accumulation also in those spoil plots that were cultivated with other plants (winter rape, winter wheat, alfalfa) (Bender, 1980).

Other plots of the Patnów spoil heap as well as plots of other spoil heaps in the Konin Brown Coal Basin (Wschodnie and Goslawice) were also studied by Bender and Gilewska (1980, 1984a, b, d) and Gilewska and Bender (1983a, b, 1984). The newly studied Patnów plots were brought into agricultural recultivation 3 or 4 years before the studies began. (The last two crops were winter wheat in some plots and alfalfa with orchard grass, *Dactylis glomerata*, in some others). Both the wheat and the alfalfa-orchard grass plots included unfertilised, 1-NPK- and 2-NPK-fertilised variants. The Wschodnie plots had been recultivated for 10 years (last crop: winter wheat) and all were 1-NPK-fertilised. The Goslawice plots were 6-year-old mine spoils without any plant cover, fertilised with the smaller NPK dosage. "Raw and fresh" (unplanted and unfertilised) Patnów spoils and an adjacent arable soil fertilised with 1 NPK and cultivated with wheat served as controls. Four activities (cellulolytic, invertase, urease, and catalase) were measured in the ploughed layer (0-25 cm) of the spoil plots and arable soil.

* Referring to these investigations and completing the analytical data obtained after 3 years of recultivation with the data registered after 10 years of recultivation, Bender (1989) and Gilewska (1991) have stated that, after 10 years, values of the four activities even in the plots fertilised only with 1 NPK became similar to those measured in the arable soil also fertilised with 1 NPK.

The lowest activities were registered in the raw spoils. The main factor determining the increase of enzymatic activity in coal mine spoils proved to be mineral fertilisation with a high NPK dosage. The presence of plants strongly enhanced the effect of fertilisers, which increased with recultivation time. Thus, after 10 years of recultivation, the enzymatic potential of the mine spoils reached that of the arable soil. The nature of the crop plant used for recultivation also had an evident influence: e.g. invertase activity of the recultivated spoils was highest under the alfalfa-orchard grass mixture; the highest urease activity was recorded in the wheat plots. The optimum moisture content for the enzyme activity was in the 40-60% range of the maximum water-holding capacity of the spoils. The recultivated spoils containing higher amounts of clay minerals showed higher catalase activity.

In parallel with increasing enzyme activity, the production of plant biomass, humification of plant residues, accumulation of microbial biomass, soil respiration (CO_2 evolution), N_2 fixation (acetylene reduction), and formation of a humus layer have intensified. The C-to-N ratio in spoil decreased from 22-26 (raw spoil) to 15 (spoil recultivated for 10 years). Owing to the agricultural recultivation techniques applied, it became possible to transform the spoils into fertile soils in a relatively short period of time (10-15 years) (Bender and Gilewska, 1980, 1984b, d, e; Golebiowska and Bender, 1980, 1983).

Bender and Gilewska (1983) have demonstrated that invertase, urease, and catalase activities are good indicators of the soil-forming processes occurring even in those Patnów spoils that had been submitted to technical recultivation only 1 year before. Samples were taken from the 0-25-, 25-50-, and 75-100-cm depths of 15 representative profiles. Invertase and urease activities were always highest in the 0-25-cm layer and decreased with depth. Catalase activity of the 0-25-cm layer did not give the highest value in each profile and showed only a trend to decrease with depth. This differentiation of the enzyme activity is a noticeable symptom of the soil profile variation and indicates that the soil-forming processes are most intense in the top layer, which is due to the microorganisms penetrating the spoils from the surrounding agro-ecosystem. These soil-forming processes are, however, limited because of the shortage of biogenic elements, especially N and P, in the spoils. Therefore, the initiated soil-forming processes would develop very slowly; the transformation of spoils into soils would take many decades. This is why the NPK fertilisation and biological recultivation of spoils are required.

Some enzymological aspects of the forest recultivation of strip mine spoils in the Konin Brown Coal Basin have also been studied (Bender and Gilewska, 1984c; Gilewska and Wójcik, 1984). Black locust was used as a pioneer forest plant on the slopes of the Goslawice and Nieslusz spoil heaps (medium-heavy loam and sandy loam). At the time of study, the locust plantations were 15 years old. The average tree height was 10.8 m at Goslawice and 11.6 m at Nieslusz. The stand density was medium on both spoil heaps. Black cherry (*Prunus*

Table 23
Enzyme activities in technogenic soils formed in some plots planted to black locust on the Goslawice and Nieslusz coal strip mine spoil heaps

Profile No.	Depth (cm)	Invertase*	Urease*	Catalase*
1	0-5 (A_1)	322	38.0	132
	5-25	156	18.2	22
	25-40	35	10.0	26
	40-70	45	9.5	20
	70-100	47	8.7	20
8	0-5 (A_1)	565	90.0	340
	5-25	240	49.4	108
	25-40	80	19.0	10
	40-70	57	9.5	2
	70-100	65	7.7	4
10	0-25	115	15.5	14
	25-40	22	9.0	14
	40-70	65	13.2	10
	70-100	75	18.2	26
15	0-25	175	20.5	40
	25-40	60	11.8	16
	40-70	55	8.7	6
	70-100	47	10.0	6
20	0-5 (A_1)	520	80.0	312
	5-25	233	41.8	68
	25-40	83	7.4	12
	40-70	70	10.0	2
	70-100	87	8.2	4

Reproduced from: M. Gilewska and A. Wójcik, Archiwum Ochrony Środowiska, 1984, No. 3-4, pp. 148-149.

* Expression of enzyme activities: invertase in mg of "glucose" 100 g^{-1} soil 24 h^{-1}; urease in mg of NH_4^+-N 100 g^{-1} soil 3 h^{-1}; and catalase in ml of O_2 100 g^{-1} soil 15 min^{-1}.

serotina), purging buckthorn (*Rhamnus catharctica*), golden elder (*Sambucus nigra*), and aspen (*Populus tremula*) occur sporadically in the plantations. Twenty profiles were examined; Nos. 1-10 are located on Goslawice and Nos. 11-20 on Nieslusz.

The results showed that invertase activity in all profiles and urease and catalase activities in most profiles were highest in the 0-25-cm layer (see Table 23, containing the analytical data obtained in profiles 1, 8, 10, 15, and 20). In general, organic matter, total N, P_2O_5, and K_2O contents were also highest in this layer, but in most cases, there was no distinct outline of the humus horizon in contrast to the agriculturally recultivated coal strip mine spoils, in which the 0-25-cm humus layer, after 10 years of recultivation, was clearly outlined.

The studies performed by Bender and his co-workers in the Konin Brown Coal Basin were amply reviewed in a synthesis work of Bender (1992) dealing with the biological recultivation in Poland, and were also referred to in Strzyszcz's review work presented at the Symposium "Minesite Recultivation" (Cottbus, Germany) in June 1994 and published in 1996.

Six enzyme activities (dehydrogenase, catalase, invertase, β-glucosidase, urease, and asparaginase) were determined in the Smolnica black coal mine spoil heap (Silesia), which had been submitted to three recultivation experiments (Osmańczyk, 1980; Osmańczyk-Krasa, 1984a, b). The spoil consisted of sandstones and, mainly, of shales; pH in H_2O was 4-5.

In the first experiment, which began in 1973, ten mixtures of grasses and legumes (*Arrhenatherum elatius, Bromus inermis, Dactylis glomerata, Festuca capillata, F. heterophylla, F. pratensis, F. rubra, Lolium multiflorum, L. perenne, L. westerwolticum, Lotus corniculatus*, and *Trifolium repens*) were used for revegetating spoil plots. The second and third experiments started in 1975, and the spoil plots were recultivated with four tree species: European ash (*Fraxinus excelsior*), red oak (*Quercus rubra*), sycamore (*Acer pseudoplatanus*), and black alder (*Alnus glutinosa*). The enzymological analyses were carried out in 1977 and 1978 (every 6 weeks during the vegetation period).

1. The plots submitted to recultivation with grass-legume mixtures were fertilised with 120 kg of N, 75 kg of P_2O_5, and 25 kg of K_2O ha^{-1} in the first year (1973), only with N and P in the next years, and with the N dosage being reduced to 60 kg ha^{-1} in the third year; later no fertilisers were applied. For enzymological analyses, samples were taken from the 0-10-cm layer of each spoil plot. The results showed that each of the six enzymatic activities determined increased in the recultivated plots as compared to the control (unreclaimed, stored

raw spoil). The increase was highest in the case of invertase and β-glucosidase activities. Of the ten plant mixtures tested, that consisting of *Bromus inermis*, *Festuca pratensis*, *F. rubra*, *Lolium perenne*, and *Lotus corniculatus* manifested the strongest positive effect on the enzymatic potential in recultivated spoil (Osmańczyk, 1980).

2. The plots submitted to recultivation with tree species were fertilised with 100 kg of N, 50 kg of P_2O_5, and 25 kg of K_2O ha^{-1} in the first year (1975), with the same N and P dosages but without K in the second and third years, and with doubled N and P dosages in the fourth year. The samples for enzyme analyses were collected from below the root system (at 30-cm depth). In the case of the control plot (not recultivated, not revegetated), the samples were taken from the 0-10-cm layer. It appeared from the analytical data that the enzyme activities, except for catalase activity, increased in the recultivated plots in comparison with those measured in the control. The highest increase occurred in dehydrogenase activity. The increase was remarkable in the invertase and β-glucosidase activities also. The trees enhanced the enzymatic potential of the recultivated spoil plots in the following order: alder > sycamore ≈ ash > oak (Osmańczyk-Krasa, 1984a).

3. Tree seedlings were planted in spoil plots in three different ways. The first tree seedlings were planted directly into the spoil. The second ones were planted in holes made on the spoil surface and filled with about 6 l of podzolic soil taken from pine forests. The third ones were planted in holes filled with about 6 l of fly ash derived from a slag dump. The fertilisers applied were 100 kg of N and 50 kg of P_2O_5 ha^{-1} in 1975 and 1976, 200 kg of N and 100 kg of P_2O_5 ha^{-1} in 1977, and 100 kg of N and 40 kg of P_2O_5 ha^{-1} in 1978. As in the second experiment, the samples were taken from below the root system (at 30-cm depth).

Comparison of enzyme activities in the untreated, soil-treated, and fly ash-treated spoils has shown that treatment of spoils with podzolic soil or fly ash led to an increased dehydrogenase activity. The effect of soil, as compared with that of the fly ash, was stronger in the plots with ash tree and oak, weaker in the plot with sycamore, and identical in the alder plot. Catalase activity also increased under the influence of soil and fly ash treatments. The increase was more pronounced in the fly ash-treated plots than in those treated with soil, except for the alder plot. Invertase and β-glucosidase activities behaved like catalase activity. Urease activity was not influenced by treating the spoils with soil or fly ash; the only exception was the soil-treated ash tree plot, in which urease activity increased. Asparaginase activity was favourably influenced by both soil and fly ash treatments in each plot, but the effect of fly ash was stronger than that of the soil. Of the four species planted, alder caused the greatest increase in the activity of the majority of enzymes in both untreated plots and those treated with soil or fly ash. The effect of alder should be attributed to the N_2-fixing capacity of its root nodules, which finally leads to improved nutritional conditions for microorganisms and, implicitly, to increased enzyme production. The enzyme activities were also higher under ash trees planted in fly ash (Osmańczyk-Krasa, 1984b).

Reviewing the results of these experiments, Osmańczyk-Krasa (1987) emphasises that the enzyme activities appearing in unreclaimed spoil after its 3- to 5-year storage indicate the initiation of the soil-forming process in it, while the increase in enzyme activities of the revegetated spoil proves further development of pedogenesis.

Enzymological research in the Russian Federation

The spoil heaps (mostly medium loams; pH ~7) at the brown coal strip mines located in the forest-steppe zone in the Nazarovo Basin (which belongs to the Kansk-Achinsk Fuel-Energetic Complex, Siberia) were studied enzymologically by two research groups.

1. Naprasnikova *et al.* (1982) and Naprasnikova (1983, 1985a, b, 1987a, 1988) determined enzyme activities in spoil heaps covered with 3-, 5-, and 15-year-old spontaneously developed plant communities, in spoil heaps recultivated with pines, larches, willows, or sweetclover for a maximum of 15 years, and in zonal (grey forest) soils that had not been affected by the strip mining. Proteinase activity was always lower in the recultivated spoils and much lower in the spontaneously revegetated spoils than in the zonal soils. Peroxidase activity remained low in both recultivated and spontaneously revegetated spoils as compared with that in the zonal soils. Invertase and acid, neutral, and alkaline phosphatase activities in the recultivated spoils and in spoils under 15-year-old spontaneous vegetation, however, approached or even exceeded those found in the zonal soils. Polyphenol oxidase and the other enzymes were most active in the rhizosphere of the dominant plant species. The results obtained underline the advantages of recultivation over spontaneous revegetation of the coal strip mine spoils.

Based on the finding that the enzyme activities increased twice in non-topsoiled spoils during their agricultural recultivation for 3.5 years, Naprasnikova (1986) and Naprasnikova and Makarova (1986) recommend this recultivation method without covering the spoils with a fertile soil layer.

The spoil heaps at the Berezovo mine were submitted to spontaneous revegetation <1, 1-3, 3-5, and 5-10 years before the studies by Snytko *et al.* (1988) began. The spoils, containing calcareous loams and sands, were not neutral; their pH ranged from 8.2-8.3 (in heaps younger than 1 year) to 7.6-8.0 (in the 5-10-year-old heaps). Enzyme activities were determined in the 0-5-cm layer of heaps, at 5-day intervals during the growing season.

Invertase, proteinase, urease, and phosphatase activities showed a trend to increase with the age of heaps. Peroxidase activity gave the highest values in the 3-5-year-old heaps, whereas polyphenol oxidase activity, which was lacking in the heaps younger than 1 year, did not significantly differ depending on the age of the other heaps. It was also found that the enzyme activities were more variable in the heaps than in the adjacent soils, the variation coefficients being 60-94% for the heaps and 12-41% for the adjacent soils. This finding was explained by structural nonhomogeneity and relatively young age of the spoil heaps. Biomass of plants, bacteria, and invertebrates as well as the humus, total N, and P_2O_5 contents in the spoils increased in parallel with the age of heaps. In contrast, the fungal biomass was most abundant in the 1-3-year-old heaps. Thickness of the humified layer in the heaps of increasing age had the following values: 0, 0.3, 0.7-1.0, and 1.0-1.5 cm, respectively.

Of the spoil heaps submitted to spontaneous revegetation at the Berezovo mine, Naprasnikova (1987b) selected, for enzymological comparison, the 1.5- and 6-year-old ones and found significantly higher invertase, urease, and phosphatase activities in the 6- than in the 1.5-year-old heaps which proves that even during this short (4.5-year) period the spoils could acquire an increased enzymatic potential.

Naprasnikova and Makarova (1992) described new microbiological and enzymological investigations, dealing with the rhizosphere of three perennial plants (red clover, white sweetclover, and meadow fescue) growing on 3-4-year-old levelled spoil heaps and on undisturbed zonal soils. The rhizosphere effect (i.e. the ratio between the numbers of micro-organisms in the rhizosphere soil and nonrhizosphere soil) was higher in the young tech-nogenic soils than in the zonal soils. Activities of hydrolases and oxidoreductases were also higher in the rhizosphere of plants growing on young technogenic soils than in the rhizo-sphere of the same plants growing on the zonal soils. The formation of humus layer was more evident in spoil heaps recultivated with the three plants mentioned above than in the spon-taneously revegetated spoil heaps.

In a 5-year experiment, described by Naprasnikova (1993), spoil heaps of different ages (<1 to 10 years) were sampled every day during the growing season for determination of their enzyme activities (sampling depth was 0-5 cm). It was found again that the enzymatic potential of spoil heaps increased with their age. Numbers and biomass of microorganisms in the 0-5-cm layer also increased with the age of spoil heaps.

2. Shugalei *et al.* (1984, 1985) and Korsunova and Shugalei (1986) studied the enzyme activity and chemical composition of the topsoil (50-cm layer) stockpiled before surface mining of coal. The stockpiles are 1.5-4 m high and, after 3-5 years of storage, they are used to cover the levelled spoils for their recultivation. Catalase activity in the stored topsoil remained nearly at the same level as in the 0-50-cm layer of the adjacent undisturbed leached chernozem, but urease activity decreased and proteinase activity increased to some extent in the stored topsoil. The quantity and quality of humic substances in the stored topsoil and in the undisturbed soil were nearly the same.

The topsoil, after its replacement on the levelled spoils and recultivation with perennial grasses, showed catalase and urease activities approaching the seasonal mean values recorded in the 0-50-cm layer of the undisturbed soil, whereas proteinase activity was three times higher in the replaced topsoil than in the undisturbed soil.

In different raw spoil samples, Shugalei and Yashikhin (1985) registered high variation coefficients of catalase, urease, and proteinase activities, respiration (CO_2 evolution) rate, organic matter, total N, NH_4^+-N and NO_3^--N contents (77, 90, 66, 58, 69, 72, 82, and 151%, respectively) which was attributed to nonhomogenous mixing of the different overburden layers. Levelling of raw spoils led to diminution of the variation coefficients, except that of the urease activity which increased.

Shugalei *et al.* (1985) also determined enzyme activities in the 1-1.5-cm humus layer that had formed in 10 years on levelled spoils not covered with stored topsoil but cultivated with Scotch pine or Siberian larch. It was found that catalase, urease, and proteinase activities in the newly formed humus layer exceeded by 5-20 times the activities measured in the 0-20-cm layer of the undisturbed soil.

By using coal mine spoils from the Nazarovo Basin, Bezkorovainaya (1994) carried out a pot experiment to study the role of earthworms in decomposition of litters produced by six tree species: aspen (*Populus tremula*), birch (*Betula fruticosa*), Siberian larch (*Larix sibirica*), Siberian pine (*Pinus sibirica*), Scotch pine (*Pinus sylvestris*), and Siberian spruce (*Picea *obovata*). Each pot contained 2.5 kg of spoils. Litter (50 g dry weight) and five earthworms (*Eisenia* sp.) were placed on the surface of spoils. Pots with spoils and litter, but without earthworms were the controls. After moistening, the pots were exposed to hydrothermal conditions similar to the natural ones. The experiment lasted 3 years (1985-1987). The litters and spoils were submitted yearly to mechanical, chemical, microbiological, and enzymological analyses.

It was found that decomposition of each litter was stimulated in the presence of earthworms and, as expected, the decomposition degree was more advanced in deciduous than in coniferous litters. In the presence of earthworms, protease and urease activities of litters increased. The increase was highest in the urease activity of birch litter: it was 1.5-3-fold in comparison with the other litters and 2.5-fold compared with the control birch litter.

According to Naplekova *et al.* (1982, 1983) and Trofimov *et al.* (1986), the soil-forming processes in the technogenic landscapes created by the coal strip mining in the Kuznetsk Basin (Kuzbass, Siberia) take place very vigorously, owing to the spontaneous revegetation and forest recultivation of spoil heaps (slightly alkaline, calcareous sandy loams and clay shales). In 18-year-old spoil heaps in the mountain taiga zone and in 8-12-year-old spoil heaps in the steppe zone the number and composition of microflora approach those in the zonal soils (pseudopodzolic soil and leached chernozem, respectively). This causes the intensification of respiration (CO_2 evolution), catalase, dehydrogenase, amylase, and phosphatase activities, gelatin and cellulose decomposition, and amino acid accumulation in the technogenic soils and leads, finally, to an increase in their fertility.

In the forest-steppe zone, Naplekova *et al.* (1985) determined the polyphenol oxidase and peroxidase activities of the technogenic soils formed during the development of spontaneous plant communities on the spoil heaps. They established that the nature of the dominant plant had a decisive influence on these activities and on their ratio. The polyphenol oxidase-peroxidase ratio, expressed as a percentage, is considered an indicator of the intensity of humification processes (humification coefficient). High values of the humification coefficient were registered in technogenic soils under perennial legumes and grasses.

Klevenskaya *et al.* (1986) and Klevenskaya (1989, 1992) emphasise that the associations between microorganisms and plants, owing to accumulation of enzymes, especially of polyphenol oxidase and peroxidase in the rhizosphere, speed up the elementary pedogenetic processes in the technogenic soils.

Coal mine spoil heaps located in the Moscow Basin were also studied enzymologically (Lukina, 1991). Experimental recultivation plots were installed on the spoil heaps at the Kimovsk coal mine belonging to the "Tulaugol" Enterprise. All plots were submitted to a basal treatment with lime and phosphorite meal, then the majority of them were fertilised in the following variants:

1. $N_{150}P_{300}K_{300}$;
2. $N_{300}P_{600}K_{600}$;
3. $N_{300}P_{900}K_{600}$;
4. $N_{150}P_{300}K_{300}$ plus farmyard manure (120 t ha^{-1});
5. $N_{150}P_{300}K_{300}$ plus a fertile soil layer (240 t ha^{-1});
6. $N_{150}P_{300}K_{300}$ in mixture with coal (6 t ha^{-1});
7. $N_{300}P_{600}K_{300}$ in mixture with coal (6 t ha^{-1}); and
8. $N_{150}P_{300}K_{300}$ in mixture with coal (6 t ha^{-1}) plus coal (60 t ha^{-1}).

The plots were sown with green manure crops. In 1988*, when the enzymological analyses were carried out, a mixture of vetch and oats was used. Untreated spoil plots and the undisturbed zonal soil (leached chernozem) were the controls. Catalase, dehydrogenase, urease, nitrate reductase, and nitrogenase activities in the 0-20-cm layer of plots and undisturbed soil were determined.

Catalase and nitrate reductase activities were very low, whereas dehydrogenase, urease, and nitrogenase activities were not detectable at all in the control heap.

Catalase activity increased 3-5 times in the plots treated with lime and phosphorite meal and in those fertilised with NPK or NPK-coal mixtures as compared to the activity of the control plot. The increase of catalase activity in the plots treated with NPK plus farmyard manure and NPK plus a fertile soil layer was much more pronounced, namely 10- and 7.5-fold, respectively. However, catalase activity remained, even in these plots, about 5 times as low as in the undisturbed soil.

A similar situation was observed in the case of dehydrogenase activity, but this activity was only twice as low in the plot treated with NPK plus farmyard manure and only 1.5 times as low in the plot treated with NPK plus a fertile soil layer as in the undisturbed soil.

Fertilisation led to the appearence and increase of urease activity. The highest increase occurred in the NPK-farmyard manure-treated plot; urease activity in this plot even exceeded

* The age of the recultivation plots in 1988 is not mentioned in Lukina's paper.

the activity measured in the undisturbed soil. This activity also increased remarkably in the plots treated with NPK plus a fertile soil layer and with NPK plus coal (6 + 60 t ha^{-1}), but it did not reach the level of urease activity in the undisturbed soil.

The recultivated plots exhibited an increased nitrate reductase activity in comparison with that of the control plot. This effect of recultivation was strongest in the plots treated with $N_{300}P_{600}K_{600}$ and $N_{300}P_{900}K_{600}$, in which this activity had higher values than in the undisturbed soil.

Nitrogenase activity behaved essentially like the nitrate reductase activity.

By generalisation of the results obtained, the conclusion may be drawn that the treatments with NPK plus farmyard manure or a fertile soil layer, applied after a basal treatment with lime and phosphorite meal, were the most efficient methods for creating an enzymatic potential in spoils submitted to agricultural recultivation; the treatments with NPK-coal mixtures were, in general, not more advantageous than the treatment with NPK alone.

Enzymological research in Hungary

In a pot experiment carried out by Sulyok and Vörös (1983), refuse soils (yellow sand, yellow clay, grey clay, and andesitic tuff) that resulted from the strip mining of lignite in Gyöngyös-Visonta and the original topsoil (brown chernozem) were fertilised with NPK (at a rate equivalent to 309 kg of N, 189 kg of P_2O_5, and 180 kg of K_2O ha^{-1}) or with NPK plus 10 t of wheat straw per ha and sown with alfalfa. The controls were: 1. unfertilised and unsown and 2. unfertilised but sown refuse soils and topsoil.

After 2 years of recultivation, invertase activity of both controls and fertilised variants was highest in the topsoil and lowest in the yellow sand. Rate of respiration (CO_2 evolution) and intensity of cellulose decomposition were also higher in the unfertilised and unsown topsoil than in the unfertilised and unsown refuse soil. In the case of the sown controls and fertilised variants, however, the largest amount of CO_2 was produced by the yellow clay. Cellulose decomposition was most intense in the yellow clay fertilised with NPK and straw, whereas the highest alfalfa yields were obtained on the yellow clay fertilised with NPK and on the grey clay fertilised with NPK and straw.

The results suggest that enzyme (invertase) accumulation in the recultivated sandy refuse soil is a very slow process. In the recultivated clayey refuse soils this process is somewhat faster but not so fast as the microbial cellulose decomposition or CO_2 production.

Therefore, enzyme accumulation should be considered a more reliable indicator of the transformation of mine spoils into soils than cellulose decomposition or CO_2 evolution.

At the same lignite mine, a recultivation field experiment was also carried out. The experimental conditions and the results obtained after 5 years of recultivation were described by Vörös and Antal (1988). In the spring of 1983, two of the refuse soils (yellow sand and yellow clay) were placed in concrete rings (having a diameter of 80 cm) buried in the soil. Plots of the original, undisturbed soil served for control. The refuse soils in rings and the control soil were fertilised in three variants: 0, NPK, and NPK plus lignite. In additional variants, the refuse soils were fertilised with sewage sludge. Rates of fertiliser application were: 390 kg of N, 200 kg of P_2O_5, 200 kg of K_2O, 1.6 t of lignite powder, 10 or 40 t (dry matter) of sludge ha^{-1}. Then, *Sorghum bicolor* was sown. In August 1983, the sorghum as a green manure crop was incorporated into the refuse soils and control soil. In the autumns of 1983 and of the next 4 years, the fertilisation with NPK and NPK plus lignite was repeated using 240 kg of N, 120 kg of P_2O_5, 120 kg of K_2O, and 1.2 t of lignite ha^{-1}; winter wheat was the test plant.

Results of the determination of invertase activity in refuse soils after 5 years of their recultivation showed that the activity was very weak in the unfertilised refuse soils and increased under the influence of fertilisation in the order: 10 t of sludge $<$ NPK $\approx$ NPK plus lignite $<$ 40 t of sludge. The activity increase due to fertilisation with NPK and NPK plus lignite was significantly higher in the sand than in the clay. Contrarily, under the influence of fertilisation with sludge (40 t ha^{-1}), the activity increased to a larger extent in the clay than in the sand. A parallelism was observed between invertase activity and humus content in the recultivated refuse soils. Although invertase activity increased in refuse soils due to fertilisation, it remained significantly lower than that recorded in the fertilised control soil. Fertilisation of refuse soils also enhanced microbial decomposition of cellulose and led to higher wheat yields.

These investigations were also referred to in Tóth's (1988) and Strzyszcz's (1996) review works.

Enzymological research in Germany

Schröder *et al.* (1985) studied, from physical, chemical, microbiological, enzymological, and micromorphological viewpoints, seven profiles of technogenic soils formed during the agricultural recultivation of spoils (loess; pH in 0.01 M $CaCl_2$ was 7.1-7.9) that resulted from the strip mining of brown coal in the Rhine region. The soils studied were located near Grefrath and Berrenrath. The recultivation started about 20 years previously through redeposition of stored loess either as dry material, followed by levelling (five soils), or as wet material (pumping of water-loess mixture into depressions, i.e. slurry poldering), followed by partial evaporation of water (two soils). Recultivation led to humus and K accumulation and to Na and Mg loss, but due to tillage, the 30-50-cm layer of soils became compacted and impermeable. In the soils formed through redeposition of dry loess, the 50-120-cm layer also frequently became compacted.

Dehydrogenase activity was detectable, in general, only from the 0-30-cm layer, but even in this layer, the activity reached only about 10% of that of the native soils. Respiration (CO_2 evolution) was more intense in the upper layers than in the deeper ones, the lowest values being recorded in the compacted layers. In each layer of the native soils, respiration rate was greater as compared with the corresponding layers of the seven technogenic soils. Similar results were also obtained in regard to cellulose decomposition. For improving the compacted soils, deep loosening and drainage were recommended.

In the same coal mine region, but in another locality (Gustorf near Grevenbroich), Lessmann and Krämer (1985) determined, in 1983, some enzymological and microbiological parameters in a levelled spoil (loess) recultivated with alfalfa for 2 years. For comparison they used a native, vegetated loamy riverside soil. This study site was located at Kirchhoven (at approximately 33 km from Gustorf). The Gustorf plot was never fertilised and the Kirchhoven plot was not treated with mineral fertilisers in the last 2 years. Both plots are serving as controls for an experiment on organic fertilisation. The studies will continue for many years, and the organically fertilised plots will also be analysed.

The unfertilised plots differed from each other in the enzymological and microbiological parameters of their 10-20-cm layer. In this layer, the loess plot had lower dehydrogenase and urease activities, contained less bacteria, respired less strongly (produced less CO_2), and degraded the cellulose more slowly than the plot of native soil. In the 35-45-cm layer, there were no significant differences between the two plots. The profile differentiation as indicated by enzymological and microbiological data was reduced in the loess plot.

Dealing with the problems of soil assessment in the recultivated brown coal mine area located in the "Niederrheinische Bucht", Schröder (1986) compared a native soil, characteristic for this area, with two representative technogenic soils on loess, the agricultural recultivation of which began in 1970. Recultivation of one of these two soils was considered good and that of the other bad. Thus, in each horizon, the bulk density was lower than 1.65 g cm^{-3} in the good recultivated soil and had a higher value in the bad recultivated soil. In the 0-40-cm layer, dehydrogenase activity, like CO_2 production and cellulose decomposition, showed the following order: native soil > good recultivated soil > bad recultivated soil.

Haubold *et al.* (1987) performed a similar comparison using 15 good and 15 bad recultivated soils on loess, also from the brown coal area in the Rhine region. They found that dehydrogenase activity, microbial biomass, and cellulose decomposition in the 0-40-cm layer were, in general, about 50-100% lower in the recultivated soils than in the native soils. At the same time, the mean values of these microbial parameters and those of the chemical parameters analysed (C, N, Na, K, Mg, and Ca contents, and cation-exchange capacity) indicated no remarkable differences between the good and bad recultivated soils. We think that this finding, although valid for dehydrogenase activity (an activity depending on the momentary proliferation of microorganisms), cannot be applied to those enzymes that are able to accumulate in soil and to be, in the accumulated state, independent of the momentary microbial proliferation. The activity of such enzymes was not assayed by these investigators.

In other studies carried out in the same coal strip mining area, Schröder *et al.* (1987a) and Schröder (1988a, b) have determined, among other things, the dehydrogenase activity in the 0-40-cm layer of 13 technogenic loamy-silty loess soils formed after slurry poldering of spoils followed by their agricultural recultivation for 6-25 years, and have found that the activity was significantly higher in the older soils than in the younger ones. The C, N, K, and Ca contents, cation-exchange capacity, and microbial biomass also increased significantly over time, while the time-dependent decrease of carbonate, Na, and Mg contents and increase of cellulose decomposition were insignificant. The significant correlations between age and microbial and chemical parameters of the technogenic soils had the following coefficients. 0.82 (dehydrogenase activity), 0.72 (microbial biomass). 0.95 (C content), 0.59 (N content), 0.87 (K content), 0.74 (Ca content), and 0.70 (cation-exchange capacity).

Generalising the results obtained in a comparative study of 29 technogenic soils formed after redeposition of loess as dry material (16 soils) or as wet material (13 soils) and of the adjacent native soils under compact plant cover, all being located in the brown coal area of the Rhine region, Schröder (1988b) drew the conclusion that dehydrogenase activity, microbial biomass, and cellulose decomposition in the ploughed layer of the technogenic soils, including the old ones, reached only about 30-50% of the values recorded in the corresponding layer of the native soils.

Müller *et al.* (1988) studied, in 1987, the biological properties, besides the chemical and physical ones, of 12 technogenic soils located in the Köln-Bergheim zone (within the brown coal area in the Rhine region). These soils were recultivated after redeposition of loess as dry material 20 years previously. Since that time, four of them were used as maple and hornbeam forest, four as pastures, and four as arable lands.

Dehydrogenase activity and microbial biomass were highest in the 0-10-cm layer of the pasture soils, intermediary in the same layer of the forest soils, and lowest in the ploughed layer (0-30 cm) of the arable soils. Accumulation of humus and N was also more evident in the 0-10-cm layer of the forest and pasture soils than in the 0-30-cm layer of the arable soils. Dehydrogenase activity and microbial biomass were also determined in the 10-30-cm layer of the pasture and forest soils and the values obtained were lower than those registered in the 0-10-cm layer. Humus and N contents decreased with sampling depth in all soils. It should be added that numerous earthworm burrows were observed in the forest and pasture soils but only a few occurred in the arable soils.

Although the biological and chemical properties of the 0-10-cm layer were better in the pasture soils than in the arable soils, the physical properties of pasture soils were unsatisfactory, because their 30-40-cm layer was compacted under the track of the grazing animals as the herbage cover was not yet resistant to mechanical disturbance. In conclusion, during the first decades the recultivated soils are not recommended for use as pastures.

In continuation of these investigations, Schumacher *et al.* (1993) sampled, at the beginning of March 1992, two young (<12 years) and four old (>25 years) recultivated spoil (loess) plots. One of the young plots contained arable soil and the other was under forest vegetation. The four old plots were used as non-manured and farmyard-manured arable lands, forest and grassland, respectively. A native arable loess soil was the control. Sampling depths were 0-15 and 15-30 cm (arable soils), Ah horizon (mostly the 5-10-cm depth), and Y1 horizon (up to 30-cm depth) (forest and grassland soils). Dehydrogenase and invertase activities, and substrate-induced respiration (CO_2 evolution from glucose-amended samples) were always highest in the old grassland soil, and lowest in the young arable soil. Both young and old forest soils were more active than the old arable soils. Surprisingly, farmyard-manuring of the old arable soil did not result in increased activities and respiration*. In concordance with these findings, organic C content was low in arable soils and showed important accumulations in the order: old forest soil > old grassland soil > young forest soil.

In a similar study, Schneider *et al.* (1995) compared about 10- and 25-year-old recultivation spoil (loess) plots used as arable lands or forests. Samplings were made in the spring months of the 1988-1993 period. Dehydrogenase activity and substrate-induced respiration determined in both whole soil aggregates (>1.5 cm) and in their outer, middle, and inner parts from the 5-10-cm layers of 10- and 25-year-old arable spoil plots did not show any significant differences depending on plot age and soil aggregate parts. Contrarily, dehydrogenase activity and respiration in the 5-10-cm layers of the forest soils were significantly higher in the 25- than in the 10-year-old plots and increased, within the soil aggregates, from the outer parts towards the inner ones, in plots of both ages. At the 20-25-cm layers of the same soils, the activity and respiration gave low values, were not age-dependent and did not show evident differences in the three parts of the soil aggregates. Organic C content was found again to accumulate in the forest soils. Abundance of different invertebrates, including earthworms, was also greater in the forest than in the arable soils. In conclusion, it was recommended that agricultural recultivation of brown coal spoils in the studied region should begin with forest recultivation, under such conditions, the arable soils will reach a "maturity stage" more rapidly.

Some of the results of the complex soil investigations, including the enzymological ones, performed in the Rhine brown coal area and reviewed above, were reiterated by Schröder and Schneider (1992) and Schneider and Schröder (1995).

In the Halle-Leipzig zone, Machulla and Hickisch (1988) analysed, enzymologically and microbiologically, the spoil heaps (predominantly brown loamy or marly overburdens) that resulted from the strip mining of brown coal at Espenhain. In 1987, samples were taken from the 0-20-cm layer of 1-, 9-, 18-, and 27-year-old heaps. The 1-year-old heap was not

* Similar results were already published by Schumacher *et al.* in 1989. Exactly 30-year-old spoil plots were studied in the Köln zone. Two plots were farmyard-manured three times and two plots were left non-manured. Samples taken from the 0-30-cm spoil layer in September 1988 and March 1989 exhibited only slightly different values of dehydrogenase, catalase, invertase, protease, and alkaline phosphatase activities, and microbial biomass and N mineralisation rate between farmyard-manured and non-manured plots. The effect of farmyard-manuring on the improvement of humic matter system was also very weak.

sown with crop plants. One of the 9-year-old heaps was cultivated with alfalfa, whereas another 9-year-old heap and the 18- and 27-year-old heaps were fertilised with diluted liquid dung and cropped to cereals. A fallow area at the margin of the mining site and a plot left fallow in a cereal field were the controls.

Catalase and dehydrogenase activities and numbers of bacteria, actinomycetes, micro-fungi, cellulolytic and phosphate-mobilising microorganisms, and *Pseudomonas fluorescens* cells were determined. The presence of proteolytic microorganisms, abundance of *Azoto-bacter chroococcum*, and intensity of cellulose decomposition were also studied.

The results showed that the enzyme activities increased and the microflora enriched both quantitatively and qualitatively in parallel with the age of spoil heaps. After 18-27 years of recultivation, catalase activity and microflora in heaps became similar to those of the control soils. Contrarily, dehydrogenase activity remained much lower even in the 27-year-old heap in comparison with the activity measured in the control soils.

In the Wetterau brown coal strip mining region, north of Frankfurt am Main, complex soil investigations were performed by Schröder and Schneider (1992) and Schneider and Schröder (1995). For recultivation of spoils (loess; pH 7.4-7.7), the dry deposition method was applied. Young (8-11 years) and old (20-25 years) spoil plots and undisturbed soil plots under agricultural use were studied.

Accumulation of organic C and total N throughout the years was very slow in the 0-30-cm layer and even much slower in the deeper layers of spoil plots. But dehydrogenase activity was markedly higher in the old spoil plots than in the young ones. However, the activity level of the undisturbed soils was not attained.

For characterisation of spoils in the Niederlausitz (Lower Lusatia) brown coal strip mining area (Cottbus region), Katzur and Haubold-Rosar (1996) and Kolk and Hüttl (1996) have applied enzymological methods, too.

These spoils are very acidic (pH 1.7-3.5) due to their iron disulphide (pyrite, marcasite) content, from which sulphuric acid is generated through chemical and bacterial (*Thiobacillus ferrooxydans*) weathering; their dumps remain barren of vegetation for decades, if not covered with a 100-cm thick cultivable soil layer or not subjected to amelioration. For amelioration, the spoil dumps are treated with lime or base-rich brown coal ash (to raise pH to 5.0 in the 0-45-cm layer which is required for reforestation) and with mineral fertilisers (at rates of 100-

150 kg of N, 25-50 kg of P, and 100-200 kg of K ha^{-1}). The lime or ash should be incorporated into the spoil to a depth of 60 cm, better 100 cm. The fertilisers are merely incorporated into the upper spoil layer. Then, the ameliorated spoil dumps are planted with forest tree species.

Katzur and Haubold-Rosar (1996) found that catalase activity and substrate (glucose)-induced respiration determined in the organic surface layers and topsoils on ameliorated and forestally recultivated spoil dumps were higher in the deciduous than in the coniferous forest stands. For example, in ash-treated spoil dumps under different forest stands of different age, values of the catalase activity and respiration increased in the following order: *Pinus sylvestris* (22 years) < *P. sylvestris* (27 years) < *Quercus rubra-Tilia cordata* (31 years) < *Populus nigra* (29 years). Accumulation of C and N was also the highest under the poplar stand and the lowest under the 22-year-old pine stand.

Kolk and Hüttl (1996) determined alkaline phosphatase activity in the 0-10-cm mineral layer of young (<5 years) spoil dumps planted with rapid-growing poplar and willow clones (at the Welzow-Süd mine). Spoil plots not covered with forest soil (at the Reichwalde mine) served for comparison. These, control plots contained no coal or only small amounts of coal powder.

The order of alkaline phosphatase activity in spoils sampled in April 1996 was the following: control plot containing no coal < control plot containing small amounts of coal powder < poplar plot ≈ willow plot.

It has also been found that alkaline phosphatase activity is an incomparably more sensitive parameter for characterisation of spoils than microbial biomass, because microbial biomass in the relatively highly phosphatase-active 0-10-cm mineral layer of the poplar and willow plots was not measurable, being very small.

Enzymological research in New Zealand

Speir *et al.* (1986) described a trial set up at Waimumu (Southland) for modelling different methods of soil replacement after completion of coal (lignite) strip mining. The soil was a silt loam. The trial comprised four treatments:

1. control, no soil stripping or replacement;

2. soil stripped by layers and replaced by layers designated arbitrarily (A layer = 0-20 cm; B layer = 20-40 cm; C layer = 40-80 cm), i.e. A/B/C;

3. A and B layers mixed and then replaced over C, i.e. (A + B)/C; and

4. A, B, and C layers each mixed with equal quantities of overburden (O) (deeper material overlying coal seams) and replaced in order, i.e. (A + O)/(B + O)/(C +O).

Within each treatment there were high and medium compaction plots. The medium and high compactions were achieved by deep ripping of soil after replacement, and not ripping, respectively. Consequently, the high compaction (unripped) control plot was essentially undisturbed soil.

In March 1983, all plots were fertilised with molybdic superphosphate (400 kg ha^{-1}) and sown with a ryegrass-clover mixture. At the end of spring 1983, urea (50 kg of N ha^{-1}) was added to plots 2, 3, and 4. In April 1984, heavy dressings of superphosphate were applied to all plots. Over 3 years, four biochemical parameters were periodically determined in the topsoil, namely invertase and sulphatase activities, net mineralisation of soil organic N (a measure of N availability), and mineral N flush* (a measure of microbial biomass). The annual herbage yields (dry matter) were also determined.

The biochemical parameters in each year and the herbage yields in the second and third years showed, in both high and medium compaction plots, the following order: treatment 1 (control) > treatment 2 > treatment 4 > treatment 3. When expressed as a percentage of treatment 1 (control) values, invertase and sulphatase activities and mineral N flush correlated significantly with herbage yields in ripped and unripped plots. These parameters, especially sulphatase activity in unripped plots, appear promising as predictors of recovery of soil fertility. In some high compaction plots, N mineralisation and herbage yield decreased in the third year, as compared to values recorded in the first and second years, which indicates N defficiency and, implicitly, the necessity of supplementary N fertilisation.

The results obtained in this trial during 5 years were described by Ross *et al.* (1992). They showed that beginning with the third year invertase activity in topsoil of treatment 1 (control) did not differ significantly from that of treatments 2-4. Sulphatase activity in treatment 1 became similar to that in treatments 2 and 4 only in the fifth year, but also remained in this year significantly higher than that in treatment 3. In other words,

* Soil samples moistened to 60% of their water-holding capacity were exposed to alcohol-free chloroform vapours in desiccator, for 24 h. Then, the chloroform was removed by repeated evacuation. In the next step, the fumigated soil was inoculated with a suspension prepared from the control (unfumigated) soil and incubated at 25°C for 14 days. Following incubation, mineral N (NH_4^+-N plus NO_3^--N plus eventual NO_2^--N) was extracted with 2 M KCl and determined. The difference between the net amount of mineral N produced by the fumigated and unfumigated samples was taken to be the mineral N flush.

accumulation of invertase in the technogenic soil was more rapid than that of the sulphatase. It is likely that plant materials contributed appreciably to accumulation of invertase, whereas accumulation of sulphatase was dependent primarily on microbial synthesis. Ripping had no effect on the enzyme activities.

Over the whole 5-year period, herbage yields correlated significantly with invertase activity in treatments 2-4 and with sulphatase activity in treatments 3 and 4, but the correlation between herbage yields and mineral N flush was significant in treatment 4 only.

Enzymological research in the United Kingdom

Harris and Birch (1989) determined, among other things, the dehydrogenase activity of technogenic soils that resulted from opencast mining of coal at five sites in England. Recultivation began 0.5-15 years previously; more precisely, the age of technogenic soils was 0.5, 1.5, 2.5, 3.5, and 4.5 years at Buckhead, 6 years at Ryefield, 7 years at Morrells, 11 years at Steadmill (fields 1 and 2), and 15 years at Salterwood. For comparison, three undisturbed soils were used: a pasture soil at Buckhead, a soil cultivated with cereals and a pasture soil at Kirk. Samples taken from the 0-30-cm depth were analysed. All the soils sampled were fine loams over clayey subsoil, and all recultivated sites had similar fertiliser and management regimes.

Dehydrogenase activity was highest in the Buckhead pasture soil and exceeded many times the activity values registered in the technogenic soils at the same site. There was no linear relationship between dehydrogenase activity and the age of technogenic soils; the activity showed only a trend to increase with the age of technogenic soils. Thus, the 6-15-year-old technogenic soils were, in general, more dehydrogenase-active than the 0.5-4.5-year-old ones, but the 11-year-old technogenic soil of field 2 at Steadmill was the least active. Based on the results obtained, the suggestion was made that, for increasing dehydrogenasic potential of technogenic soils, the seed mixtures used for recultivation should contain more perennial grasses.

Harris and Birch (1989) also determined dehydrogenase activity in a 5-year-old, 3-m high topsoil stockpile at Buckhead. Dehydrogenase activity in different depths of the pile presented the order: 0-30 cm > 180-210 cm > 90-120 cm. However, the activity, even in the 0-30-cm layer, was only ~15% of that measured in the 0-30-cm layer of the Buckhead pasture soil.

For the investigations described by Birch *et al.* (1991) and Bentham *et al.* (1992), 12 sites were selected. Four of them had been restored after opencast coal mining operations, and eight were nature reserves managed by traditional methods, in which the soils had remained undisturbed. Of the restored sites, two were 8-year-old woodlands (a poplar wood and a willow and poplar mixed wood, respectively) at Morrells, and the other two were 5- and 10-year-old grasslands at Shilo. The eight undisturbed sites comprised two productive flood-meadows, a chalk grassland, a rough grassland, a sandy heathland, and three established woodlands. All sites were sampled between December 1988 and March 1989, excepting the two restored grasslands which were sampled in December 1989. The samples taken from the 0-30-cm soil depth were analysed for determination of their dehydrogenase activity (considered as an indicator of total soil microbial activity), ATP content (an indicator of soil microbial biomass), and ergosterol content (a measure of the size of soil fungi). Many physico-chemical soil characteristics (total C, total N, moisture content, bulk density, pH, sand, silt, and clay contents) were also analysed.

Plotting of ATP content versus dehydrogenase activity and three-dimensional ordination of ATP content, dehydrogenase activity, and ergosterol content in undisturbed and restored soils have shown that the two floodmeadows clustered together as did the established woodlands and both clusters were distinct from the other sites. The two restored (8-year-old) woodlands and the younger (5-year-old) restored grassland clustered together. This means that soil microbiology of these 8-year-old woodlands was closest to that of the 5-year-old grassland. At the same time, the 10-year-old restored grassland clustered together with the rough grassland. This suggests that, 10 years after restoration, the microbiology of this grassland soil was consistent with a low-productivity undisturbed grassland soil. Position of the sandy heathland was close to the cluster of the restored woodlands and younger restored grassland, whereas the chalk grassland, having very low soil dehydrogenase activity, differed from the other grasslands. A clustering similar to that based on the three microbiological indices specified above was not possible using physical and chemical soil characteristics alone. The findings led to the conclusion that the microbiological indices could be used to assess the restoration potential and progress of reclaimed soils.

These investigations were also referred to in a review work of Harris and Birch (1992).

Edgerton *et al.* (1995) sampled soils from three restored field areas at the Erin opencast coal mining site located near Denby in Derbyshire. The areas were recultivated using stored

Table 24
Physical, chemical, and microbiological properties of soils in restored areas at the Erin opencast coal mining site located near Denby in Derbyshire (values represent means ± standard errors; all results are expressed on an oven-dried weight basis)

Age of areas	Number of replicates	Textural class	Bulk density (g cm^{-3})	pH in H_2O	Total C (mg g^{-1})	Total N (mg g^{-1})	DHA (μg TPF g^{-1})	ATP (ng g^{-1})
11 years	15	Loam	1.30±0.03	6.64±0.10	21.9±1.5	1.61±0.11	159.1±29.7	520±90
6 years	5	Silt loam	1.31±0.01	6.99±0.02	18.9±0.9	1.75±0.08	144.6±23.6	347±60
6 years (flooded)	5	Silt loam	1.34±0.01	6.76±0.01	18.5±2.2	1.45±0.01	74.9±8.8	123±30
6 years (compacted)	5	Silt loam	1.52±0.03	7.72±0.12	23 8±2.6	1.52±0.14	65.4±13.0	95±30
5 years	15	Silt loam	1.32±0.02	7.94±0.06	30.4±2.8	1.65±0.13	83.9±12.8	221±70

Reproduced from: D.L. Edgerton, J.A. Harris, P. Birch and P. Bullock, Soil Biology & Biochemistry, 1995, Vol. 27, p. 1500, with kind permission from Elsevier Science.

DHA – Dehydrogenase activity. TPF – Triphenylformazan.

subsoil material due to a lack of original topsoil. At sampling time (early April 1991), the three areas were 5-, 6-, and 11-year-old, respectively. The 5- and 11-year-old areas were allowed to revegetate naturally. The 6-year-old area was managed as a low input-low output grass-clover ley for the first 3 years and then an attempt was made to create a diverse grassland. As water-logging and compaction are common problems in restoration of mine soils, a part of the 6-year-old area was subdivided into an area subjected to periodical flooding and another in which the soil was artificially compacted during restoration. The sampling depth was 0-30 cm.

Besides several physical and chemical soil properties, dehydrogenase activity (as an indicator of total soil microbial activity) and ATP concentration (taken as microbial biomass C after multiplying by the factor 168.5) were also determined. Table 24 shows that the textural class is loam in the 11-year-old area and silt loam in the other areas. Dehydrogenase activity and ATP concentration increased with increasing age of the three areas, not considering the flooded and compacted 6-year-old areas. However, for both properties the increase was only significant ($P<0.05$) between the 5- and 11-year-old areas. Flooding and compaction had a significantly ($P<0.05$) negative influence on both properties, whose values in the flooded and compacted 6-year-old areas were even lower than those measured in the 5-year-old area.

A high linear correlation was found between the amount of water-stable aggregates (>2 mm) and microbial biomass C with both values increasing with age in all areas. The highest values were recorded, as expected, in the 11-year-old area, whereas the compacted 6-year-old area gave the lowest ones.

Enzymological research in Spain

González-Sangregorio *et al.* (1991) have studied one of the spoil benches of the opencast lignite mine at Meirama, Galicia. The spoil bench grew yearly by an area of 0.5 ha. The spoil had a sandy-loam texture (56% sand, 12% clay) and contained almost no pyrite and coal remains. The recultivation trial began in 1984 on an 0.5-ha plot formed in 1983, and a new trial started every year on the 0.5-ha plot formed during the preceding year. The original topsoil was not used for recultivation. Each plot was recultivated by the following technology: ploughing and harrowing; liming with coarse agricultural lime at a rate of approximately 1 t ha^{-1}; fertilising with NPK 15-15-15 (170 kg ha^{-1}); sowing of seeds (150 kg ha^{-1}) of four species: perennial ryegrass (*Lolium perenne*, 50%), red fescue (*Festuca rubra*, 25%), Kentucky bluegrass (*Poa pratensis*, 15%), and white clover (*Trifolium repens*, 10%); and mulching with cellulose and stabilising agents (300 kg ha^{-1}). The application of fertilisers was repeated yearly in early spring. For increasing the organic matter content in spoil plots, the plants were not harvested.

In the spring of 1987, before fertilisers were spread, soil samples were taken from the 0-10-cm layer of the plots planted in 1986 (1-year-old), 1985 (2-year-old), and 1984 (3-year-old). A raw spoil sample was taken from a recently built bench. Samples were also collected from the 0-10-cm layer of two pasture soils near the lignite mine; these native soils served for comparison. The samples were analysed enzymologically and chemically.

Phosphatase activity increased with the age of mine soils, without attaining the values recorded in the native soils*. At the same time, pH optimum of activity changed from 5 (1-year-old plot) to 5.5 (2-year-old plot) and 6 (3-year-old plot). Only acid phosphatase activity was found in the 1-year-old mine soil. This activity was predominant in the 2- and 3-year-old mine soils, in which a slight alkaline phosphatase activity was also detectable. This may be explained by the lime-induced pH increase from 4.32 (raw spoil) to 5.02, 5.04, and 5.71 in the 1-, 2-, and 3-year-old mine soils, respectively.

The kinetic parameters K_m (Michaelis constant) and V_{max} (maximum velocity) of phosphatase activity gave very high values in the 1-year-old mine soil, whereas the values recorded in the 2- and 3-year-old mine soils were similar to those obtained for native soils. The thermodynamic parameters E_a (activation energy), ΔH^* (activation enthalpy), and Q_{10} (temperature coefficient) of phosphatase activity in mine soils exhibited the same kind of age dependence as the kinetic parameters, except that all values were in this case relatively close to those recorded in native soils.

Total C, N, and P contents increased, both inorganic and organic N and P accumulated as the mine soil aged.

The conclusion was drawn that evolution of pedogenetic processes in the mine soils studied was rapid, which could be attributed to favourable properties of the parent material, the soil management practices applied, and the favourable climatic conditions in the region of the Meirama mine.

---

* See also Blanco *et al.* (1991).

Continuation of these investigations was described by Gil-Sotres *et al.* (1992). In the spring of 1989, samples were taken from the 0-10-cm depth of 1-5-year-old plots to be analysed for pH, total C and N contents, and C and N mineralisation capacity. In the spring of 1990, the plots sampled from their 0-10-cm layer were aged 1, 2, 4, 5, 6, and 7 years (i.e. recultivation of the 7-year-old plot began in 1983, not in 1984). The 1990 samples were used for determination of four enzyme activities (phosphatase, urease, and protease towards two substrates: casein and Nα-benzoyl-N-argininamide), ATP and total C contents. Raw spoil and adjacent native soils were analysed for comparison.

The results showed again that, owing to liming, the pH had increased values in all mine soils. Total C and N contents increased gradually with the age of mine soils. However, their values were considerably lower than those of the adjacent native soils.

Mineralisation of C, expressed as the amount of CO_2 produced (respiration), manifested a rising trend as a function of soil age, although even in the oldest mine soils it was far below the levels of native soils. Mineralisation of N, expressed as the amount of NH_4^+-N plus NO_3^--N produced, increased during the first 4 years of recultivation, and decreased sharply in the fifth year. NH_4^+-N predominated over NO_3^--N, which means that ammonification predominated over nitrification.

The enzyme activities and ATP content were very low in the raw spoil (protease activity towards casein was even lacking) and tended to increase with the age of mine soils. However, their values, except for protease activity towards casein, were much smaller even after 7 years of recultivation than those measured in a native grassland soil. All the enzyme activities correlated very significantly both with each other and with ATP content*. They also correlated significantly with total C content. However, the correlation coefficients between activities of phosphatase, urease, and protease towards benzoylargininamide and total C content were smaller than the correlation coefficients between these three activities and ATP content. This finding suggests that the organic matter present was not yet able to stabilise extracellular enzyme molecules which reflects the immaturity of these soils.

The conclusion was drawn that in the mine soils studied establishment of C and N cycles progressed satisfactorily, although less clearly for N than for C.

* Results of this correlation analysis were previously presented at scientific congresses held in Spain and Italy (Trasar-Cepeda *et al.*, 1990a, b).

The phosphorus cycle in the Meirama lignite mine soils was also studied (Trasar-Cepeda *et al.*, 1993).

In the spring of 1989, samples were taken from the 0-10-cm layer of 1-, 3-, and 5-year-old recultivation plots. In the spring of 1990, raw spoil and the 0-10-cm layer of 4- and 7-year-old plots were sampled. Native Galician soils served for comparison. Phosphatase activity was determined in the 1989 samples, while those collected in 1990 were used for analysis of P. As Figure 12 shows, phosphatase activity was found again to increase with the

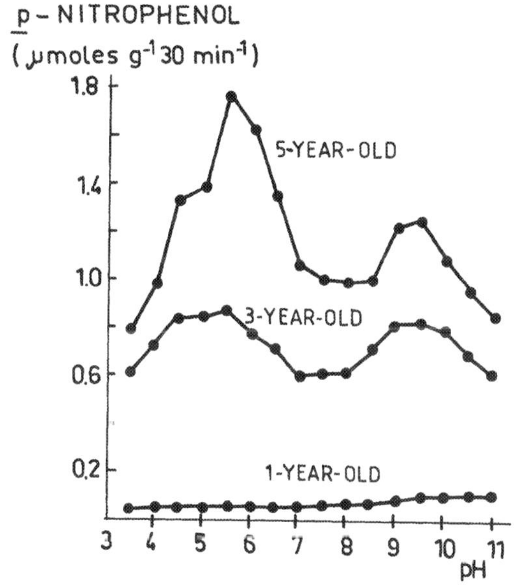

Figure 12. pH dependence of the phosphatase activity in mine soils from Meirama.

Redrawn from: M.C. Trasar-Cepeda, E. de Blas, I. Garcia-Rodeja, M.C. Leirós and F. Gil-Sotres, Land Degradation and Rehabilitation, 1993, Vol. 4 (3), p. 137, with permission from John Wiley & Sons, Ltd.

mine soil age. In contrast to another 1-year-old plot in which only acid phosphatase activity was detectable (see page 199), the soil in the 1-year-old plot studied by Trasar-Cepeda *et al.* (1993) manifested a slightly higher alkaline than acid phosphatase activity. The reverse was true for the older plots. Thus, the ratio of peak acid to peak alkaline phosphatase activity was 0.55, 1.05, and 1.41 in the 1-, 3-, and 5-year-old plots, respectively*.

The total P content was low in the raw spoil (253 µg of P g^{-1}) and the majority (77%) corresponded to non-extractable forms. In the 4- and 7-year-old mine soils the total P contents were very similar and with about 100 µg g^{-1} higher than in the spoil; this can be attributed to the P fertiliser applied at the start of recultivation. Total extractable inorganic and organic P contents increased with the age of mine soils. However, although inorganic P levels were similar to those found in native Galician soils, organic P levels were extremely low (even in the 7-year-old soil) compared with native soils of similar pH and organic matter content. The ratios C-to-organic P and N-to-organic P were much lower in the 4- than in the 7-year-old mine soil, but much higher than in the native soils. In other words, accumulation of organic P in mine soils was slow, more precisely it was much slower than accumulation of C and N. These findings clearly indicate that the P cycle has not been successfully established in the studied mine soils. This may, in the long-term, lead to serious deficiencies in plant-available P.

In a laboratory experiment for studying the possibility to replace inorganic fertilisers by cattle slurry in recultivation of lignite mine spoils, Saá *et al.* (1994) mixed samples of raw spoil from the Meirama lignite mine with cattle slurry (containing 9% dry matter; 44.20% total C, 3.04% total N, 1.14% P_2O_5, 3.95% K_2O) in two doses (175 ml kg^{-1} of spoil – mixture P and 350 ml kg^{-1} of spoil – mixture G). The mixtures were moistened to 80% of field capacity, then incubated at 28°C for 5 months. During incubation, every second day the moisture content was restored to 80% field capacity. After 0, 2, and 5 months of incubation, the mixtures were analysed to determine several enzyme activities (dehydrogenase, β-glucosidase, urease, protease on casein, protease on benzoylargininamide, phosphatase), total C and N contents, amounts of monosaccharides and phenols in aqueous extract as well as amounts of humic substances and organo-mineral complexes.

Each enzyme activity was extremely low in the raw spoil and very high in the cattle slurry. This is why the mixtures at time zero exhibited high activities; the activity increase was more pronounced in mixture G than in mixture P, excepting activities of β-glucosidase and protease on benzoylargininamide, the increase of which was not dependent on the slurry dose. During the first 2 months of incubation, dehydrogenase and urease activities decreased,

* See also González-Sangregorio *et al.* (1993).

while the other activities increased in both mixtures. During the next 3 months, the activities slightly increased or remained unchanged in mixture P and markedly decreased in mixture G. Nevertheless, at the end of the 5-month incubation period, each activity in each mixture exceeded many times the activity values recorded in the raw spoil before incubation. This means that the cattle slurry amendment led to the formation of an enzymatic potential in the spoil.

During incubation of the spoil-cattle slurry mixtures, the total C contents steadily decreased indicating mineralisation processes, whereas the total N contents remained practically constant indicating that there was no N loss due to volatilisation. The monosaccharide contents increased continuously, while the phenol contents decreased. Accelerated formation of clay-bound humic substances also occurred. In conclusion, fertilisation with cattle slurry was considered as an attractive option for rapid recultivation of lignite mine spoils.

Leirós *et al.* (1996) compared, under field conditions, the effectiveness of cattle slurry and inorganic fertilisers in recultivation of the Meirama lignite mine spoils. Three types of raw spoil plots were installed:

plots N – slurry-fertilised at the time of seeding*;

plots R – slurry-fertilised at the time of seeding and in the springs (prior to the growing season) of 3 consecutive years; and

plots I – NPK-fertilised at the time of seeding

In all cases, dose of slurry was 200 m^3 ha^{-1}.

Plant yield was found to be lower in the slurry- than in the NPK-fertilised plots (e.g. the third-year green matter yields, t ha^{-1}, were 0.5 in plots N, 6.5 in plots R, and 8.2 in plots I), but, more relevantly, the slurry-fertilised plots had higher vegetative cover, a higher proportion of native plants (largely *Festuca* spp.), and a lower proportion of bryophytes. Enzyme activities determined in the 3-year-old plots showed the following order: urease and phosphomonoesterase I < N < R, and phosphodiesterase N < I < R, i.e. the yearly slurry fertilisation had the strongest enhancing effect on the formation of an enzymatic potential in spoils.

These observations and other findings (such as the improvement of pollution-neutralising properties in slurry-fertilised spoils) support the conclusion reiterated by Leirós *et al.* (1996) that the use of cattle slurry for rapid recultivation of lignite mine spoils is an attractive alternative to the use of inorganic fertilisers.

---

* Composition and rate of the seed mix applied are the same as those specified on pages 198-199.

Enzymological and other biochemical properties as well as physical, chemical, and pollution-neutralising properties of the Meirama opencast lignite mine spoils and soils have been reviewed by Gil-Sotres *et al.* (1995a) and are also referred to in three other review articles (Gil-Sotres *et al.*, 1995b; Leirós *et al.*, 1995; Seoane *et al.*, 1995).

Enzymological research in the Czech Republic

Technogenic soils in the North Czech Brown Coal District were submitted to enzymological investigations in the March 1984-September 1987 period. The investigations were described by Šíša in two unpublished reports in 1985 and 1988, by Šíša and Belatková in a poster communication in 1991 and by Šíša in an article published in 1993. The mining sites selected for investigations are typical representatives of the entire Brown Coal District as regards both opencast mining and recultivation technologies applied.

Three enzyme activities (catalase, invertase, and amylase), basal respiration (CO_2 evolution from unamended samples), potential respiration (CO_2 evolution from ammonium sulphate- and glucose-amended samples), ratio between potential respiration and basal respiration (taken as an indicator of humus stability), humus content, and pH were determined in overburdens and adjacent native soils.

The investigations carried out at the selected mining sites will be summarised below.

At Vršany, before opencast mining of coal, the removed topsoil and subsoil were dumped as a mixture. Enzyme activities in the 0-20- and 20-40-cm layers of dump and native soil (chernozem on loess) presented the following order: dump 20-40-cm < dump 0-20 cm < native soil 20-40 cm < native soil 0-20 cm. In other words, dumping of overburdens led to degradation of enzyme activities. Basal respiration showed a rather reverse trend, but potential respiration and potential per basal respiration ratio as well as humus content behaved like enzyme activities. pH was about 8.5 in the dump and nearly 8 in the native soil.

At Mariánské Radčice, where the topsoil and subsoil were also dumped together, the differences between the 0-20- and 20-40-cm layers of dump and native soil (chernozem on clay) were less pronounced than at Vršany. The chernozem at Mariánské Radčice was somewhat less active biologically than the chernozem at Vršany.

At Užin, some overburden (grey clay) plots not covered with topsoil were submitted to agricultural or sylvicultural recultivation. Other plots, covered with topsoil, were used only for agricultural recultivation. In the 0-40-cm layer analysed, each enzyme activity was highest in the topsoiled plots, whereas the lowest values were registered in the non-topsoiled agriculturally recultivated plots (catalase) or in the sylviculturally recultivated ones (invertase and amylase). This means that topsoiling of overburdens had a beneficial effect on their enzyme activities, and agricultural recultivation was a better technology than the sylvicultural one. Humus content and basal respiration were lowest in the topsoiled plots. Contrarily, potential respiration and potential per basal respiration ratio were highest in the topsoiled plots and, at the same time, they were lowest in the sylviculturally recultivated plots. Mean values of pH were 6.15 (non-topsoiled agriculturally recultivated plots), 6.60 (topsoiled plots), and 6.00 (sylviculturally recultivated plots).

At the Lotta Marie mine, the agricultural recultivation of the non-topsoiled overburdens (loess mixed with sand) was also more efficient than the sylvicultural recultivation. Both 0-20- and 20-40-cm layers of the agriculturally recultivated plots were more catalase-active, showed higher basal and potential respiration and potential per basal respiration ratio and contained more humus than the corresponding layers of the sylviculturally recultivated plots. pH in the two layers was higher in the agriculturally than in the sylviculturally recultivated plots (7.80 and 7.81, and 5.04 and 5.39, respectively).

At Střimice, toxic sands (containing large amounts of H^+, Al^{3+}, Fe^{3+}; pH 4.79) were recultivated agriculturally on two areas. Previously, both areas were covered with an isolating layer of bentonite followed by overlayering of original topsoil. The recultivation began in 1973/1974 on area 1 (called Svoboda) and in 1982 on area 2 (experimental field). Thickness of bentonite layer was 20 cm on both areas, whereas that of topsoil was 50 cm on area 1 and 30 cm on area 2. An area with non-covered and non-recultivated toxic sands served as control. The 0-20- and 20-40-cm layers on areas 1 and 2 and the 0-15-cm layer of the control sands were analysed in 1986 and 1987. The results have shown that invertase and amylase activities, potential respiration, potential per basal respiration ratio as well as humus content were highest on area 1, while catalase activity and basal respiration gave the highest values on area 2. All parameters, excepting amylase activity, were lowest in the control sands. Due to recultivation, the pH increased to 7.07-7.48 on area 1 and to 8.21-8.53 on area 2.

In continuation of these investigations, Šíša *et al.* (1997) have selected eight localities within the North Czech Brown Coal District for collecting samples from recultivated overburdens and undisturbed soils, 5 times during the vegetation period of 1995. The samples were analysed for determination of their invertase, phosphatase, and catalase activities, basal and potential respiration, ammonification and nitrification capacities, biomass C content and some chemical parameters.

Comparison of the results registered for the eight localities has made it possible to establish that the topsoiled, then agriculturally recultivated overburdens at Užin have the highest enzymatic and microbial potential: invertase activity and a part of microbial parameters were highest, phosphatase and catalase activities and the other part of microbial parameters had high values in these overburdens. At the same time, the recultivated loessoid soil in a vineyard at Čepirohy exhibited the lowest enzymatic and microbial potential; for example, invertase and phosphatase activities were lowest in this soil.

Chapter 8. TECHNOGENIC SOILS FROM POWER PLANT WASTES

Enzymological research in Poland

Field experiments were carried out over 10 years to cultivate different crop species on the ash-containing wastes of the Halemba power plant. Plots fertilised and cultivated with potato, barley, field bean, and flax and an unfertilised and uncultivated plot were selected by Balicka and Wegrzyn (1984) for microbiological and enzymological analyses. For this purpose, the 0-10- and 10-20-cm layers were sampled two or three times yearly.

Owing to cultivation, the bacterial numbers increased in each plot; the highest increase occurred in the potato plot. The increase of dehydrogenase and catalase activities and of nitrification potential was similar in plots with different crop plants. It has also been found that the low biological activity in the unfertilised and uncultivated plot is the result of the lack of N; these wastes are not toxic for soil microorganisms.

Enzymological research in Hungary

The field experiment started in 1983 for recultivation of refuse soils at the Gyöngyös-Visonta lignite strip mine was referred to on page 187. Under similar conditions, Vörös and Antal (1988) also dealt with the cultivation of ash at the power plant in Visonta. The fertilisation comprised only three variants: 0, NPK, and NPK plus lignite.

In the fifth year of cultivation, invertase activity (expressed as mg of glucose g^{-1} ash 24 h^{-1}), very low in the unfertilised ash (0.1), increased very significantly to 0.7 in the NPK-fertilised ash and to 5.0 in the NPK plus lignite-fertilised ash. Nevertheless, these values are much lower than those recorded in the unfertilised, NPK- and NPK plus lignite-fertilised control soil (24.5, 52.6, and 52.0, respectively). Fertilisation of ash also resulted in intensification of microbial cellulose decomposition and in increase of wheat yield, especially when the ash was covered with a 20- or 60-cm soil layer.

Enzymological research in Romania

Păunescu and Ştefanic (1989) described microbiological and enzymological investigations carried out in 1980 on ash heaps that originated from the Işalniţa-Craiova power plant and were cultivated with different crop plants. The ash heaps had pH values of 7.15-8.15; their mobile P and K contents were satisfactory. They were fertilised with farmyard manure and mineral N and frequently irrigated at a low rate. Total numbers of viable microorganisms and dehydrogenase, catalase, invertase, urease, and phosphatase activities were determined in the 0-20-cm layer of ash plots cultivated with rye, alfalfa, kidney bean, soybean, potato, Jerusalem artichoke, or sugarbeet. An unfertilised plot under spontaneous vegetation was used for comparison.

The microorganisms were found to be most numerous in the plots cultivated with Jerusalem artichoke or alfalfa, whereas the lowest microbial numbers were recorded in the rye and potato plots. Based on the measured enzymatic activity values, a synthetic indicator of enzymatic potential was calculated for each plot. Depending on the nature of the plants, this indicator showed the following order: Jerusalem artichoke ≈ kidney bean > alfalfa ≈ rye > spontaneous vegetation ≈ sugarbeet > soybean > potato. It is evident from this order that the enzymatic potential of technogenic soil formed from power plant wastes was enhanced to a larger extent by some crop plant than by the spontaneous vegetation. Dehydrogenase activity correlated positively and significantly with invertase and urease activities, whereas its correlation with catalase activity was negative, but insignificant. No correlation was found between dehydrogenase and phosphatase activities.

Enzymological research in the United States of America

For enzymological, microbiological, and chemical analyses, Klubek *et al.* (1992) collected samples from three coal ash basins* located in the area of the coal-fired power plant at the Savannah River Site in Barnwell County, South Carolina. Soil samples taken near the ash basins were used as a control. Sampling depth was always 0-15 cm. All samplings were carried out on a single date (June 5, 1989). Two of the ash basins were old sites, abandoned in 1964 and 1974, respectively. The 1964 abandoned site received both cyclone and bottom ash in dry form, whereas the 1974 abandoned site received bottom ash and was reused as a

* As Klubek *et al.* (1992) point out, the usual method of coal ash disposal is the transport of fly ash as a slurry to a settling basin with the removal of water via surface drainage.

settling basin for sluiced ash. The third basin is currently used for settling of sluiced ash (in use since mid-1970s). The control site is a floodplain.

On the 1964 abandoned (i.e. 25-year-old) site the vegetative cover consisted of sweet gum (*Liquidambar styraciflua*), sycamore (*Platanus occidentalis*), loblolly pine (*Pinus taeda*), longleaf pine (*Pinus palustris*), one or two wax myrtle (*Myrica cerifera*); little to no understory was present. The 1974 abandoned (i.e. 15-year-old) site was dominated by sycamore (*Platanus occidentalis*) and hybrid poplar (*Populus* sp.); wax myrtle (*Myrica cerifera*) dominated in the understory; other species present included red maple (*Acer rubrum*), willow (*Salix* sp.), box elder (*Acer negundo*), loblolly pine (*Pinus taeda*). On the current use site, cattail (*Typha latifolia*), wax myrtle (*Myrica cerifera*), and willow (*Salix* sp.) grew; no understory was present. The vegetative cover on the control site was formed of sweet gum (*Liquidambar styraciflua*), loblolly pine (*Pinus taeda*), red maple (*Acer rubrum*), willow oak (*Quercus phellos*), and blackgum (*Nyssa sylvatica*).

For determination of dehydrogenase activity, the ash and soil samples were not treated or were treated with 0.5% glucose (on dry weight basis) and 0.5% glucose plus 0.025% ammonium nitrate, respectively. The activity was found to be significantly lower in the untreated ash samples from both abandoned sites than in the soil from the control site, whereas the activity was lacking at the current use site. Glucose treatment brought about significant activity increases at the 1974 abandoned and control sites, but not at the 1964 abandoned and current use sites. Treatment with both glucose and ammonium nitrate led to significantly increased dehydrogenase activity at all sites, except for the current use site. The significant differences in dehydrogenase activity by site and treatment indicate a deficiency in the availability of C and N at the studied sites.

Respiration (cumulative CO_2 evolution throughout a 21-day incubation period) was also enhanced by treatment of ash and soil samples with glucose and glucose plus ammonium nitrate. As in the case of dehydrogenase activity, the effect of glucose plus ammonium nitrate was more marked than that of glucose alone.

The microbiological analyses showed that abundance and diversity of microorganisms increased as the age of ash basin increased and their plant community matured. Total numbers of heterotrophic bacteria, actinomycetes, and fungi correlated significantly with the % total N content, but not with the % total C content in ash and soil samples.

The results suggest that the reclamation and utilisation of coal ash basins will require the addition of organic residues (characterised by a low C-to-N ratio) to assure the availability of C and N, and the enhancement of an active microflora. The establishment of a mature plant community can then be more readily achieved.

Chapter 9. TECHNOGENIC SOILS ON EXHAUSTED PEATLANDS

Enzymological research in the Russian Federation

Artem'eva *et al.* (1980) and Shiroķikh *et al.* (1987) have studied the low moor peatland Gadovo boloto (Kirov, formerly Vyatka, region) which was dried through open deep draining ditches, and then its peat was extracted by means of rotary cutter. The extraction ended in 1965, but a peat layer remained on the bottom (sand) of the exhausted peatland, on different areas of which the residual peat had the following characteristics: thickness = 10-50 cm (however, in some places there was no peat and so the sand appeared at the surface); decomposition degree = 20-45%; pH = 4.75-6.01; mediocre total N and low plant-available N, P, and K contents.

Some plots of the exhausted peatland were submitted to cultivation in 1970, and other plots were reclaimed in the next years. The plots were seeded to awnless bromegrass (*Bromus inermis*) and fertilised annually with N (NH_4NO_3), P (superphosphate), K (KCl), NP, NK, PK, or NPK. Each plot was treated with Cu as a trace element at the rate of 7 kg ha^{-1}. Plots not seeded and not fertilised and plots seeded but not fertilised served for comparison. Another peatland, never used for peat extraction, but cultivated with agricultural crops since 1935 and fertilised annually with NPK, also served for comparison.

The peat layer of some plots cultivated since the spring of 1973 on the exhausted peatland was sampled by Artem'eva *et al.* (1980) in 1974 and 1975, i.e. in the second and third years of cultivation, for determining invertase and catalase activities. The results presented in Table 25 show that the activities remained low even after 2-3 years of cultivation, excepting the plots fertilised with NPK, in which they increased to some extent in the second and third years and in the third year, respectively. The hay yield was also highest in the NPK-fertilised plots.

In preliminary reports on these investigations, Nekrasova and Krylova (1975) and Nekrasova *et al.* (1977, 1979) have emphasised that in the NPK-fertilised plots the increase in enzyme activities and hay yield was associated with an increase in numbers of algae.

These investigations were also referred to in the book of Zverkov (1982).

Table 25
Enzyme activities in technogenic soils of plots cultivated with awnless bromegrass and fertilised annually for two and three years on an exhausted peatland

Fertilisers and their rates ($kg\ ha^{-1}$)	Invertase*		Catalase*	
	Year 2	Year 3	Year 2	Year 3
Control (unfertilised)	16.60	N.D.	3.2	4.6
N 60	6.50	8.90	3.0	5.6
P 90	7.14	2.86	2.9	5.0
K 120	5.90	1.59	3.8	6.2
N 60, P 90	7.74	6.29	3.2	6.6
N 60, K 120	8.70	9.22	2.5	7.2
P 90, K 120	9.60	13.07	3.4	7.2
N 60, P 90, K 120	15.10	27.53	3.9	7.6
N 90, P 90, K 120	17.64	22.57	3.3	7.8
N 120, P 90, K 120	12.84	32.11	4.0	8.0

Reproduced from: T.I. Artem'eva, K.A. Nekrasova, I.A. Vertogradskaya, Yu.V. Zverkov, T.M. Borisovich and L.N. Krylova, Pochvennaya Fauna i Biologicheskaya Aktivnost' Osushennykh i Rekul'tiviruemykh Torfyanikov, Moscow, 1980, pp. 86-87.

* Expression of enzyme activities: invertase in mg of "glucose" g^{-1} soil 24 h^{-1} and catalase in ml of O_2 g^{-1} soil min^{-1}.

N.D. – Not determined.

In 1984 and 1985, Shirokikh *et al.* (1987) analysed enzymologically the 0-30-cm layer of some plots cultivated on the exhausted peatland. The plots had a 30-50-cm peat layer over sand and were annually fertilised with NPK. In 1985, their ages were 3, 13, and 16 years, respectively. The analyses indicated that catalase, urease, proteinase, and dehydrogenase activities measured in the 3-year-old plots and in the uncultivated ones were similarly low. The activities were much more pronounced in the 13- and 16-year-old plots and in the control peatland cropped and fertilised since 1935, i.e. for 50 years. Catalase and urease activities were highest in the 13-year-old plots and proteinase and dehydrogenase activities gave the highest values in the peatland cultivated for 50 years.

However, other results, published by Shirokikh and Zimenka (1989), indicate that long-term mineral fertilisation of the exhausted peatlands may lead to a general decrease of their enzymatic potential. In the experiments of these authors, the control plots were fertilised only once (at the beginning of the experiments), whereas the other plots were fertilised annually during 16 years with $N_{120}P_{60}$, $N_{120}K_{120}$, and $N_{120}P_{60}K_{120}$, respectively, all fertilisers having been applied in mineral form. Although in the first years mineral fertilisation had a beneficial effect on the enzymatic potential of the exhausted peatland, in the 16th year a decrease was found in almost each of the activities (catalase, urease, proteinase, nitrogenase, polyphenol oxidase, and peroxidase) determined in the annually fertilised plots as compared to the control plots. It should be added that in the 16th year the hay yields in the annually fertilised plots were at the level of the yields obtained in the control plots.

In a short report, Shirokikh and Ulanov (1986) described some biological effects of using sand (800 m^3 ha^{-1}) to improve the technogenic soil formed from residual peat on exhausted low moor peatlands. The studies, that were carried out in the 1981-1985 period, showed that, owing to sand treatment, productivity of perennial and annual grasses increased to a large extent; the total number of soil microorganisms manifested a tendency to increase, whereas soil respiration (CO_2 evolution), catalase, urease, proteinase, and dehydrogenase activities surprisingly decreased. However, according to another short report, by Shirokikh (1988), activity of oxidoreductase and hydrolase enzymes became higher when the sand or clay treatment of the residual peat was accompanied by organic fertilisation (farmyard manure, green manure, or sewage sludge) and application of mineral fertilisers.

Plots recultivated with perennial grasses for 20 years were compared with spontaneously revegetated plots and it was found (Shirokikh and Vertogradskaya, 1992) that accumulation of humus was more pronounced and number and biomass of microorganisms were much higher in the peat under perennial grasses (8-10.10^6 cells ml^{-1} peat and 2.4 t of microbial biomass ha^{-1}) than in that under spontaneous vegetation (0.4-1.10^6 cells ml^{-1} peat and 1.3 t of microbial biomass ha^{-1}). The cellulolytic microflora consisted of fungi, streptomycetes, and myxobacteria in the recultivated peat, whereas in the peat under spontaneous vegetation cellulose was degraded predominantly by fungi. In the recultivated peat peroxidase activity increased 5 times, urease and proteinase activities showed 2-2.5-fold increases, but polyphenol oxidase activity decreased 1.5-2 times as compared to activities measured in the spontaneously revegetated peat.

In plots submitted to forest recultivation for 20 years, it was characteristic that the microbial biomass was more abundant, the number of microorganisms and the enzyme (polyphenol oxidase) activity were 2-3 times higher in the litter than in the subjacent peat layer. Shirokikh (1993) found that peroxidase activity behaved like polyphenol oxidase activity.

The enzymological investigations carried out by Shirokikh and co-workers were also referred to in two brief review articles (Shirokikh and Shirokikh, 1993; Starodumova and Shirokikh, 1993).

Enzymological research related to recultivation of exhausted peatlands was also carried out at the Tolmachevo-Krivodanov peatland (Siberia, Novosibirsk region) (Naplekova and Gorskikh, 1986; Potapova, 1989).

Naplekova and Gorskikh (1986) conducted experiments, under field conditions and in vegetation pots, for studying the effect of liquid dung on catalase and proteinase activities, on a lot of microbiological parameters (number of saprophytic microorganisms, respiration, assimilation of organic and mineral N, ammonification, nitrification, N_2 fixation, accumulation of amino acids, and aerobic decomposition of cellulose), and on herbage yield in the exhausted peatland cultivated with fodder plants.

The liquid dung, applied at high rates, equivalent to 180 and 240 kg of N ha^{-1}, had, immediately after its application, a negative effect on the enzymatic activities and microbiological parameters of peat. After 2 months, the negative effect diminished; the diminution was more marked under field conditions than in the vegetation pots. When applied at lower rates, equivalent to 60 and 90 kg of N ha^{-1}, the liquid dung activated the microbiological processes in peat. The yield of fodder plants reflected the changes that occurred in the biological activity of peat.

The experiment described by Potapova (1989) was carried out in microplots, in which the exhausted peat was fertilised with $P_{90}K_{90}$ plus the trace element molybdenum or without Mo, or with $N_{60}P_{90}K_{90}$ plus Mo or without Mo, and cultivated with oats. Unfertilised microplots were the controls. During the growing season, the peat was sampled from the 0-10- and 10-30-cm depths and analysed for determining cellulase, proteinase, nitrogenase, and catalase activities as well as respiration (CO_2 evolution). The determinations showed that, under the influence of fertilisation, the enzyme activities and respiration increased in both peat layers sampled, except for catalase activity which remained unaffected in the 0-10-cm layer, increasing only in the 10-30-cm layer. Cellulase activity and respiration were highest in the microplots fertilised with NPK plus Mo, whereas proteinase and nitrogenase activities gave the highest values in the microplots fertilised with PK plus Mo. Green matter yield of oats was also higher in the microplots that received mineral fertilisers together with molybdenum.

Enzymological research in Belorussia

Polyakov *et al.* (1987) performed enzymological investigations related to the influence of forecrops on the efficiency of using perennial grasses for the recultivation of an exhausted peatland belonging to the "Berdovka" farm (Lida district, Grodno region). Thickness of the residual peat layer on the bottom (fine sand) was 50-70 cm. Decomposition degree of peat was 35-40% and pH in KCl solution 5.3-5.6. The following plants were used as forecrops: a pea-oats mixture; a foxtail clover-timothy grass-meadow fescue mixture; an awnless brome-grass-timothy grass-meadow fescue mixture; annual grasses; barley; lupine; and potato. The plots were installed in 1982. Some of them were seeded to perennial grasses in the same year and others were sown in 1982 with forecrops which were replaced by perennial grasses after 1 year (in 1983), 2 years (in 1984), and 3 years (in 1985). The plots were fertilised with NPK. Plots kept free of plants served for comparison. Yields of perennial grasses and activity of some enzymes in plots were estimated in 1985.

Yields of perennial grasses were lowest in plots cultivated only with these plants for 4 years, increased in plots cultivated with forecrops for 1, 2, and 3 years and then sown with perennial grasses. The highest yield was recorded in plots cultivated with forecrops in the first year and then with perennial grasses for 3 years. Of the forecrops, the pea-oats mixture, barley, and lupine proved to be the best.

Catalase, urease, and dehydrogenase activities were lower in the control plots than in the cultivated ones. Catalase activity was highest in the plots cultivated with perennial grasses for 4 years, whereas urease and dehydrogenase activities exhibited the highest values in plots cultivated with forecrops for 2 years followed by cultivation of perennial grasses also for 2 years.

The conclusion has been drawn that cultivation of forecrops is also recommended for enhancing the enzymatic potential of technogenic soils formed on exhausted peatlands.

Chapter 10. TECHNOGENIC SOILS FROM RETORTED OIL SHALE

Enzymological research in the United States of America

Hersman and Klein (1979) studied, under laboratory conditions, the effects of retorted oil shale additions on some enzymological and microbiological characteristics of surface soils. The retorted oil shale used was produced by the Paraho process. Samples of this retorted oil shale transported to the study site (the Piceance Basin, northwestern Colorado) and samples of surface soil collected from the study site were sieved and mixed in different proportions by weight, obtaining five variants: 100% soil (control); 95% soil plus 5% shale; 90% soil plus 10% shale; 75% soil plus 25% shale; and 100% shale (control). The mixtures were stored at 23-25°C with the soil water content maintained at 10% weight and sampled and analysed at 2-week intervals over a 10-week period.

The results showed that retorted shale addition caused, for all sampling times, significant decreases in acetylene reduction (N_2 fixation) rate, dehydrogenase activity, ATP content, mineralisation rate of uniformly ^{14}C-labelled glucose, respiration (O_2 uptake) rate, and fungal numbers, although numbers of bacteria and actinomycetes were not significantly affected by the presence of up to 25% retorted shale.

In another experiment (Klein *et al.*, 1982), retorted oil shale produced by the Lurgi process was used. Its addition to soil samples in 5, 10, and 25% proportions also brought about significant diminutions in dehydrogenase and phosphatase activities as compared to those in the control soil.

These results have a great importance for the revegetation programmes in areas disturbed by oil shale processing. If retorted oil shales are covered with stored surface soils, it may be necessary to insure that little physical mixing of the surface soil and retorted oil shale occurs. It may even be necessary to consider the construction of capillary barriers below the surface soil column to minimise upward movement of soluble fractions from the retorted oil shale.

These suggestions were verified by Sorensen *et al.* (1981) and Klein *et al.* (1982, 1985), under field conditions using experimental plots installed in the Piceance Basin in the autumn of 1977 and designed to allow the study of vegetation succession on surface soil (a fine loam) of various depths (thicknesses) overlying Paraho retorted oil shale and on surface soil separated from retorted shale by a gravel barrier. The profile configuration for five panels included panels with soil depths of 30, 61, and 91 cm over compacted shale, a capillary barrier panel (61 cm of soil over 30 cm of fine and coarse gravel), and a control panel (soil).

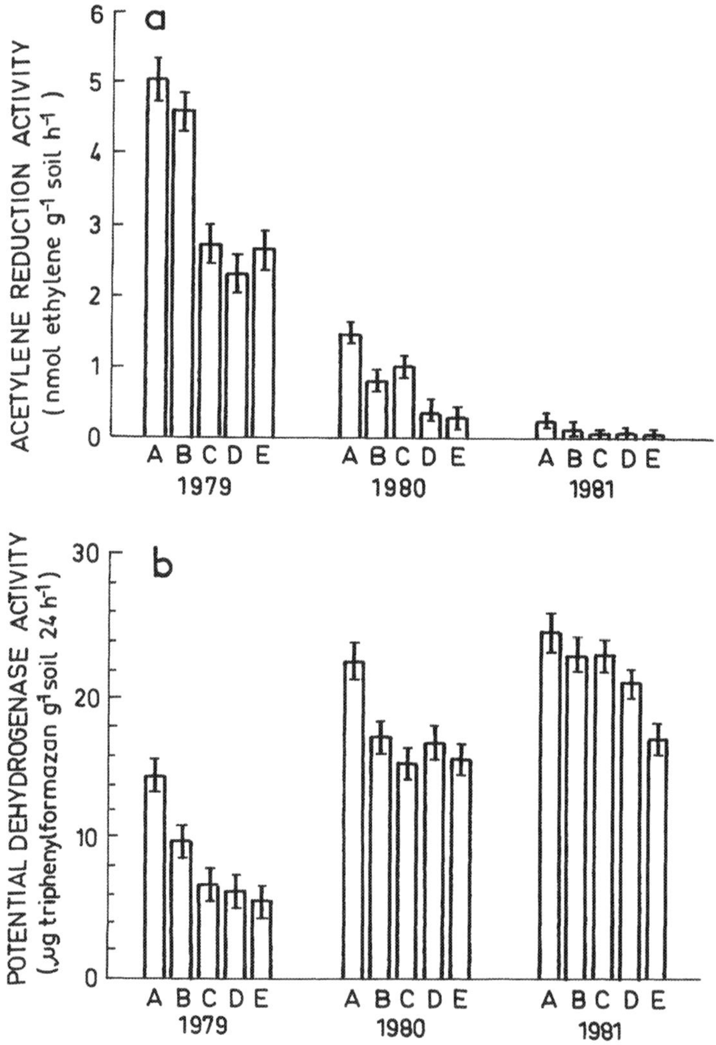

The surface of each panel was divided into three replicate plots. Within each of these plots nine possible combinations of three seed mixtures (native species, introduced species, and native plus introduced species) and three fertilisation rates (112 kg of N plus 56 kg of P ha^{-1}, 56 kg of N plus 28 kg of P ha^{-1}, and no N and P) were applied randomly to subplots.

In the summers of 1979, 1980, and 1981, samples were taken from the 5-10-cm depth after removal of the 0-5-cm layer. The samples were analysed to determine their acetylene reduction, dehydrogenase, and phosphatase activities.

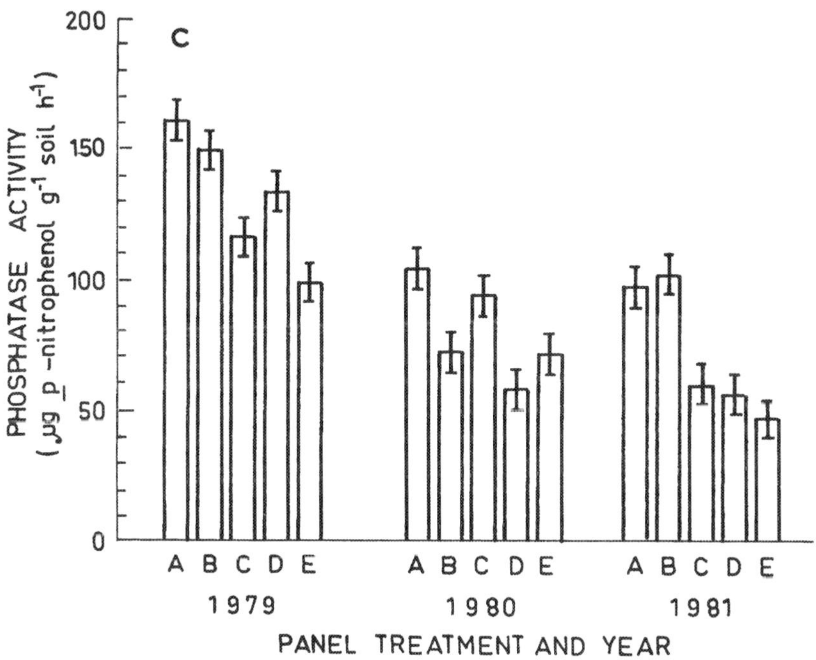

Figure 13. Acetylene reduction (*a*), potential dehydrogenase (*b*), and phosphatase (*c*) activities of soil of various depths over retorted oil shale and capillary barrier.

Redrawn from: D.A. Klein, D.L. Sorensen and E.F. Redente, Soil Reclamation Processes: Microbiological Analyses and Applications, 1985, pp. 153-155, by courtesy of Marcel Dekker, Inc.

A – Control soil. B – Soil (30 cm) placed over capillary barrier (30 cm of fine and coarse gravel) overlying retorted oil shale. C, D, E – Soil placed directly over retorted oil shale; depth (thickness) of soil: 91 cm (C), 61 cm (D), and 30 cm (E).

The error bars indicate least significant differences between means.

The average yearly activities across the panels from 1979 to 1981 are shown in Figure 13, from which one can see that over the 3 years acetylene reduction decreased to a large extent in all experimental variants. In each year, this activity was most intense in the control soil and weakest in the 30- or 61-cm deep soil over retorted shale. In 1979, this activity in soil overlying the capillary barrier was not significantly lower than that in the control soil.

Potential (zymogenous) dehydrogenase activity and phosphatase activity behaved contrarily over the 3-year period: potential dehydrogenase activity increased, whereas phosphatase activity showed a consistent downward trend since 1979 in all variants. In each year, both activities had the highest values in the control soil and in the soil overlying the capillary barrier, whereas the soil 30 deep over shale was, in general, the least active.

The acetylene reduction data manifested a significant interaction between seed mixture (or the ensuing plant community) and fertilisation in 1979, although in 1980 and 1981 only a simple effect of fertiliser on acetylene reduction occurred. In 1981, potential dehydrogenase activity showed a significant three-way interaction between seed mixture, fertilisation, and soil-shale arrangements. In the case of phosphatase activity, a significant interaction between seed mixture, fertilisation, and soil depth was observed when the activity values from 1979, 1980, and 1981 were analysed together.

Measurements of percentage mean plant cover over the 3-year period in the control plot, corresponding capillary barrier plot with 61 cm of soil, and plots with 91, 61, and 30 cm of soil directly over shale indicated no significant decrease in percentage mean plant cover when comparing the control plot with only 30 cm of soil. These measurements also suggested an improved plant development in the plot with the capillary barrier. In comparison, acetylene reduction, potential dehydrogenase, and phosphatase activities decreased to a greater extent than the corresponding values for percentage mean plant cover on the plots with less surface soil overlying the retorted shale.

Thus, this field study confirms the laboratory results of Hersman and Klein (1979). Appreciable diminution of microbial and enzymatic activities occurred in soil up to 91 cm in depth when it was placed directly over retorted oil shale. Suppression of these activities in such soils might lead to a long-term reduction of productivity. A capillary barrier composed of fine and coarse gravel helped maintain enzyme activities in soils placed over retorted oil shale during revegetation.

In 1981, the retorted shale over which soil had been placed was also analysed enzymologically. Potential dehydrogenase and phosphatase activities were essentially absent in the shale. In contrast, actual dehydrogenase activity in the shale showed values equivalent to those found in the control soil (Klein *et al.*, 1982), but the possibility that this activity was due to nonenzymatic factors is not excluded.

Also in 1981, Klein *et al.* (1982) analysed enzymologically a stored topsoil pile. The planted north end of the soil pile in comparison with its unplanted south end manifested higher dehydrogenase and phosphatase activities. Contrarily, numbers of actinomycetes, fungi, and bacteria and the soil moisture content were higher in the unplanted south end of the pile.

In several industry-constructed reclamation plots in which the surface soil was placed over shale materials retorted by different processes (TOSCO II, USBM, Union B, Union Decarbonized), dehydrogenase activity decreased with soil depth and the shale-to-surface had the lowest activity. The TOSCO II shale-to-surface was less dehydrogenase-active than the USBM shale-to-surface. The plots were constructed in western Colorado during 1970-1975 and analysed enzymologically in 1981 (Klein *et al.*, 1982, 1985).

Biondini *et al.* (1985) described a detailed study concerning the relations between plant succession and soil biological activity in 8 of the 45 subplots installed in the Piceance Basin in 1977 (see page 218-219). These eight subplots which were studied not only in 1979, 1980, and 1981, but also in 1982, i.e. not only in the second, third, and fourth years, but also in the fifth year after their installation, correspond to the variants "introduced species" and "native species". Each variant comprised four treatments:

1. 30 cm of topsoil over 60 cm of retorted shale;
2. 30 cm of topsoil over 60 cm of retorted shale plus 112 kg of N and 56 kg of P ha^{-1};
3. 60 cm of topsoil and 30 cm of capillary barrier over 60 cm of retorted shale; and
4. 60 cm of topsoil and 30 cm of capillary barrier over 60 cm of retorted shale plus 112 kg of N and 56 kg of P ha^{-1}.

The subplots were designated 30-F, 30+F, 60C-F, and 60C+F, respectively (F = Fertilisers; C = Capillary barrier).

It should be reiterated that the seed mixture of introduced species, like that of the native ones, was sown in 1977. At the end of the growing seasons of 1979 to 1982, the species composition was established, the percentage of the participation of species in the plant cover was calculated, and the potential dehydrogenase and phosphatase activities, acetylene reduction (N_2 fixation) capacity, and organic matter content in the 5-10-cm layer of soil were determined.

The *introduced species*, 16 in number, were: *Agropyron desertorum, A. sibericum, A. elongatum, A. trichophorum, A. intermedium, Bromus inermis, B. biebersteinii, Elymus junceus, Medicago sativa, Melilotus officinalis, Astragalus cicer, Onobrychis viciaefolia, Saponaria officinalis, Sanguisorba minor, Caragana arborescens,* and *Elaeagnus angustifolia*. Total weight of seeds was 17.8 kg ha^{-1}.

Proportion of the perennial grasses in plant cover during the 5-year period was significantly smaller in subplot 30-F (e.g. 47% in year 2 and 44% in year 5) than in 30+F (65 and 85%), 60C-F (84 and 98%), and 60C+F (95 and 95%). In 30-F, the dominant species after 2 years was *Melilotus officinalis* (24%) which disappeared after 5 years, when *Medicago sativa* became the dominant species (9% in year 2 and 51% in year 5). The other perennial forbs taken together had the highest proportion (20%) in 30-F in the second year. In the other subplots, the dominant species were *Agropyron desertorum, A. intermedium*, and *A. trichophorum*. *Agropyron elongatum* disappeared from all subplots after 3 years.

The statistical analysis showed that in year 2 the species composition in 30-F and 30+F was not significantly different; it was significantly different in 60C-F and 60C+F, and in 30-F and 30+F as a group related to 60C-F and 60C+F as a group. In the fifth year, significant differences were found only between the species composition of 30-F and those of the other subplots. In other words, there was a certain degree of convergence in the species composition of 30+F, 60C-F, and 60C+F as time elapsed. These subplots were dominated by grasses, whereas in 30-F the dominant plant was *Medicago sativa*. The rate of succession, as calculated from the changes in proportion of total grasses, *Melilotus officinalis, Medicago sativa* and the other forbs, was similar in 30-F, 30+F, and 60C-F, but significantly higher than in 60C+F. This means that the grasses became dominant most rapidly when the retorted shale was covered with a 30-cm capillary barrier and a 60-cm fertilised topsoil layer.

The yearly percentage proportion of perennial grasses in all subplots and that of *Medicago sativa* in 30-F correlated significantly with dehydrogenase activity but not with phosphatase activity, acetylene reduction capacity, and organic matter content in soil. Dehydrogenase activity after 2 years was significantly lower in 30-F and 30+F than in 60C-F and 60C+F and increased in all subplots in the next years. In year 5, values of dehydrogenase activity in the four subplots were convergent.

Phosphatase activity and acetylene reduction capacity, which significantly correlated with each other, decreased during the 5-year period in each subplot. However, the order of phosphatase activity in soil of subplots after 2 and 5 years remained the same: 30-F < 30+F < 60C-F < 60C+F; this order was also registered in the acetylene reduction capacity after year 2 and another order was found in year 5 (30-F ≈ 30+F ≈ 60C-F < 60C+F). The organic matter content in soil did not correlate significantly with the enzyme activities.

The *native species* also comprised 16 species: *Agropyron smithii, A. riparium, A. inerme, Oryzopsis hymenoides, Stipa viridula, Festuca ovina duriuscula, Poa ampla, Sporobolus airoides, Sphaeralcea munroana, Hedysarum boreale, Penstemon palmeri, Cowania mexicana stansburiana, Ephedra viridis, Atriplex canescens, Ceratoides lanata*, and *Purshia tridentata*. Total weight of seeds was 16.7 kg ha^{-1}.

Proportion of *Agropyron inerme* and *Poa ampla* in plant cover tended to increase with time in all the subplots. In years 4 and 5, *Agropyron inerme* had the highest proportion (53 and 47%, respectively) in 60C-F and *Poa ampla* in 30+F (44 and 28%, respectively). During the 5-year period, *Agropyron riparium* had relatively constant proportions in 30-F (27-31%) and gradually smaller proportions in the other subplots. Proportion of *Agropyron smithii* did not show any particular trend to change with time; its highest value (19%) was registered in 60C+F in year 5. Participation of the other species in plant cover showed a decreasing tendency with time.

In year 2, species composition did not differ significantly in 30-F, 30+F, and 60C-F, whereas the difference between these subplots and 60C+F was significant. In the fifth year, significant differences in species composition appeared between 30-F and 30+F and between these subplots and 60C-F and 60C+F; the difference between 60C-F and 60C+F was, however, not significant. The successional rate, calculated from the changes in the proportion of *Agropyron inerme, A. riparium, A. smithii, Poa ampla*, and *Ceratoides lanata*, had no significantly different values in the four subplots.

Soil dehydrogenase activity correlated significantly with the yearly percentage proportion of perennial grasses in plant cover. In year 2, dehydrogenase activity in 30-F and 30+F was lower than in 60C-F and 60C+F. In the next years, higher values of this activity were recorded in all subplots. In year 5, the subplots 30-F, 30+F, and 60C-F did not differ significantly from each other in regard to their dehydrogenase activity. At the same time, dehydrogenase activity in these three subplots was significantly lower than that measured in 60C+F.

The findings concerning phosphatase activity, acetylene reduction capacity, and organic matter content in soil were similar to those made in the subplots sown with introduced species.

It was concluded that in both subplots sown with introduced species and those sown with native species the potential dehydrogenase activity in soil was affected by plant species composition, the increase of participation of perennial grasses in plant cover having been accompanied by increase in activity, whereas phosphatase activity and acetylene reduction capacity were related not to plant species composition but to topsoil (its thickness on retorted shale), fertilisation, and time.

The investigations of Biondini *et al.* (1985) were also amply referred to in Harris and Hill's (1995) review paper.

Chapter 11. TECHNOGENIC SOILS FROM IRON MINE SPOILS

Enzymological research in Romania

Blaga *et al.* (1979, 1981) compared dehydrogenase, catalase, and invertase activities in spoils (calcareous sandy loams or clays; pH in H_2O was 7.7-8.3) levelled with the aim of their agricultural recultivation in the northern zone of the iron strip mine in Căpuş (Cluj county) and in adjacent soils. In the soils the activities decreased with depth (0-70 or 0-80), whereas in the spoils the activities were approximately the same in the 0-20- and 50-80-cm layers. In the 0-20-cm layer each activity was many times lower in the spoils than in the soils*. In the 50-80-cm layer the differences between spoils and soils in dehydrogenase activity were great, but differences were not so pronounced in the case of their catalase and invertase activities.

In another study carried out in the same zone by Bunescu and Blaga (1980), similar results (i.e. very low and higher activities, respectively) were registered in different spoil and soil profiles, except for a spoil profile that showed relatively high dehydrogenase and catalase activities. The activities correlated with the total N content of spoils and soils.

On the levelled spoils in the southern zone of the Căpuş iron strip mine, recultivation plots were installed. Some plots were recultivated with sainfoin (*Onobrychis viciaefolia*) and others with orchard grass (*Dactylis glomerata*). After 3 years of recultivation the spoils of these plots were analysed enzymologically by Drăgan-Bularda *et al.* (1983). For comparison, the 0-15-cm layer of an adjacent native soil (rendzina) and the same layer of a spoil plot not submitted to recultivation were also analysed. Some of the results are presented in Figure 14. They show that recultivation led to increased enzyme activities in iron mine spoils during their transformation into technogenic soils. Potential dehydrogenase activity increased to a lesser extent than phosphatase activity, which reached values similar to that of the native soil. Both activities were higher in the sainfoin plots than in those recultivated with orchard grass. In each case, the 0-20-cm layer was more active than the 20-40-cm one.

---

* See also the paper of Blaga *et al.* (1988).

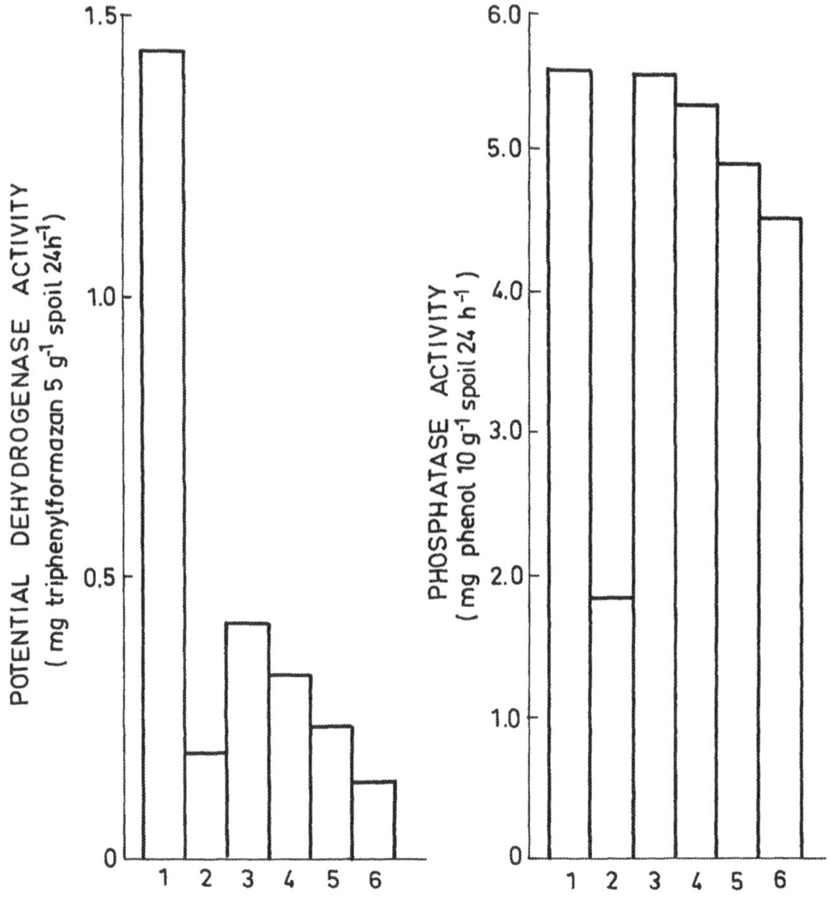

Figure 14. Potential dehydrogenase and phosphatase activities in recultivated iron strip mine spoils.

Redrawn from: M. Drăgan-Bularda, S. Kiss, D. Paşca and V. Olar-Gherghel, Publicaţiile Societăţii Naţionale Române pentru Ştiinţa Solului (Bucharest), 1983, No. 21B, pp. 111-112.

1 – Adjacent native soil. 2 – Spoil not submitted to recultivation. 3 – Spoil recultivated with sainfoin (depth: 0-20 cm). 4 – Spoil recultivated with sainfoin (depth: 20-40 cm). 5 – Spoil recultivated with orchard grass (depth: 0-20 cm). 6. Spoil recultivated with orchard grass (depth: 20-40 cm).

In the Şatra zone of the Căpuş iron strip mine, Blaga *et al.* (1984) recorded very low values of dehydrogenase, catalase, and invertase activities in the 0-20- and 50-80-cm layers of three profiles of spoils levelled for their recultivation.

In the autumn of 1985, the ninth year of a fertilisation and crop rotation experiment on the southern zone of the Căpuş iron strip mine, samples were taken for enzymological analyses from the 5-20-cm depth of spoil plots representing five variants:

V_1: unfertilised (control);

V_2: fertilised with farmyard manure (40 t ha^{-1});

V_3: complexly fertilised (40 t of farmyard manure plus 100 kg of N plus 60 kg of P plus 40 kg of K ha^{-1});

V_4: lightly NPK-fertilised (100 kg of N plus 60 kg of P plus 40 kg of K ha^{-1}); and

V_5: heavily NPK-fertilised (300 kg of N plus 180 kg of P plus 120 kg of K ha^{-1}).

The mineral fertilisers applied were NH_4NO_3, single superphosphate, and potash salt. Each variant included plots that in 1985 were planted in maize, oats, or sainfoin. The enzymatic and nonenzymatic catalytic potential (actual and potential dehydrogenase, invertase, phosphatase, urease, and nonenzymatic H_2O_2-splitting capacity), like the yield of crops, was highest in the complexly fertilised spoil plots and lowest in the unfertilised ones. Significant correlations were found between invertase activity and corn yield and between phosphatase activity and sainfoin yield. But under the influence of long-term fertilisation, the crop production capacity of the studied technogenic soil increased to a larger extent than its biological

Table 26
Enzymatic indicators and relative crop yields in plots of a technogenic soil in the ninth year of a long-term fertilisation and crop rotation experiment at the Căpuş iron strip mine

Fertilisation variant	Enzymatic indicator	Relative crop yield
V_1	302.596 a	71.60 a
V_2	375.684 ab	198.23 b
V_3	440.374 b	300.00 c
V_4	349.742 ab	244.92 bcd
V_5	364.946 ab	266.43 bce

Reproduced from: M. Drăgan-Bularda, G. Blaga, S. Kiss, D. Paşca, V. Gherasim and R. Vulcan, Studia Universitatis Babeş-Bolyai, Biologia, 1987, Vol. 32 (2), p. 51.

Numbers followed by the same letter in a column are not significantly different at $P = 0.05$.

potential reflected by its enzymatic activities (Table 26). This means that long-term fertilisation is able to greatly enhance the crop production capacity of the technogenic soil, but the increasing effect of fertilisation on the biological potential of the technogenic soil, as reflected by its enzymatic activities, is the result of much slower processes (Drăgan-Bularda *et al.*, 1987).

Blaga *et al.* (1991) described an experiment in which fermented and dried sewage sludge was used as a fertiliser. At the Căpuş iron strip mine, plots were installed in 1986 on a spoil area which in 1983 had been levelled for agricultural recultivation. The sludge was applied at rates of 20-140 t ha^{-1}. Untreated plots and plots fertilised with NPK (150 kg of N ha^{-1} as NH_4NO_3 plus 80 kg of P ha^{-1} as single superphosphate plus 60 kg of K ha^{-1} as potash salt) served for comparison. The sludge, superphosphate, and potash salt were administered in the autumn of 1986, whereas NH_4NO_3 was applied in the springs of 1987, 1988, and 1989. The test plant was maize. The corn yield was recorded in each year. In 1989, at the end of the growing season the plots were sampled for determination of dehydrogenase, catalase, and invertase activities and for other analyses.

In each year, the corn yield was smallest in the untreated (control) plots, whereas the highest yields were recorded in the plots treated with 80 or 100 t of sludge ha^{-1} (in 1987), in those treated with 120 t of sludge ha^{-1} or fertilised with NPK (in 1988), and in those fertilised with NPK (in 1989). In 1989, there was a parallelism between the rate of sludge application and corn yield.

The enzymatic activities were lowest in the untreated plots and increased, like the corn yield, with the rate of sludge application. They also increased in the NPK-fertilised plots but to a lesser extent than in the sludge-treated ones.

Taking into consideration these results as well as the findings showing that sludge application led to humus, N, and P accumulation in the technogenic soil and the amount of heavy metals in this soil remained under the permissible threshold values, the conclusion was drawn that for fertilisation of the technogenic soil in Căpuş the sewage sludge can be applied at rates of up to 120 t ha^{-1}, once at 3-year intervals.

Enzymological research in the Russian Federation

Catalase and invertase activities in the revegetated spoil heaps around the Lebedin strip mine (located in the Starooskol iron ore zone within the Kursk Magnetic Anomaly region) were determined by Sviridova and Panozishvili (1979). The spoil heaps here were of three types: sandy, loamy, and cretaceous-marly. They were revegetated with spontaneous and introduced grasses or with forest plants (sea buckthorn, black locust, and oleaster). After 8-10 years the heaps revegetated with grasses were covered with a 1.5-2-cm sod, protecting them against erosion. In the 8-10-year-old forest stands the mass of litter reached densities of 0.5-0.9 t ha^{-1}. Catalase activity was measurable in the soil of all revegetated heaps. This activity, like respiration (CO_2 evolution) and humus accumulation, was more intense under herbaceous vegetation than under forest. Invertase activity was influenced not only by the nature of vegetation but also by the nature of spoil heaps. This activity was highest in the rhizosphere of sea buckthorn growing on the cretaceous-marly heap.

Zasorina (1985a, b) has studied, enzymologically, the spoil heaps near the Stoilensk iron strip mine (located within the Kursk Magnetic Anomaly region). These spoil heaps are of two types: sandy and cretaceous. The age of natural vegetation growing on heaps varied between 3 and 20 years. Invertase, urease, and catalase activities in the 0-5-cm layer of spoil heaps increased in parallel with the age of vegetation. The increase was more pronounced in the sandy spoils than in the cretaceous ones. During the growing season, the maximum activity values were registered in midsummer. In both young and old spoil heaps, after their revegetation with a mixture of six grasses and legumes (bromegrass, fescue, wheatgrass, alfalfa, clover, and sainfoin), the enzymatic activities increased 1.5-3 times.

The investigations described by Shcherbakov *et al.* (1991) were related to the revegetation of the spoil heap in "Berezovyi log" (Broad gorge with birch) at the Lebedin mine. The spoil plots were covered with topsoil (chernozem) redeposited as dry material (plots I) or as wet material (pumping of soil and water mixtures) (plots II). Alfalfa was used for revegetation. In the third year of revegetation, samples were taken from the 0-10-, 10-20-, 20-30-, and 30-40-cm depths of plots I and II and of the zonal soil (typical chernozem) that was not affected by the mining operations. The soil samples were analysed chemically (humus, total N, etc.), enzymologically (urease activity), and microbiologically (respiration, measured as CO_2 evolution, and nitrification capacity). The samples collected from the centre and periphery of plots II were examined separately. The alfalfa yield in plots was also recorded.

All the parameters mentioned above tended to decrease with sampling depth. Urease activity, respiration, and nitrification capacity gave much higher mean values in plots I than in plots II. At the same time, the mean values in plots I were similar to those obtained in the zonal soil. Consequently, redeposition of topsoil as dry material was, from a biological viewpoint, a better method than its redeposition as wet material. Within plots II no marked differences were observed between the central and peripheric places. It was also established that urease activity and nitrification capacity of plots I and II correlated significantly with the total N content; urease activity correlated with the alfalfa yield, too.

Germanova *et al.* (1995) determined urease and catalase activities in the 0-5-cm layer of raw and 5-year-old spoil heaps at the Kostomuksha strip iron mine located in the northern subzone of the taiga zone (northern Karelia). The adjacent, undisturbed podzolic soil under pine forest served for comparison.

Due to the unfavourable climatic conditions (number of days without frost: 80-90 per year; medium monthly temperature in July: 14°C; annual amount of precipitations: 600 mm), development of natural vegetation was very poor; microbial numbers and diversity were small and enzyme activities were similarly low in both raw and 5-year-old spoils as compared to values registered in the undisturbed soil. As the first step in recultivation, levelling of spoils is recommended to improve their water regime.

Chapter 12. TECHNOGENIC SOILS FROM MANGANESE MINE SPOILS

Enzymological research in Ukraine

According to Keleberda (1973), the 0-20-cm layer of the spoil heap (consisting of medium loams) at the Aleksandrov manganese quarry (Dnepropetrovsk region) contains more humus and shows higher H_2O_2-splitting (enzymatic, i.e. catalasic plus nonenzymatic) capacity than the 20-40-cm layer. In another study, carried out in the same area, Keleberda (1978) has found that recultivation of spoil plots with oleaster for 11 years resulted in the formation of a technogenic soil with increased humus and N contents and invertase, urease, and proteinase activities in the 0-5-, 5-10-, and 10-20-cm layers as compared with the uncultivated control plot.

On the same spoil heap, plots planted with Scotch pine (*Pinus sylvestris*) or with Scotch pine in mixture with black alder (*Alnus glutinosa*) were also installed. The observations and analyses carried out 8 years after planting showed that the pines grew higher and their needles contained more N, P, K, and chlorophyll in the mixed stand than in the pure one. The alder also improved the chemical properties of technogenic soil. Thus, humus and N contents increased in the following order: soil under pine < soil at the middle of the 2.5-m distance between the pine and alder plants < soil under alder. Invertase, urease, and proteinase activities were highest in the soil under alder (Dan'ko *et al.*, 1980).

At the same manganese quarry, Uzbek (1986) determined several enzyme activities in different layers of a 20-year-old spontaneously revegetated spoil plot, now covered with a stable phytocoenosis made up of meadow grass (*Poa angustifolia*) and sagebrush (*Artemisia austriaca*) and found that the activities were much higher (catalase 1.5, phosphatase 13, urease 36, invertase 46, and dehydrogenase 72 times) in the 1-cm-deep surface layer rich in roots than at 6-cm depth and in deeper layers of the spoil plot. Microbial numbers as well as humus, total N, mobile P, and exchangeable K contents were also highest in the surface layer.

Continuing the enzymological investigations in the Nikopol manganiferous basin, Uzbek *et al.* (1988) determined the invertase, urease, and phosphatase activities in raw overburdens (loess-like loam, red-brown clay, and grey-green clay) from the walls of quarry as well as in the 0-20-cm layer of experimental plots that had been formed of topsoil or of raw overburdens and submitted to spontaneous revegetation 17 years previously. A plot on the uncultivated adjacent native soil (chernozem) served for comparison.

The enzyme activities in the raw overburdens from the quarry walls were very low or lacking: each activity was detectable in the loess-like loam, but in the two clays only urease activity was measurable.

The activities were significantly higher in the revegetated overburden plots and gave the highest values in the topsoil plot. However, none of the overburdens reached the activity values recorded in the native soil. Thus, activity of invertase, urease, and phosphatase, respectively, represented, depending on the nature of overburdens, 39.4, 34.5, and 28.4% (topsoil), 23.2, 14.8, and 9.3% (loess-like loam), 34.0, 22.6, and 7.6% (red-brown clay), and 37.7, 27.3, and 5.4% (grey-green clay) of the activities measured in the native soil. These data prove that, in comparison with the activity levels in the native soil, accumulation of enzymes in the overburdens, irrespective of their nature, occurred in the order: invertase > urease > phosphatase.

Uzbek (1989, 1991) found that in spontaneously revegetated 18-year-old overburden plots there was an increase not only in the activity of hydrolases (invertase, urease, and phosphatase) but also in the activity of oxidoreductases (catalase and dehydrogenase). Catalase activity increased more markedly than dehydrogenase activity. Other 18-year-old overburden plots located in the Nikopol basin were also studied enzymologically. These plots were not covered with spontaneous vegetation since they were recultivated with alfalfa and sainfoin. Owing to the influence of alfalfa and sainfoin, the processes leading to accumulation of hydrolases in the overburdens became more intense in all plots. The activity of oxidoreductases also increased but, in contrast to spontaneous revegetation, recultivation with alfalfa and sainfoin led to a higher increase in dehydrogenase than in catalase activity.

Enzymological research in Georgia

The enzymological properties of the recultivated strip mine spoils in the Chiatura manganese ore zone (situated in the Kvirila Basin, western Georgia) were described by Daraseliya and Kalatozova (1973, 1976) and Daraseliya (1979). The experimental variants and the results obtained are specified in Table 27. One can see from this table that the native soil had much higher enzyme activities than the uncultivated spoils (calcareous sands and clays; pH in H_2O was 8.4). The activities increased in the recultivated spoils but did not reach the values registered in the native soil. The perennial grass-legume mixture was more efficient

Table 27
Effect of the measures applied for recultivation of strip mine spoils in the Chiatura manganese ore zone on the enzyme activities (mean values of analytical data over 3 years: 1969-1971)

Experimental variants	Depth (cm)	Humus (%)	pH in H_2O	Invertase*	Phosphatase*
Adjacent native brown forest soil	0-20	4.20	5.5	6.5	6.8
	20-40	1.65	5.2	4.3	3.0
Mine spoils without plants	0-20	0.05	8.4	0.8	1.9
	20-40	0.05	8.4	0.6	1.2
Perennial grass with legume (ryegrass-alfalfa) on mine spoils	0-20	2.23	7.0	2.4	4.1
	20-40	0.10	7.9	1.4	1.6
Grapevines on mine spoils	0-20	0.36	8.1	0.7	1.1
	20-40	0.21	8.5	0.8	0.5
Grapevines on stored topsoil (40-45-cm layer) reapplied on mine spoils	0-20	0.59	7.9	1.5	4.4
	20-40	0.36	8.1	1.1	2.8

Reproduced from: N.A. Daraseliya, Biologicheskaya Aktivnost' Osnovnykh Pochv Zapadnoi Gruzii, Tbilisi, 1979, p. 258.

* Expression of enzyme activities: invertase in mg of "glucose" g^{-1} soil 24 h^{-1} and phosphatase in mg of P_2O_5 10 g^{-1} soil h^{-1}

than grapevines. Reapplying the stored topsoil on the surface of spoils had beneficial effects on the accumulation of enzymes (invertase, phosphatase) in the grapevine plots. The 0-20-cm layer was more active and richer in humus and microorganisms than the 20-40-cm one. No relationship was found between catalase activity in spoils and the recultivation measures applied.

Chapter 13. TECHNOGENIC SOILS FROM LEAD AND ZINC MINE WASTES

Enzymological research in the United Kingdom

Studying the decomposition of vegetation growing on metal mine waste, Williams *et al.* (1977) also carried out enzymological analyses. The waste studied was located around the disused mine at Y Fan (Powys, Wales) and contained high concentrations of lead and zinc. After the abandonment of the mine (1928), the waste was partially colonised naturally by metal-tolerant *Agrostis tenuis*. An evenly colonised area was selected for study. A similar but uncontaminated area was also selected on a pasture situated about 500 m from the mine. The vegetation on this site consisted primarily of *A. tenuis* and *Festuca ovina*. Urease activity in soil, microbial populations in litter and soil, and microfauna in litter from both sites were compared. Accumulation of litter was greater on the waste, which also contained significantly less humic and fulvic acids in the soil immediately beneath the litter layer. Urease activity was also significantly lower in the mine soil than in the nearby pasture soil (Table 28).

Table 28
Urease activity in a mine soil and a pasture soil at Y Fan, Powys, Wales

Reaction mixture	mg of NH_4^+-N released from 100 g of soil (on air-dry basis) at 37°C in 3 h	
	Mine soil*	Pasture soil*
Soil + urea solution	2.55±0.14	42.62±2.38
Soil + water	1.97±0.25	3.32±0.58
Urease activity	0.58±0.19	39.31±2.17

Reproduced from: S.T. Williams, T. McNeilly and E.M.H. Wellington, Soil Biology & Biochemistry, 1977, Vol. 9, p. 273, with kind permission from Elsevier Science.

* Means of two soils significantly different at $P = 0.05$.

Microbial numbers from litter at the two sites were not markedly different, although numbers of fungi were lower on litter from the mine waste, while those of bacteria and actinomycetes were higher. In contrast, numbers of all groups in the mine soil were considerably lower than those in the pasture soil. Similarly, there were fewer animals in the litter on the waste. The low biological activity in the litter and soil of the studied mine waste, caused by the high Pb and Zn concentrations, explains the retarded decomposition of vegetation growing on this site.

Clark and Clark (1981) have applied soil enzymological methods, in order to determine the reasons for the differences in the floras of adjacent species-poor and species-rich areas of a limestone terrace in the lead-mining complex on Grassington Moor, in Yorkshire Pennines (England).

The northern half of the terrace received drainage water and fine-textured, Pb- and Zn-containing mine waste from abandoned mine workings up slope, and the vegetation there was sparse, floristically impoverished, and composed of species typical of heavy-metal mine areas in the British Isles, i.e. *Minuartia verna, Agrostis tenuis*, and *Festuca ovina*. There was no direct input of mine waste on the southern half of the terrace, and there the vegetation was floristically rich and continuous, except where the limestone cropped out. The mean number of species per 0.25 m^2 on the species-poor area was 2.4, in contrast to the species-rich area, where it was 10.1.

The soils of the species-poor area had lower pH values and contained less humus, NO_3^--N, NH_4^+-N, available P, and exchangeable K, compared with those of the species-rich area. The total lead content averaged in the soils of the species-poor and species-rich areas 78,000 and 8,000 $\mu g\ g^{-1}$ soil, respectively, far above the 350 $\mu g\ g^{-1}$ threshold value above which lead levels are anomalously high. The soils of both these areas would therefore normally be expected to be toxic to all but tolerant races. The average level of ammonium acetate-extractable Pb was 21,800 $\mu g\ g^{-1}$ soil on the species-poor area and only 311 $\mu g\ g^{-1}$ soil on the other area. Zinc levels were mainly lower than those of lead on both areas and the difference in the levels of total and ammonium acetate-extractable Zn between the areas was less marked than for Pb.

Acid phosphatase, dehydrogenase, and urease activities and respiration (CO_2 evolution) were measured in soil samples taken in the root zone, 2-9 cm below the surface. When expressed on an air-dry soil basis, they were higher in the species-rich soil. However, when expressed on an air-dry organic matter basis, the differences were reduced or eliminated (Table 29). In each half of the terrace there were significant correlations between density of species, amounts of plant nutrients, and enzyme activities, and all were related inversely to the

Table 29
Enzyme activities and respiration in soils from a limestone terrace contaminated by Pb- and Zn-containing mine waste on Grassington Moor, Yorkshire

Activity or respiration	Area					
	Species-poor (SP)		Species-rich (SR)		Ratio (SR/SP)	
	S*	OM**	S	OM	S	OM
Acid phosphatase (μg of *p*-nitrophenol)	9.9 (2.7)***	67.4 -	18.1 (2.2)	79.5 -	1.8	1.2
Dehydrogenase (μg of triphenylformazan)	38.3 (21.5)	260.4 -	2503.0 -	11026.4 -	65.4	42.3
Urease (mg of urea)	1.5 (0.4)	10.0 -	6.9 (2.7)	30.4 -	4.6	3.0
Respiration at current moisture content***** (mg of C)	0.085****	0.578	0.093	0.409	1.1	0.7
Respiration at field capacity (mg of C)	0.067	0.456	0.090	0.396	1.3	0.9

Reproduced from: R.K. Clark and S.C. Clark, The New Phytologist, 1981, Vol. 87, p. 808.

* Activity or respiration registered in 1 g of air-dry soil in 24 h.
** Activity or respiration reported for 1 g of air-dry organic matter in 24 h.
*** For activity values, standard deviations are given in parentheses.
**** For respiration values, the least significant difference is 0.016 at $P < 0.05$.
***** Considerably less than field capacity.

levels of extractable lead. The conclusion has been drawn that nutrient enrichment is involved in the formation of the species-rich area on Grassington Moor; the higher enzyme activities in the species-rich area indicated that metal detoxification was taking place there, and the higher organic matter content of this area is related to the enzyme activities.

Enzymological research in Romania

The raw and the revegetated wastes at the Săsar mine (Baia Mare, Maramureş county), the ores of which contain Pb and Zn as well as Cu, Cd, and some other heavy metals, were studied enzymologically by Soreanu (1983). An adjacent native meadow soil served for comparison. The revegetation experiment started in 1975 and comprised unfertilised and NPK-fertilised mine waste plots seeded with a mixture of perennial grasses and legumes or with individual grass and legume species or sunflower. In 1980, samples were taken from the 0-20-cm depth of each plot and native soil for determining invertase, dehydrogenase, phosphatase, and urease activities in wastes and soil, respectively. It was found that revegetation caused an increase in each activity as compared with those measured in the raw waste, but except for urease activity, the other activities did not reach the values recorded in the native soil. The fertilised plots revegetated with the grass-legume mixture gave the best results in respect of plant cover percentage, herbage yield, and enzyme activities of wastes.

Biological recultivation experiments involving enzymological analyses were also carried out at the Rodna mine (Bistriţa-Năsăud county) (Kiss *et al.*, 1989b, 1990). The spoils submitted to recultivation had resulted from underground mining of lead and zinc ores, their concentration by flotation, and decantation in a pond.

In June 1987, small recultivation plots ($7\ m^2 = 2 \times 3.5$ m) were installed on three terraces (VIII, V, and III) from the nine terraces of the spoil dump. Terrace VIII was 2 years old, whereas terraces V and III were 7 and 10 years old, respectively. The level difference between terraces was approximately 3.5 m. On each terrace there were both plots with south-western aspect and plots with south-eastern aspect. In all, 14 spoil plots were installed in the following variants:

1. covering with soil plus fertilising with farmyard manure (FYM) plus NPK plus seeding with Italian ryegrass and meadow clover (RC);
2. FYM plus NPK plus RC;
3. NPK plus RC; and
4. NPK.

The soil used was a loamy sand of low fertility. The thickness of soil cover on spoil plots was 10 cm. The farmyard manure was applied at a rate of 40 t ha^{-1}. The mineral fertilisers used were NH_4NO_3, single superphosphate, and potash salt (KCl), applied at rates of 150 kg of N, 200 kg of P_2O_5, and 250 kg of K_2O ha^{-1}. Italian ryegrass (*Lolium multiflorum*) and meadow clover (*Trifolium pratense*) were sown at rates of 25 and 60 kg of seeds ha^{-1}, respectively. On each terrace, untreated places in the vicinity of the experimental plots served as controls (in all, six places). All treatments were performed on the same day (June 24, 1987). Then, each plot was moistened with water from the Someşul Mare River which flows at a 25-m distance from the foot of the spoil dump. Systematic watering of the plots was assured from the same source. The plots were supplementarily fertilised with NH_4NO_3 (150 kg of N ha^{-1}) in October 1987.

Two large plots ($50\ m^2 = 20 \times 2.5$ m) were installed on terrace VI. Plot I was covered with 10 cm of soil, fertilised with NPK, and sown with Italian ryegrass and meadow clover. Plot II was also covered with 10 cm of soil and fertilised with NPK, but it was not sown with ryegrass and clover, instead, it was treated with about 50 kg of spoil taken from the surface layer of terrace I (the lowest and oldest, 15-year-old, terrace). This spoil contained seeds of plants and microorganisms from the spontaneous flora and microflora, respectively. For covering plots I and II the same soil was used as in the case of the small plots. The NPK fertilisers and their rates and the seed rates were also the same. All treatments and the first watering were carried out on July 5, 1988. The large plots, like the small ones, were moistened systematically with water from the Someşul Mare River. A supplementary NH_4NO_3 dosage (150 kg of N ha^{-1}) was administered in October, 1988. Controls of the large plots were also untreated places in their vicinity.

For enzymological analyses, six series of samples were collected from the 0-10-cm depth of the small plots and their controls in the June 1987-October 1989 period, and four series of samples from the same depth of the large plots in the July 1988-October 1989 period. Phosphatase, catalase, and actual and potential dehydrogenase activities were determined. The so-called nonenzymatic catalytic activity (i.e. nonenzymatic H_2O_2-splitting capacity) was also assayed. Based on the activity values, the enzymatic indicator was calculated for each plot and control. This indicator is considered as an index of the biological quality of soil (and spoil).

The results have shown that the enzymatic activities (but not the nonenzymatic catalytic activity), the enzymatic indicators calculated for the whole period of experiments (1987-1989), and the herbage yields in 1989 gave the highest values in the spoil plots covered with soil, treated with organic and mineral fertilisers and sown with ryegrass and clover or with plants from the spontaneous flora. Complex treatment of the small plots was more efficient in creating an enzymatic and herbage production potential in spoils than was mineral fertilisation alone or mineral fertilisation plus seeding of ryegrass and clover. Covering of spoils with soil proved to be a more important recultivation measure even in comparison with organic fertilisation.

The enzymatic indicators of control places showed a tendency to increase with the age of terraces. Nevertheless, the control places even on the 10-year-old terrace III were much less enzyme-active than the soil-covered small and large plots aged only ~2.5 and ~1.5 years, respectively.

It has also been found that the south-western or south-eastern aspect of the plots subjected to the same treatment had no significant effect on the enzymatic indicators and herbage yields.

In conclusion, for the recultivation of raw and young spoils at lead and zinc mines, covering with soil, fertilisation with NPK, and sowing a grass-legume mixture were recommended, and for the recultivation of old spoils NPK fertilisation was recommended as the minimum treatment.

This conclusion was confirmed by enzymological analyses of spoils and native soil carried out in the next years. Thus, in 1992 (May 25 and August 28) samples were taken from three depths: 0-10, 10-20, and 20-30 cm. The enzymatic indicator showed again the highest values in the soil-covered small and large spoil plots and in the native soil and, as expected, the lowest values were recorded in the youngest (untreated control and only NPK-fertilised) spoils (Kiss *et al.*, 1992, 1994).

Evolution of the enzymatic potential in the 0-10-cm layer of spoil plots at the Rodna mine during the 1987 (1988)-1994 period was also evaluated (Paşca *et al.*, 1994; Cristea *et al.*, 1995). Figure 15 was selected to illustrate the evolution of enzymatic potential in soil-covered spoil plots (plots 1, I, and II) as compared to plots 2 and 3 (not covered with soil), untreated spoils (controls 1-3, I, and II), and native soil. It is evident from this figure that the favourable effect of a soil cover on the enzymatic potential of spoils is long-lasting.

Results of the analyses carried out during 1996 (i.e. 9 and 8 years after the installation of small and large plots, respectively, at the Rodna mine) have confirmed the long-lasting effect of soil cover on the enzymatic potential of spoils (0-30-cm layer), as the enzymatic indicators of soil (spoil) quality could be ranked in the order: plot I > plot II > native soil > plot 4 > plot 1 >> untreated spoils (control 1-3). The humus content showed the same order. But due to the fertilisation of plots, their total N, available P, and exchangeable K contents were higher and much higher than those of the native soil and untreated spoils, respectively. pH and carbonate content decreased in the soil-covered plots (Paşca *et al.* 1997a, b, 1998).

The soil cover also resulted in increased nematode abundance, diversity, and maturity index (Popovici, 1993, 1995; Popovici *et al.*, 1995; Paşca *et al.*, 1997b, 1998). Development of the mites and collembolans was also favoured in the soil-covered plots (Paşca *et al.*, 1997b).

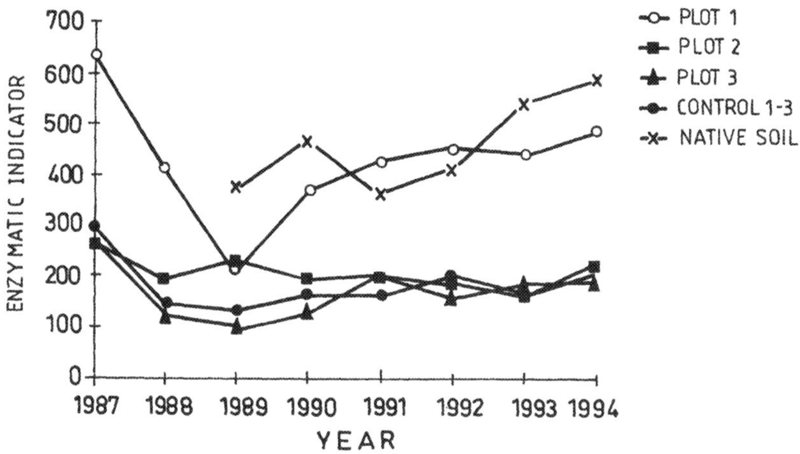

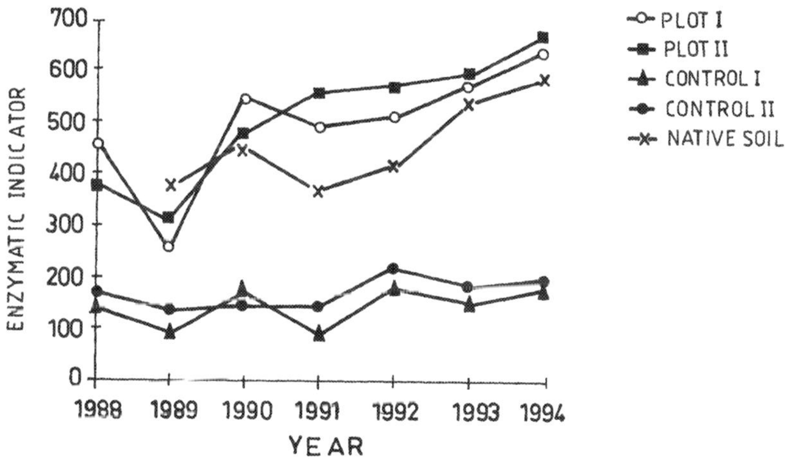

Figure 15. Evolution of the enzymatic potential in some spoil plots installed at the lead and zinc mine in Rodna.

Redrawn from: D. Paşca, S. Kiss, M. Drăgan-Bularda, R. Crişan and V. Muntean, Studia Universitatis Babeş-Bolyai, Biologia, 1994, Vol. 39 (1), pp. 97 and 101; V. Cristea, S. Kiss, D. Paşca, M. Drăgan-Bularda, R. Crişan and V. Muntean, Colloques Phytosociologiques, 1995, Vol. 24, pp. 174 and 176.

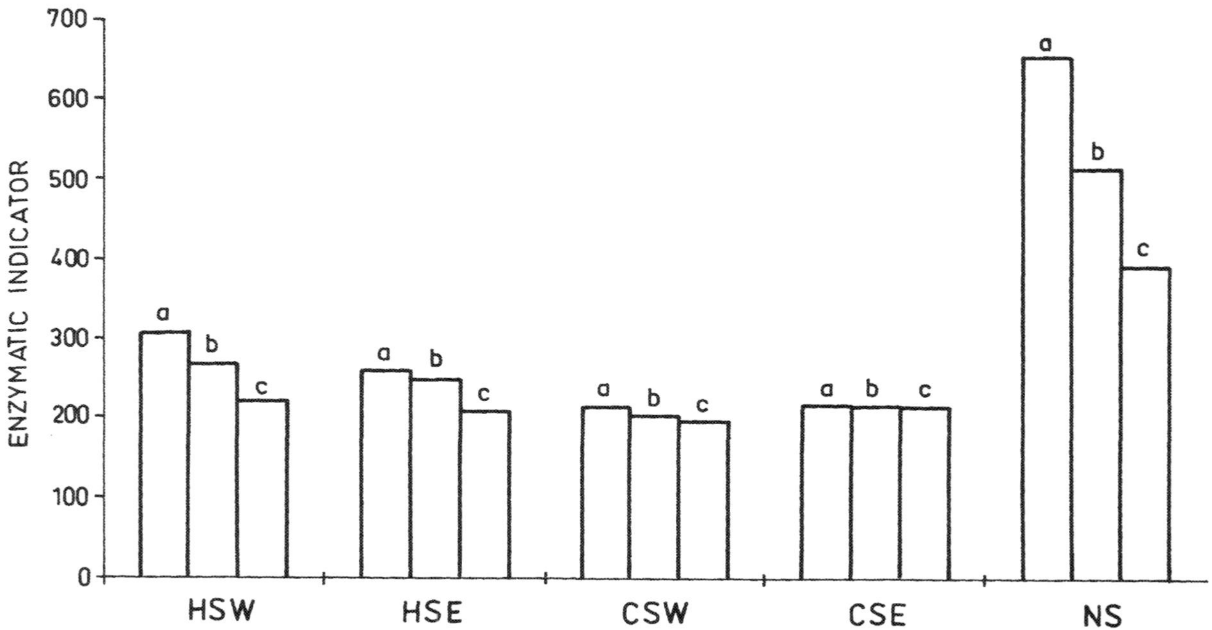

Figure 16. Enzymatic indicators of soil (spoil) quality calculated for the May 1994-September 1995 period.

Redrawn from: D. Paşca, V. Cristea, R. Crişan, V. Muntean, S. Kiss and M. Drăgan-Bularda. Studia Universitatis Babeş-Bolyai, Biologia, 1996, Vol. 41 (1-2), p. 192.

Sampling depths: 0-10 cm (*a*), 10-20 cm (*b*), and 20-30 cm (*c*). H – Spoil from the root zone of *Hippophaë rhamnoides*. C – Spoil from the control places. SW – South-western aspect. SE – South-eastern aspect. NS – Native soil.

The spoils and native soil at the Rodna mine were subjected to microbiological analyses, too. Some of the results obtained were briefly described by Paşca *et al.* (1998). Global numbers of the aerobic heterotrophic, ammonifying, sulphate-reducing, and iron-reducing bacteria determined in the 0-10-cm layer of spoils and native soil in summers of 4 years (1993-1996) were of the same level in the soil-covered plots I and 1 and in the native soil and always exceeded the numbers recorded in the untreated spoils.

At the Rodna mine, an experiment using ligneous plants for revegetation of spoils was also carried out (Cristea *et al.*, 1995; Paşca *et al.*, 1996). Young trees and bushes were planted on the 4-year-old terrace VIII (on its both parts having south-western and south-eastern aspect, respectively), in October 1989. *Hippophaë rhamnoides, Colutea arborescens, Salix gracilis, Robinia pseudacacia,* and *Pinus nigra* succeeded in surviving. Growth of *Hippophaë rhamnoides* (sea buckthorn) was especially luxuriant.

In the May 1994-September 1995 period, six series of spoil samples were taken from the root zone of *H. rhamnoides* and from the control (unplanted) places. The native soil was also sampled. Sampling depths were 0-10, 10-20, and 20-30 cm. The same enzymatic and nonenzymatic catalytic activities were determined in all samples as those specified above for the small and large experimental plots, and the enzymatic indicators of soil (spoil) quality were calculated.

Figure 16 shows that the enzymatic indicators of the 0-10- and 10-20-cm layers present the order: control spoil < spoil from the root zone of *H. rhamnoides* < native soil. In the 20-30-cm layer the enzymatic potential is similar in both spoils, and lower than in the native soil. The aspect-dependent variation of this potential is very low.

The finding that *H. rhamnoides* increased the enzymatic potential of spoils supports the opinion, according to which fixation of spoils by ligneous species supplements the coverage by herbaceous plants and contributes to a better and faster ecological and economic reintegration of spoil dumps.

In 1996, the spoil from the root zone of *H. rhamnoides* was again found to be more enzyme-active than that from the control places, but it did not reach the activity level of the soil-covered plots or native soil (Paşca *et al.*, 1997a).

Chapter 14. TECHNOGENIC SOILS FROM GOLD MINE WASTES

Enzymological research in New Zealand

Enzymological and other biochemical as well as chemical and physical properties of such soils were dealt with by Ross *et al.* (1994). For rehabilitation of gold mine wastes, a field experiment was carried out at the Golden Cross mine located in Waitekauri Valley, at the southern end of the Coromandel Range, North Island. Plots (6 × 6 m) were installed on a waste rock stack that had been levelled and consolidated. Weathered andesite overburden was subsequently spread on this rock to provide a 60-cm depth of "subsoil", that was then covered with silt loam topsoil to 10-, 15-, or 20-cm depth or left non-covered (i.e. no topsoil was added). Each of these four plots was installed in three replicates. Control plots on undisturbed silt loam were selected in a level pasture within 150 m of the experimental plots.

Before sowing of new pasture species, the existing herbage on the control plots was killed by spraying with glyphosate, and rotary-hoed 3 weeks later to a depth of 15 cm. All plots, including the control ones, received limestone (6 t ha^{-1}) and KCl (250 kg ha^{-1}). Each of the experimental plots was divided into three subplots (6 × 2 m), to which superphosphate was added to raise available phosphate in 0-7.5-cm depth to Olsen P ($NaHCO_3$-extractable P) levels of 7, 13, and 20 ppm, respectively. The plots were then rotary-hoed to 7.5-cm depth and sown by hand in April 1989 with the following seed mixture (kg ha^{-1}), the legume seeds being inoculated with appropriate rhizobia: *Lolium perenne* (20), *Lolium x hybridum hausslen* (8), *Dactylis glomerata* (5), *Trifolium repens* (4), and *Lotus pedunculatus* (3). Potassium (312 kg ha^{-1}, as KCl and K_2SO_4) was added periodically to all plots during the experiment; urea-N (91 kg ha^{-1}, in three split dressings) was included in the first year, to stimulate grass growth. Additional P (187 kg ha^{-1}, as superphosphate) was also added to the experimental plots only.

For enzymological and other studies, Ross *et al.* (1994) used the plots with no topsoil addition, those covered with 15-cm topsoil, and the control plots. Samples were taken from the 0-7.5- and 7.5-15-cm depths of plots, in October 1989, March 1990, January 1991, and

March 1992. Sampling was restricted to subplots with the highest phosphate level (20 ppm Olsen P in the 0-7.5-cm layer) to ensure minimal P limitation to plant growth.

The samples were analysed for determining invertase activity, a series of other biochemical properties: basal and substrate-induced respiration (CO_2 evolution from un-amended and glucose-amended samples), microbial C and N contents, aerobic and anaerobic N mineralisation, nitrification as well as some chemical and physical properties (total C and N contents, pH, moisture content). Dry matter yields and botanical composition of herbage were also recorded.

Invertase activity in both 0-7.5- and 7.5-15-cm layers in each year was higher in the topsoil than in the no-topsoil treatment, but never reached the values registered in the control plots. The activity increased significantly during the experiment only at the 0-7.5-cm depth, the increase being higher in the topsoiled plots than in those not covered with topsoil. The other biochemical properties were also higher in the 0-7.5- than in the 7.5-15-cm layer. Herbage dry matter annual yields in the last two years presented, in the experimental and control plots, the following order: no topsoil < 15-cm topsoil < control.

The enzymological and other biochemical data were consistent with the conclusion drawn from physical, chemical, and agronomic data that for rehabilitation of the studied mining site the wastes should be covered with an at least 15-cm-thick topsoil layer.

Chapter 15. TECHNOGENIC SOILS FROM APATITE AND PHOSPHORITE MINE WASTES

Enzymological research in the Russian Federation

Pereverzev and Podlesnaya (1986) described complex investigations aimed at the biological recultivation of wastes generated in the apatite enrichment process at the "Apatit" Enterprise located in the vicinity of the city of Kirovsk, formerly Ibinogorsk (Murmansk region, Kola Peninsula). The wastes were sands rich in nepheline; their initial pH was 7.0-7.2 in H_2O and ~7.0 in KCl solution; they contained plant nutrients (P, K, Ca, etc.) extractable with diluted acids.

Table 30
Agrochemical characteristics and urease activity of technogenic soils formed on apatite mine wastes

Depth (cm)	Organic C (%)	pH		Phosphate in saline extract*	Water-soluble K**	Urease activity***
		in H_2O	in KCl solution			
				Fescue plot		
0-1	15.50	6.8	5.8	29.6	100	27.50
1-2	0.49	7.2	6.0	7.2	10	0.13
2-10	0.33	7.8	6.9	4.6	10	0.06
10-16	0.28	8.4	8.0	6.9	14	0.04
16-23	0.33	8.4	8.0	5.8	16	0.04
23-30	0.22	8.2	7.8	9.9	14	0.04
30-40	0.23	8.3	7.7	9.7	15	0.02
				Bromegrass plot		
0-1	1.80	6.9	6.0	12.0	18	5.23
1-6	3.75	6.3	5.6	4.9	14	3.32
6-12	0.22	7.2	6.3	8.6	7	0.04
12-20	0.26	8.0	7.3	13.4	13	0.04

Reproduced from: V.N. Pereverzev and N.I. Podlesnaya, Biologicheskaya Rekul'tivatsiya Promyshlennykh Otvalov na Krainem Severe, Apatity, 1986, p. 27.

* Given as mg of P_2O_5 kg^{-1} soil.
** Given as mg of K_2O 100 g^{-1} soil.
*** Expressed in mg of NH_3-N g^{-1} soil 24 h^{-1}.

In an experiment started in 1965, a plot was seeded to meadow fescue (*Festuca pratensis*) and another to awnless bromegrass (*Bromus inermis*). Before seeding, the second plot was covered with a peat layer. The plots were not treated with mineral fertilisers. The biomass produced annually by the two plants was low (about 100 kg of dry matter ha^{-1}).

In 1974, both plots were submitted to chemical, enzymological, and microbiological analyses. The data in Table 30 show that in the fescue plots only the 0-1-cm layer accumulated organic matter, mobile P and K in significant amounts and manifested a high urease activity. In the bromegrass plot which, as stated above, was covered with peat in 1965, both the 0-1- and the 1-6-cm layers contained more organic matter and were more urease-active than the deeper layers. The pH decreased in the upper layers and increased in the deeper ones. The microbial numbers paralleled the organic matter contents.

Other experiments proved that for a more efficient recultivation of the nephelinic sands, application of mineral fertilisers (NP or NPK) is necessary.

Baeva *et al.* (1992) have studied the spoils that resulted from the opencast phosphorite mining at the "Fosforit" Enterprise in Kingisepp, located in the western part of the Sankt-Petersburg region. For enzymological analyses, the southern, central, and northern zones around a quarry were selected. They were 2-3-, 5-6-, and 10-12-year-old spoil areas, respectively.

The raw spoils consisted of loose, mostly silty and sandy materials in mixture with lime and dolomite lumps; they were rich in mobile phosphorus and their reaction was neutral or slightly alkaline. After levelling, they were covered in bands with peat and planted to forest species (spruces, larches, pines). Later, other three species (birch, alder, aspen, willow) and a herbaceous cover (common coltsfoot, fireweed, clover, sweetclover, etc.) also developed.

On the 2-3-year-old spoil area, the vegetation was formed of the planted forest species and willows; the herbaceous cover was lacking. On the older spoil areas, more tree species developed. The herbaceous cover was 30-40% on the 5-6-year-old area and 100% on the 10-12-year-old one. On the latter area, the spruce and birch trees reached 2-3 m in height.

The enzymological analyses have shown that urease activity was lacking in the 2-3- and 5-6-year-old spoils, but traces of the activity were detectable in the 10-12-year-old ones. Invertase activity (expressed in mg of "glucose" produced by 1 g of spoil in 24 h) increased with the age of the three spoil areas, having values of 3.14, 3.62, and 5.20, respectively. However, these values were lower than those recorded in soddy-podzolic soils not affected by phosphorite mining and processing. Dehydrogenase activity, which also increased with the age of spoils, gave higher values in spoil samples collected in summer than in the spring samples.

Cellulose decomposition, nitrate accumulation, respiration (CO_2 evolution), numbers of bacteria, actinomycetes, and microfungi, and humus accumulation also manifested an increasing trend during aging of spoils.

Chapter 16. TECHNOGENIC SOILS FROM SULPHUR MINE SPOILS

Enzymological research in Ukraine

Peterson *et al.* (1976, 1979) determined the actual dehydrogenase and catalase activities, numbers of microorganisms, and respiration (CO_2 evolution) in sulphur strip mine spoils (the Podorozhnen mine, Rozdol), submitted to agricultural recultivation. Spring wheat, pea, a vetch-oats mixture, spring barley, sweetclover, and trefoil were used for recultivation in unfertilised and NPK-fertilised plots. Unfertilised spoil heaps (clays and sandy loams; pH in KCl solution was 4.6-5.6) under ruderal vegetation served as controls. The published analytical data were obtained with spoil samples collected in spring, summer, and autumn during the first 3 years of recultivation (1974-1976).

In the first year, dehydrogenase activity was lacking in the control heaps. In the recultivated plots, the activity was measurable in samples collected in summer. In the second and third years, the activity in control heaps appeared in summer and autumn, whereas it was present in the recultivated plots in spring also, the highest values being found in summer. In general, the fertilisers applied in spring caused a decrease of activity in spring and an increase in summer and autumn. Of the crop plants tested, sweetclover gave the best results in increasing the dehydrogenase activity of spoils. This activity was strongly related to the number of heterotrophic microorganisms growing on starch-ammonium-agar and of oligotrophic ones growing on soil extract-agar. There was no significant correlation between dehydrogenase activity and respiration rate of the spoils and between their catalase activity and plant species or fertilisation rate.

Chapter 17. TECHNOGENIC SOILS FROM LIME AND DOLOMITE MINE SPOILS

Enzymological research in Ukraine

Keleberda (1973) collected samples from the 0-20- and 20-40-cm depths of the spoil heaps at the Ol'gin lime and dolomite quarry (Novotroitsk, Donetsk region). The spoils were sandy loams mixed with lime and dolomite in form of rubbles. Both humus content and H_2O_2-splitting (enzymatic, i.e. catalasic plus nonenzymatic) capacity were higher in the upper than in the deeper layer.

Chapter 18. TECHNOGENIC SOILS FROM REFRACTORY CLAY MINE SPOILS

Enzymological research in Ukraine

Keleberda and Dan'ko (1975) studied the Dneprov spoil heap that resulted from the strip mining of the Chasov Yar refractory clays (Donetsk region). The spoils were loamy sands. This spoil heap was recultivated with sweetclover (*Melilotus volgicus*) as a green manure plant. Uncultivated plots served as controls. It was found that invertase, urease, and

Table 31
Humus and N contents and enzyme activities in primitive soils under forest vegetation developed on a spoil heap resulting from the strip mining of the Chasov Yar refractory clays

Plant stand and its age	Depth (cm)	Humus (%)	N (%)	N in humus (%)	Invertase*	Urease*	Proteinase*
Black locust (11 years)	0-5	2.23	0.23	10.3	26.40	2.02	0.76
	5-10	0.87	0.11	12.6	10.00	0.82	0.35
	10-20	0.60	0.04	6.6	7.15	0.49	0.26
	20-30	0.54	0.03	5.4	4.20	0.42	0.20
Oleaster (11 years)	0-5	2.61	0.41	15.7	39.30	3.92	1.09
	5-10	1.03	0.17	17.0	16.70	1.60	0.81
	10-20	0.46	0.05	10.8	8.31	0.77	0.22
	20-30	0.33	0.03	9.0	2.32	0.73	0.14
Control (uncultivated)	0-5	0.98	0.08	8.1	10.80	1.06	0.36
	5-10	0.67	0.06	8.9	5.30	0.57	0.15
	10-20	0.57	0.02	3.5	3.25	0.34	0.04
	20-30	0.49	0.02	4.0	0.20	0.16	0.00

Reproduced from: T.N. Keleberda, Pochvovedenie, 1978, No. 9, p. 111.

* Expression of enzyme activities: invertase in mg of inverted sugar g^{-1} soil 40 h^{-1}; urease in mg of NH_4^+-N g^{-1} soil 40 h^{-1}; and proteinase in mg of NH_2-N g^{-1} soil 72 h^{-1}.

catalase activities and respiration (CO_2 evolution), like humus and total N contents, increased significantly in the 0-5- and 10-20-cm layers of the recultivated spoil heap as compared with the control plots. Invertase activity of the 20-30-cm layer was also higher in the sweetclover plots than in the controls (Keleberda, 1976), and proteinase activity also increased in the top layer of the sweetclover plots (Keleberda, 1977).

Afforestation of some spoil plots in this area was performed with black locust and oleaster. It has been established (Keleberda, 1978) that after 11 years the spoil was transformed into a primitive soil, characterised by increased humus and N contents and invertase, urease, and proteinase activities in its 0-20-cm layer as compared with the uncultivated control plot (Table 31).

In the same area, Keleberda (1979), Dan'ko *et al.* (1980), and Keleberda and Drugov (1984) have also studied the influence of black locust on the development of other tree species: green ash (*Fraxinus viridis*), small-leaf linden (*Tilia cordata*), and elm (*Ulmus pinnato-ramosa*). When these species were planted in rows having contact with locust, they developed better, even in the first years of their plantation, than the plants having no contact with locust. Their better development, which became very evident 9 years after their planting, was accompanied by increased total N and chlorophyll contents in their leaves and by increased invertase, urease, and proteinase activities (Table 32); humus levels; amounts of total and hydrolysable N; and mobile P and K contents of their soils (especially in the 0-5-cm layer).

Keleberda and Drugov (1984) also described another experiment, in which black alder (*Alnus glutinosa*) was used as a symbiotic N_2-fixing plant instead of black locust, whereas weeping birch (*Betula verrucosa*) served as the test tree. The results obtained in the seventh year of the black alder and weeping birch plantations were similar to those registered in the experiment with black locust and green ash, small-leaf linden, or elm.

Table 32
Enzyme activities in technogenic soils resulting from the recultivation of the Chasov Yar refractory clay strip mine spoils

Tree species	Contact with black locust	Depth (cm)	Invertase*	Urease*	Proteinase*
Green ash	+	0-5	24.8	1.97	0.69
		5-10	5.1	0.78	0.32
		10-20	1.3	0.52	0.09
	-	0-5	2.9	0.72	0.02
		5-10	2.3	0.56	0.07
		10-20	2.1	0.47	0.02
Small-leaf linden	+	0-5	28.9	1.79	0.68
		5-10	5.5	0.66	0.32
		10-20	5.4	0.63	0.24
	-	0-5	17.9	1.10	0.51
		5-10	7.2	0.64	0.10
		10-20	6.3	0.64	0.03
Elm	+	0-5	21.9	1.79	0.68
		5-10	14.7	0.71	0.31
		10-20	11.7	0.65	0.09
	-	0-5	19.8	1.14	0.45
		5-10	10.0	0.67	0.33
		10-20	9.6	0.63	0.09
Black locust	Pure stand	0-5	26.4	1.89	0.76
		5-10	10.0	0.74	0.26
		10-20	7.2	0.50	0.35

Reproduced from: T.N. Keleberda, Vestnik Sel'skokhozyaistvennoi Nauki, 1979, No. 2, p. 89.

* Expression of enzyme activities: invertase in mg of inverted sugar g^{-1} soil 40 h^{-1}; urease in mg of NH_4^+-N g^{-1} soil 40 h^{-1}; and proteinase in mg of NH_2-N g^{-1} soil 72 h^{-1}.

Chapter 19. TECHNOGENIC SOILS FROM BENTONITIC CLAY MINE SPOILS

Enzymological research in Georgia

Daraseliya *et al.* (1978) applied enzymological and microbiological methods to evaluate the efficiency of the recultivation of spoils that resulted from the strip mining of bentonitic clays (gumbrine, askanite, etc.) at Gumbra (Tskhaltubo district). The spoil heaps were recultivated with maize (unfertilised or fertilised with 300 kg of N, 240 kg of P, and 90 kg of K ha^{-1}) or with a ryegrass-alfalfa mixture. Naturally revegetated spoil heaps and an adjacent native yellow soil under forest were used for comparison.

Table 33 shows that the spoils, in comparison with the native soil, are characterised by lower invertase and phosphatase activities, which are related to the reduced humus content of spoils. Owing to recultivation, the activities increased. Fertilisation of spoil heaps recultivated with maize had a beneficial effect on the enzyme activities. The highest activities were found in the spoil heaps recultivated with the ryegrass-alfalfa mixture. The activities in the 0-20-cm layer of these spoil heaps approached those measured in the native soil. In each variant, the 0-20-cm layer was more active than the 20-40-cm one. Recultivation also led to substantial increases in the number of the main groups of microorganisms in the spoil heaps.

Table 33
Effect of the measures applied for recultivation of bentonitic clay mine spoils at Gumbra, Tskhaltubo district (mean values of analytical data over 2 years)

Experimental variants	Depth (cm)	Humus (%)	pH in H_2O	Invertase*	Phosphatase*
Adjacent native	0-20	2.05	4.4	14.2	13.9
yellow soil	20-40	1.88	4.3	6.4	8.7
Mine spoils under	0-20	0.29	7.6	0.6	5.8
natural vegetation	20-40	0.11	7.4	0.5	4.4
Maize on unfertilised	0-20	0.32	7.4	2.6	9.0
mine spoils	20-40	0.21	7.6	0.6	2.9
Maize on NPK-fertilised	0-20	0.34	7.8	4.0	10.8
mine spoils	20-40	0.24	7.9	1.1	5.8
Ryegrass-alfalfa	0-20	0.32	7.6	14.0	12.1
on mine spoils	20-40	0.16	7.7	2.7	4.5

Reproduced from: N.A. Daraseliya, G.B. Kalatozova and E.F. Lapanashvili, Soobshcheniya Akademii Nauk Gruz.SSR, 1978, Vol. 92, p. 426.

* Expression of enzyme activities: invertase in mg of "glucose" g^{-1} soil 24 h^{-1} and phosphatase in mg of P_2O_5 10 g^{-1} soil h^{-1}.

Studying the same spoil heaps, Rtskhiladze *et al.* (1981) found that dehydrogenase activity behaved like invertase activity.

Chapter 20. TECHNOGENIC SOILS FROM PUMICE MINE SPOILS

Enzymological research in Germany

The spoils of the pumice stone mining in the Neuwid Basin are loamy sands in which the sand content strongly increases with depth. For their study, Schröder and Schneider (1992) have selected eight young and three old recultivation plots. A series of age-dependent changes were registered. Thus, in the 0-30-cm layer of plots, pH decreased (from 6.6 to 5.4), organic C and total N contents slightly increased, whereas microbial biomass nearly doubled and dehydrogenase activity exhibited a nearly 7-fold increase during aging of spoil plots.

Chapter 21. TECHNOGENIC SOILS ON SAND OPENCAST MINE FLOOR DRIFT AND SPOILS

Enzymological research in Poland

Hazuk (1967) was the first to utilise enzymological methods for studying the recultivation of the floor drift of sand opencasts formerly used for the mining of filling sand (necessary for the hydraulic filling of the workings in coal mines). The study area, the Szczakowa filling sand quarry, is situated in the western part of the Little Bledowska Desert (Cracow region). The alluvial "soil" of this former sand opencast consists of deep (thick), medium-grain loose sands (pH in H_2O was 8.3). Only the top layer (1-10 cm) of these sands contains nutrients (P, K), but in very small amounts. For recultivation, a fertilisation experiment was carried out, in five variants. It was established that fertilisation with a sorbent based on bentonite and NPK without or with the addition of peat led to increased enzyme activities in the "soil": invertase, urease, and asparaginase activities approximately doubled, whereas catalase activity increased 2 to 4 times as compared with the unfertilised control.

These investigations were continued, developed, and described in detail by Greszta and Olszowski (1974). They studied seven experimental variants (each in four plots):

1. controls (no fertilisation);
2. mineral fertilisers;
3. humus (60 t ha^{-1}) plus mineral fertilisers;
4. sorbent fertiliser (30 t ha^{-1});
5. sorbent fertiliser (60 t ha^{-1});
6. sorbent fertiliser (60 t ha^{-1}) plus peat dust (12 t ha^{-1}); and
7. cinders (60 t ha^{-1}) plus mineral fertilisers.

The dosage of mineral fertilisers in variants 2, 3, and 7 was: N 20, P_2O_5 65, and K_2O 60 kg ha^{-1}. The humus used came from the caprock of the sand pit hole and consisted of the material from the humus-mineral horizon mixed with forest leafmould. The sorbent was produced from bentonite containing 73-74% clayey components (montmorillonite, kaolin, illite). The sorbent fertiliser consisted of bentonite to which lime was added in the proportion 1 part of lime per each 20 parts of clay. It was subsequently mixed with mineral fertilisers (N 4, P_2O_5 16.2, and K_2O 24 kg ha^{-1}). The cinders came from a power plant.

Following these treatments, a seed mixture of perennial plants, predominantly legumes, were sown in all the plots. Prior to sowing, the seeds of leguminous plants were treated with "Nitragina" (specific *Rhizobium* culture). The mixture consisted of: *Lupinus luteus* 100, *L. polyphyllus* 10, *Melilotus albus* 10, *Lotus corniculatus* 5, *Trifolium repens* 5, *Anthyllis vulneraria* 3, *Festuca ovina* 1, and *Bromus inermis* 1 kg ha^{-1}.

For 2 years (1966-1967), soil samples were collected throughout the vegetative season at 3-week intervals for determination of enzyme activities (invertase, β-glucosidase, urease, asparaginase, and catalase) and respiration (CO_2 evolution) rate. Numbers of bacteria, actinomycetes, and microfungi, as well as several physical and chemical properties, were also determined periodically.

The results have shown that the activity of enzymes was highest in the soil of the plots with a sorbent fertiliser, especially when this was applied with added peat. In plots fertilised with cinders the activity recorded during the first year did not differ from the activity values determined from plots treated with the sorbent fertiliser, but it appeared to be lower in the second year, showing a tendency to decrease steadily. Treatment with humus containing an addition of mineral fertilisers and with mineral fertilisers alone did not cause any significant change in the activity of enzymes.

The highest enzyme activities were found in the 1-5-cm layer. At the 6-10-cm depth a marked decrease in activities was observed.

The results obtained for the plots with the sorbent fertiliser containing added peat indicate that it is advisable to combine mineral and organic fertilisation.

The results also indicate that there exists a causative relationship between the activity of soil enzymes and herbage yield of the recultivated plots. Thus, the highest enzyme activities and the best yields were recorded for plots treated with sorbent fertiliser containing added peat. This correlation seems to be the result of a stimulating effect of the applied fertilisers on the metabolism of soil microorganisms expressed by the activity of the enzymes studied, due to which the growth of the plants was better. Another possible explanation is that the sorbent fertiliser retained more available nutrients, so the plants grew better and their residues stimulated the microorganisms.

The highest rate of CO_2 evolution and the largest number of microorganisms, especially bacteria and fungi, were also recorded in the soil of plots treated with the sorbent fertiliser containing an addition of peat.

In a recultivation experiment started in 1977 in the area of an exhausted sand mine in Choszczówka, dehydrogenase and urease activities in the technogenic soil formed were analysed in 1984, and it was found (Żukowska-Wieszczek *et al.*, 1985) that the enzyme activities like the biomass of grasses were highest in the experimental variants treated with clay plus sewage sludge, fly ash plus sewage sludge, and waste fungal mycelia from a pharmaceutical plant. The variants treated with sewage sludge alone, municipal refuse, and compost were less efficient.

Enzymological research in Georgia

Enzymological and microbiological study of the spoils resulted from the opencast mining of quartz sand in the Chiatura district (western Georgia) and recultivated with grapevine, fruit trees, forest trees, and perennial grasses has proved (Rtskhiladze *et al.*, 1981) that dehydrogenase and invertase activities and numbers of microorganisms from different physiological groups can be used as indicators of the efficiency of the recultivation measures applied.

Chapter 22. TECHNOGENIC SOILS ON EXHAUSTED GRAVEL QUARRIES

Enzymological research in Lithuania

Petrov *et al.* (1994) conducted recultivation experiments on the exhausted gravel quarry located in a large plain in the vicinity of the city of Vilnius and described the results obtained in the 1986-1991 period.

The soils covering the exhausted quarry belong to three groups:

I. 35-40-cm layer from B1 horizon;

II. 20-25-cm layer from B1 horizon covered with a 15-20-cm-thick mixture of arable layer, A2B and B1 horizons; and

III. 20-25 cm of mixed horizons from beneath the arable layer, the mixture being covered with a 15-cm arable layer.

The soils of all groups are of sandy loam texture.

In 1986, the physico-chemical analyses showed that the pH was slightly alkaline (7.1-7.9) in soils of group I, nearly neutral (6.8-7.4) in soils of group II, and slightly acid (5.9-6.6) in soils of group III. Humus content in soils of the three groups varied between 0.41 and 1.42%, total N content between 0.032 and 0.107%, P and K contents between 13.3 and 33.9 mg of P_2O_5 100 g^{-1}, and between 4.8 and 21.7 mg of K_2O 100 g^{-1}, respectively.

In the spring of 1987, a part of the experimental plots of each soil group were seeded with a perennial *legume*-grass mixture (alfalfa 50%, birdsfoot trefoil 15%, Hungarian sainfoin 15% plus grass species). The other part of plots received a perennial *grass*-legume mixture (red fescue 20%, birdsfoot trefoil 20%, Hungarian sainfoin 20% plus other species). Before sowing of the *legume*-grass mixture, the plots were fertilised with NH_4NO_3 (30 kg of N ha^{-1}). Before sowing of the *grass*-legume mixture, the plots also received 30 kg of N ha^{-1}, but they were fertilised again, after the first and second harvests, with NH_4NO_3 at a rate of 45 kg of N ha^{-1}. Before the sowing of both seed mixtures, the plots were fertilised with mineral P (90 kg ha^{-1}) and K (120 kg ha^{-1}) and some of them received organic fertilisers, too, namely farmyard manure or calcareous sapropel (120 t ha^{-1}). The control plots were not treated with fertilisers. Each fertilisation variant was carried out in four repetitions, on 24-m^2 plots.

It was found that the two seed mixtures were equally efficient, in term of herbage dry matter yields, whose mean values for the 1987-1991 period being 2.9, 4.7, and 4.1 t ha^{-1} $year^{-1}$ on soils of groups I, II, and III, respectively. In the same period, the highest dry matter yield was recorded in the farmyard-manured plots on soils of group II (5.6 t ha^{-1} $year^{-1}$). The root biomass dry matter reached 12.9-49.6 t ha^{-1}.

Both mineral and organic fertilisation had beneficial effects on physico-chemical and biological properties of soils, especially in their 0-10-cm layer. The pH in slightly alkaline and nearly neutral soils changed to a little extent, whereas in the slightly acid soils it increased up to 6.8-7.1. Humus accumulation in the farmyard-manured plots increased 2.3-3 times (in soils of group I), 1.9-2.6 times (in soils of group II), and 1.5 times (in soils of group III); the corresponding increases in the sapropel-treated soils of groups I-III were 2-, 1.7-, and 1.2-fold, respectively. Total N content increased to a large extent. Increases occurred in the P and K contents, too. Invertase and protease activities and N_2-fixation capacity measured in 1991 were nearly the same in the unfertilised control plots and in those fertilised with farmyard manure or sapropel, but in the previous 4-year period (1987-1990) biological activity showed 25-32-fold increases in the control plots and only 3.0-3.8-fold increases in the organically fertilised plots.

The investigations also showed that for agricultural recultivation covering of exhausted gravel quarries with topsoil or subsoil-topsoil mixture is obligatory.

Chapter 23. TECHNOGENIC SOILS FROM OVERBURDENS REMAINING AFTER PIPELINE CONSTRUCTION

Enzymological research in the Russian Federation

The enzymological study of the recutivation of overburdens remaining after pipeline construction was initiated by Idrisova (1984). In 1981, she modelled, in laboratory and field, a recultivation technology based on the spreading of overburdens on the surface of adjacent agricultural fields, in different proportions: overburdens (calcareous loam) remaining after the construction of 100 m of pipeline (diameter 1420 mm) were spread on 0.5, 1, 1.5, and 2 ha of agricultural field (leached chernozem on the forest-steppe zone of the Bashkirian Pre-Urals). The control field was not treated with overburdens.

Analysis of the ploughed (0-30-cm) layer of soil showed that catalase activity was practically unaffected by overburdens. In contrast, invertase and phosphatase activities and respiration (CO_2 evolution) rate of soil decreased in parallel with the diminution of the field surface on which the overburdens were spread. Thus, the maximum values were found in the control soil and the minimum ones in the field with the smallest surface (0.5 ha) on which the overburdens were spread. The contents of humus, mineral N, mobile P, and exchangeable K in soil as well as crop yields (in 1981: vetch-oats mixture; in 1982: winter rye) decreased in a similar manner. The decrease in crop yields was lower in the plots fertilised with 40 t of farmyard manure plus 90 kg of N as NH_4NO_3 plus 60 kg of P as double superphosphate plus 45 kg of K as KCl ha^{-1} than in the unfertilised plots.

In addition to the technology of spreading overburdens on soil surface, in 1981 Ishem'yarov *et al.* (1984) and Idrisova *et al.* (1986a, b, 1987) applied, under both laboratory and field conditions, another technology: mixing of the ploughed soil layer with overburdens in the following proportions: 87.5, 75.0, 62.5, 50.0, 37.5, 25.0, and 12.5% soil plus overburdens up to 100%.

As expected, in the soil-overburden mixtures, in comparison with the control soil, catalase, invertase, phosphatase, and urease activities and the contents of humus, mineral N, mobile P, and exchangeable K decreased with increasing proportions of overburdens. In this case, too, some plots were fertilised with the same amounts of farmyard manure plus NPK as specified above. Other, unfertilised plots served for comparison. In the 1981-1984 period, the crop yields were estimated in a rotation composed of vetch-oats mixture, winter rye, spring wheat, and maize. In the soil-overburden mixtures, the crop yields decreased, mainly in the unfertilised plots. Comparison of crop yields after application of the two recultivation technologies (spreading or mixing of overburdens on or with soil, respectively) indicated that diminution of crop yields was lowest with the spreading technology applied in association with organic and mineral fertilisation. In addition, the cost of recultivation was also lower when this technology was used. At the same time, it has been emphasised that the ratio between the amounts of overburden and soil (ploughed layer) should never exceed 1:1.

At this ratio, the organic and mineral fertilisation at rates specified above and the crop rotation composed of vetch-oats mixture, winter rye, spring wheat, maize, and spring wheat resulted in markedly increased catalase and urease activities in the soil-overburden mixture (1.1 ml of O_2 g^{-1} mixture min^{-1} and 2.3 mg of NH_3 g^{-1} mixture, respectively) and these values approached those registered in the control soil (1.4 ml of O_2 g^{-1} soil min^{-1} and 2.5 mg of NH_3 g^{-1} soil, respectively) (Idrisova, 1990).

Geltser *et al.* (1988) studied 5- and 20-year-old plots covering pipeline on a sloping territory in the Podolsk area (Moscow region). Initially, the overburdens were planted with spruce and seeded with herbs. But within a few years the vegetation has changed: spruce remained dominant on the superior part of the slope, and it was suppressed by herbaceous vegetation on the middle part of slope and replaced by white alder on the inferior part of slope. The 0-5-cm and deeper soil layers of the plots and of an undisturbed pine forest were analysed.

Due to the accumulation of plant residues under the herbaceous vegetation, cellulolytic activity in the soil layers was about 4 times more intense in the 20-year-old plots than in the 5-year-old ones and exceeded the activity values registered in the undisturbed forest soil. Protease activity was higher in both young and old plots than in the undisturbed soil, but, excepting the 0-5-cm soil layer, the deeper layers were more protease-active in the young than in the old plots.

Naprasnikova (1995) has determined ureolytic activity in 21 soils along the future, more than 400-km-long pipeline planned to be constructed between the cities of Kavykta and Angarsk, in valleys of the rivers Kuda, Kulenga, and Tutura (southern Siberia). The activity values measured at present will serve for assessing the soil enzymological effect of the disturbance that will be caused by the pipeline construction and other engineering-technical operations.

Enzymological research in the United States of America

The work edited by Redente and Cook (1986) contains information concerning revegetation of soils disturbed by pipeline construction in the Piceance Basin, northwestern Colorado. In 1985, three high-elevation sites (2250 m) and a low-elevation site (2040 m) disturbed by pipeline construction 2-27 years previously were compared with undisturbed native controls. The soils were sampled from the 5-10-cm depth. The data included in Table 34 indicate that in the technogenic soils the pH was higher, whereas the organic matter and

Table 34
Influence of revegetation on some parameters of sites disturbed by pipeline construction as compared to native sites in the Piceance Basin, Colorado

Parameters	High-elevation sites				Low-elevation sites	
	Disturbed			Native	Disturbed	Native
	2	4	27		23	
	years previously				years previously	
pH	8.36	7.72	7.35	6.70	8.54	7.13
Organic matter (%)	0.95	1.49	1.04	2.66	0.53	0.93
Phosphatase activity (μg of *p*-nitrophenol g^{-1} soil h^{-1})	193	467	690	781	70	377
Mineralisable N (μg of NH_4^+-N g^{-1} soil)	47.4	63.7	62.7	91.9	29.2	46.1
Dehydrogenase activity (μg of triphenylformazan g^{-1} soil 24 h^{-1})						
Actual (autochthonous)	37.1	17.7	11.5	13.6	13.7	3.8
Potential (zymogenous)	25.3	25.9	9.8	15.4	13.9	8.0
Litter cover (%)	2	0	47	63	6	4
Total plant cover (%)	19	20	60	67	38	20

Reproduced from: E.F. Redente and C.W. Cook, Eds., Structural and Functional Changes in Early Successional Stages of a Semiarid Ecosystem, Fort Collins, Colorado, 1986, p. 19.

mineralisable N contents, and phosphatase activity were lower than in the native soils. In contrast, the technogenic soils, except for the 27-year-old one, exhibited higher actual and potential dehydrogenase activities than the native soils. Diminution of dehydrogenase activities in the 27-year-old technogenic soil, in comparison with the activity values measured in the younger ones, may be due to the fact that this area has been heavily grazed. In the 2-, 4-, and 27-year-old technogenic soils, phosphatase activity increased and dehydrogenase activities decreased with their age and total plant cover percentage.

Enzymological research in Ukraine

In the 1985-1987 period, Stefurak (1989) analysed, enzymologically, microbiologically, and chemically, the soils covering the pipelines constructed in the Carpathians, more precisely, on the territories of the Lyudvikov and Vyshkov forestry districts, Ivano-Frankovsk region. The plots of disturbed soils selected for analyses had been submitted to spontaneous revegetation 1, 10, and 20 years previously. Adjacent undisturbed dark-brown mountain soils under forest vegetation were the controls.

The data of Table 35 show that catalase, urease, cellulolytic, and proteolytic activities and respiration of soils increased with the age of vegetation. However, they did not reach, even after 20 years, the values recorded in the undisturbed soils. In other words, spontaneous revegetation of soils disturbed by pipeline construction did not lead, even after 20 years, to complete regeneration of the biological potential of these soils.

Similar results were also obtained in the analyses for total number and biomass of bacteria, total numbers of actinomycetes and microfungi, numbers of oligonitrophilic, ammonifying, denitrifying microorganisms, spore-forming bacteria, presence of *Clostridium pasteurianum* and of different *Bacillus* species, ammonification and nitrification capacities as well as humus content.

Table 35

Biological activity of soils covering the pipelines constructed on territories of the Lyudvikov and Vyshkov forestry districts, Ivano-Frankovsk region

Soils	Age of vegetation on disturbed soils (years)	Catalase*	Urease*	Cellulolysis*	Proteolysis*	Respiration**
Disturbed, spontaneously revegetating soils	1	2.7±1.3	0.8±0.4	2.3±0.6	2.1±0.7	0.7±0.3
	10	8.5±1.9	3.9±1.1	3.1±0.9	7.4±1.3	1.4±0.9
	20	14.2±0.9	9.6±1.4	13.1±0.3	18.3±1.4	2.9±0.2
Undisturbed, control soils under forest vegetation	-	26.3±0.3	23.9±2.0	29.0±1.7	34.0±2.1	6.3±0.4

Reproduced from: V.P. Stefurak, Mikrobiologicheskii Zhurnal, 1989, No. 2, p. 16.

* Expression of activities: catalase in ml of O_2 5 ml^{-1} soil 5 min^{-1}; urease in mg of urea 10 g^{-1} soil h^{-1}; and cellulolysis and proteolysis in percentages.

** Expressed in kg of CO_2 ha^{-1} soil h^{-1}.

Chapter 24. RECULTIVATION OF SOILS REMAINING AFTER TOPSOIL "MINING"

Enzymological research in New Zealand

Topsoil is "mined", i.e. removed, around many urban areas for use in landscape improvement. Recultivation of the remaining soil for restoring its fertility is required with the same emphasis as the recultivation of wastelands resulting from mining or other industrial activities. This problem was enzymologically studied by Ross *et al.* (1982). The soil studied, the Judgeford silt loam (Wellington area), was originally under grazed grass-clover pasture, then used for topsoil "mining", removal to depths of 10 cm (S_{10} plots) and 20 cm (S_{20} plots) in March 1978. The remaining soil was treated with lime (2.5 t ha^{-1}) and urea (120 kg ha^{-1}) and with high amounts of P and K fertilisers (1.5 t of molybdic superphosphate and 300 kg of potassium sulphate ha^{-1}), and resown in pasture species. Plots without topsoil removal (S_0) served for comparison. Between March 1978 and March 1981, seven enzymatic activities and several other biochemical as well as chemical and physical properties were determined periodically in the 0-10-cm layer of all the plots. Herbage yield was also recorded and taken as the criterion for soil fertility.

The herbage yield and enzyme activity values obtained are presented in Figure 17. They show that these values increased rapidly in the S_{10} and S_{20} plots but had not reached S_0 levels after 3 years. Similar results were obtained in the measurements of CO_2 and mineral N production, biomass C, mineral N flush* and ATP. Organic C and total N contents increased only slowly. Enzymatic and other biochemical activities were, in general, significantly correlated with herbage yields in the S_{10} or S_{20} plots, but not in the S_0 plots; organic C and total N contents were not generally correlated significantly with yields over the first 2 years. Overall, invertase, and then sulphatase, activity appeared to be the best indicators of soil fertility status in the stripped ("mined") soil studied.

---

* See the footnote on page 194.

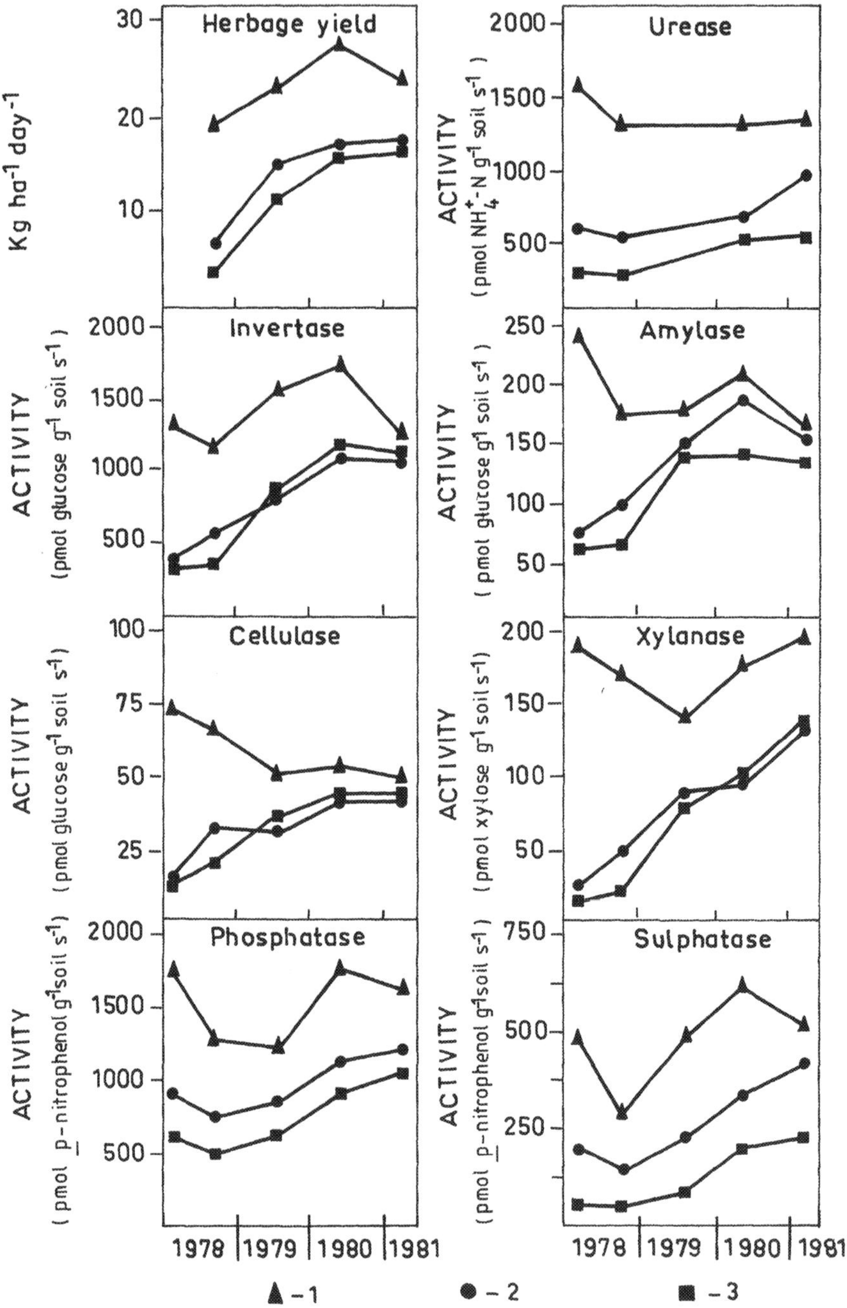
Herbage yield
Kg ha-1 day-1
Urease
ACTIVITY (pmol NH4+-N g-1 soil s-1)
Invertase
ACTIVITY (pmol glucose g-1 soil s-1)
Amylase
ACTIVITY (pmol glucose g-1 soil s-1)
Cellulase
ACTIVITY (pmol glucose g-1 soil s-1)
Xylanase
ACTIVITY (pmol xylose g-1 soil s-1)
Phosphatase
ACTIVITY (pmol p-nitrophenol g-1 soil s-1)
Sulphatase
ACTIVITY (pmol p-nitrophenol g-1 soil s-1)
1978 1979 1980 1981
▲ – 1
● – 2
■ – 3

Ross *et al.* (1984) have established that the enzyme activities and the other biochemical properties in the S_{10} and S_{20} plots remained, even in the fourth and fifth years after recultivation, under the level of the values recorded in the S_0 plots. The only exceptions were cellulase activity in the fourth year and amylase activity in the fifth year. At the same time, if in the first 3 years the correlation between herbage yield and enzymatic activities and other biochemical properties was most significant in the case of invertase and sulphatase activities, in the last 2 years the correlation became most significant in the case of xylanase activity.

These investigations were also referred to in the review works of Tate *et al.* (1985), Speir *et al.* (1986), and Smith and Paul (1990).

Figure 17. Herbage yield and enzyme activities from plots of Judgeford soil that had been stripped of 0, 10, and 20 cm of the original topsoil in March 1978, reestablished in a grass-clover pasture, and sampled periodically over the following three years.

Redrawn from: D.J. Ross, T.W. Speir, K.R. Tate, A. Cairns, K.F. Meyrick and E.A. Pansier, Soil Biology & Biochemistry, 1982, Vol. 14, p. 577, with kind permission from Elsevier Science.

1 – S_0 plots (no topsoil removed). 2 – S_{10} plots (10 cm of topsoil removed). 3 – S_{20} plots (20 cm of topsoil removed).

Herbage yield is expressed on a dry matter basis.

Chapter 25. REVEGETATION OF SOILS USED FOR COVERING RUBBLE DUMPS

Enzymological research in Germany

The Siedenburg rubble dump, studied by Weidemann (1985) and Schulz *et al.* (1985a, b), is located at the periphery of the city of Bremen. It occupies a surface of approximately 4 ha and is 21 m high. It was formed mostly of rubbles in mixture with earth excavated from foundation holes, with sludges and small amounts of domestic refuse, the deposition of which was prohibited in 1977. The dump was covered with about 60-100 cm of excavation earth, on which a spontaneous vegetation later developed. In the winter of 1979/1980, the unflat parts of the dump plateau were levelled and, thus, they became an area of ~0.8 ha free of vegetation. The central, flat part of the plateau, not being affected by levelling, retained its spontaneous vegetation. The levelled area was divided into plots subjected to spontaneous revegetation and plots recultivated in June 1980 by sowing the seeds of eight plant species. The plants and their percentage in the seed mixture were the following: *Festuca ovina* 20, *F. rubra* 20, *Poa pratensis* 20, *Agrostis tenuis* 15, *Lolium perenne* 15, *Bromus erectus* 5, *Lotus corniculatus* 4, and *Achillea millefolium* 1.

In the spring of 1982, about 120 m^3 of rubble-containing earth was removed from an about 100-m^2 surface at the margin of the dump plateau. The removed earth was replaced by a weakly humous soil mixed with sand (thickness of soil-sand mixture on the rubbles was 1 m). The new surface created in such a manner was divided into two plots, one left to revegetate spontaneously and the other recultivated in May 1982 by sowing the seed mixture specified above. In September 1982, *Chenopodium album* and *Sinapis arvensis* were the dominant plants in the unsown plot and *Lolium perenne* in the sown one.

In the May 1982-May 1983 period, the 0-20-cm soil layer was analysed 18 times for determining, among other things, the microbial net mineralisation of N, nitrification, NH_4^+ and NO_3^- contents, protease activity, and dehydrogenase activity. Microbial net mineralisation of N and percentage of NH_4^+ in total mineral N content had higher values, whereas nitrification and NO_3^- content had lower values in the recultivated plot than in that under spontaneous vegetation. The plants growing on the recultivated plot took up more NO_3^- than those in the spontaneously revegetated plot.

Considering that the difference between protease activity and microbial net mineralisation of N is equal to microbial immobilisation of N and eliminating two extreme values from the 18 values measured, the coefficient of linear correlation between microbial immobilisation of N and dehydrogenase activity was calculated. The coefficient was higher for the recultivated plot ($r = 0.631$) than for that with the spontaneous vegetation ($r = 0.513$).

The results of this field experiment led to the conclusion that the recultivated plot was more efficient in retaining N compounds than was the plot with spontaneous vegetation. It is well known that the existence of bioelement reserves constitutes an essential attribute of stable ecosystems.

A laboratory experiment was also performed. Soil sampled from the plot under spontaneous vegetation was used. The soil samples were treated with cellulose and peptone in different ratios (8 variants), incubated at 25°C for 14 days and then analysed to determine microbial net mineralisation of N, protease activity, and dehydrogenase activity. The results showed that microbial immobilisation of N (= protease activity minus microbial net mineralisation of N) correlated very significantly with dehydrogenase activity ($r = 0.979$). The conclusion drawn from this experiment was that the global mineralisation of soil N can be evaluated by determining protease activity.

Chapter 26. REVEGETATION OF LANDFILL SOILS

Enzymological research in the United Kingdom

Wigfull *et al.* (1987a, b) and Wigfull and Birch (1987) have studied the soils used as covers in the restoration of completed domestic refuse landfill areas at St. Osyth (Colchester,

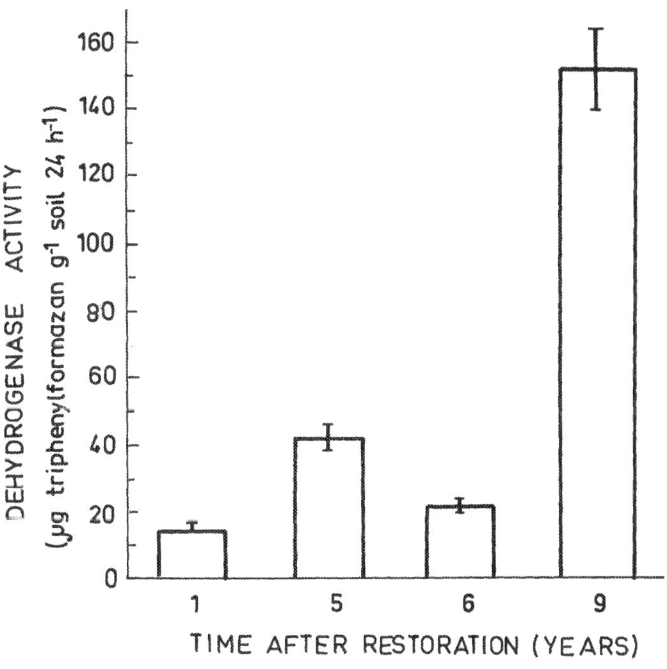

Figure 18. Soil dehydrogenase activity against time after restoration of four completed landfill areas.
Redrawn from: S.D. Wigfull, J.A. Harris and P. Birch, Conference on "Reclamation and Restoration of Soils" (London, 1987), Graph 1.

The bars indicate standard deviations.

Essex, England). For enzymological, microbiological, and chemical analyses, samples were taken, in April 1986, from the 0-30-cm layer of soils that had been placed on the completed landfill areas 1-9 years previously. After being used for covering the landfills, these soils were not fertilised and sown, but were sparsely or abundantly revegetated with grasses from their seed-bank.

Figure 18 shows the results obtained in the determination of dehydrogenase activity in soils of four landfill areas under spontaneous revegetation for 1, 5, 6, and 9 years. The 1-year-old area was restored with poor-quality topsoil, whereas for the restoration of the older areas subsoils of different qualities were used. The vegetation cover was sparse on the 1- and 6-year-old areas and well developed on the 5- and 9-year-old ones.

One can deduce from Figure 18 that dehydrogenase activity increased with time after restoration, except for the sparsely vegetated 6-year-old area, the soil of which was less dehydrogenase-active than the soil of the 5-year-old area. It has also been established that, within the 0-30-cm soil layer, dehydrogenase activity decreased with increasing depth, except for the 9-year-old area, in which the activity was low in the 0-5-cm depth, and increased and became constant in the 5-30-cm depth. This particular behaviour of the 9-year-old area might be caused by the development of anaerobiosis within the higher zone of the soil profile due to surface pooling above compacted and drier lower zones.

Dehydrogenase activity correlated significantly with total organic C and N contents (Table 36) as well as with microbial biomass (ATP content) in the soils of the four areas.

Table 36
Correlation matrix concerning the age of restored landfill areas, soil physico-chemical properties and dehydrogenase activity (the figures denote correlation coefficients)

	Age	pH	Total N content	Total organic C content	Moisture content	Dehydrogenase activity
pH	-0.357					
Total N content	0.596	-0.440				
Total organic C content	0.657	-0.494	0.995*			
Moisture content	-0.135	-0.474	0.682	0.640		
Dehydrogenase activity	0.813	-0.580	0.957*	0.966*	0.467	
C-to-N ratio	0.823	-0.621	0.232	0.324	-0.282	0.556

Reproduced from: S.D. Wigfull, J.A. Harris and P. Birch, Conference on "Reclamation and Restoration of Soils" (London, 1987), Table 3.

* The asterisks indicate significance at >95%.

These soils, in comparison with native undisturbed soils, contained less ATP and their dehydrogenase activity was lower, although they contained similar numbers of viable microorganisms.

Wigfull and co-workers drew the conclusion that the measurement of soil dehydrogenase activity could be useful in estimating the restoration potential of soils and as an index of the success of landfill and other restoration programmes.

Chapter 27. TRANSFORMATION OF HARBOUR MUDS INTO AGRICULTURAL SOILS

Enzymological research in Germany

Schröder *et al.* (1987b) and Schröder (1988a) have studied three silty-clayey soils derived from harbour mud, on the former low moor Riepster Hammrich area located between the cities of Emden and Riepe.

The mud was dragged from the basin of the Emden harbour and from the estuary of the Ems River and the slurry was pumped, through a tube system, to the about 18-ha polder (on the former low moor). The pumping, that was interrupted periodically for allowing the mud particles to settle, lasted 3 years and, as a result, the deposited mud attained a height of 2 m on the polder. During the next 3 years, partial drying and subsidence of the mud reduced its thickness to about 1 m ("mature soil"). In the sixth year, the excess water was removed through ditches and the mud soil was sown with spring barley. Thereafter, different cereals as well as rape and legumes were cultivated. About 12 years after the commencement of slurry poldering, the soil was drained for removing the infiltration water that resulted from precipitations.

The three soils studied were 6, 17, and 28 years old, respectively. For analyses, soil samples were collected from depths of 0-30, 30-50, 50-70, and 70-90 cm , in six repetitions. Three biological parameters (dehydrogenase activity, cellulose decomposition, and microbial biomass) and a series of chemical and physical parameters were analysed. It was found that, in dependence of soil age, dehydrogenase activity increased in the upper layers and decreased in the deeper ones. Thus, the correlation between dehydrogenase activity and age of soils was very significant in the 0-30-cm layer ($r = 0.91$). Cellulose decomposition in soils did not show age-dependent variation, but decreased with depth. The microbial biomass gave similar values in each layer of each soil. It is evident from these findings that dehydrogenase activity was a more sensitive biological parameter than cellulose decomposition or microbial biomass.

The correlations between soil age and chemical parameters in the 0-30-cm layer were insignificant for C and N contents and cation-exchange capacity and significant for contents of $CaCO_3$ ($r = -0.72$), Ca^{2+} ($r = 0.79$), Mg^{2+} ($r = -0.96$), K^+ ($r = -0.63$), and Na^+ ($r = -0.75$) as well as for electrical conductivity ($r = -0.76$). These data suggest that, during the aging of soils, a part of the salts originating from harbour mud migrated from the 0-30-cm layer into deeper soil layers. Improvements in the physical parameters of the 0-30-cm soil layer were also registered.

In newer investigations, described by Schumacher *et al.* (1989), Schröder and Schneider (1992) and Schneider and Schröder (1995), harbour mud-derived soils in 21 recultivation plots, namely in five young plots (3-4 years), eight plots of medium age (13-15 years), and eight old plots (30-35 years), were compared 1. with three undisturbed soils and 2. with each other, all soils being under agricultural use. Several soil parameters were determined. Some of the results registered in the ploughed soil layer are cited below.

The first comparison has shown that mean values of both dehydrogenase activity and microbial biomass were higher in each age group of recultivation plots than in the undisturbed soils.

According to the second comparison, dehydrogenase activity was highest in the young plots and lowest in plots of medium age, microbial biomass showed a tendency to slightly decrease, catalase, invertase, and alkaline phosphatase activities rather increased, protease activity and N mineralisation rate rather decreased with the age of recultivation plots.

Chapter 28. RECULTIVATION OF SOILS DISTURBED BY DIFFERENT INDUSTRIAL ACTIVITIES

Enzymological research in Canada

The investigations described by Rowell and Florence (1993) could not be included in any separate chapter of our review, due to the fact that the industrial disturbance of the investigated soils was of different nature (being caused by surface mining of coal, heavy oil extraction, pipeline construction, oil well drilling, and hydrocarbon contamination), but the results were evaluated together, irrespective of the cause of disturbance.

Industrially-disturbed soils, taken as a whole, were compared with undisturbed soils from similar areas. Biological and conventional soil assay methods were used to determine their usefulness in reclamation management. Five soil groups were studied comprising the following soil types according to the Canadian and FAO systems of classification:

group 1 – Black Chernozem (chernozem, rendzina);

group 2 – Dark Brown Chernozem (kastanozem, typic);

group 3 – Brown Chernozem (kastanozem, aridic);

group 4 – Gray Brown Luvisol (albic luvisol) and Gray Luvisol (albic luvisol, podzo-luvisol); and

group 5 – Eutric Brunisol (eutric cambisol) and Dystric Brunisol (dystric cambisol).

Disturbed and undisturbed surface soils were collected between September and November 1989 from southern and central Alberta (68 in total). Description of the samples is given in Table 37.

The biological methods used served for measuring enzyme activities (dehydrogenase, invertase, protease, arginine deaminase, alkaline phosphatase, and arylsulphatase), microbial biomass C and basal respiration rate, extractable organic C content and extractable colour index (an index of humification), mineralisation and nitrification potentials.

Table 37

Distribution of samples by soil groups and nature of disturbance

Sample locations and nature of reclamation practices	Number of samples in each group						
	Soil groups						
	1	2	3	4	5	D*	UD*
Samples from surface mining areas							
Paintearth coal mine: reclamation since 1983 has involved topsoiling, fertilisation, and seeding to grain, forage, and pasture crops. Undisturbed soils were from areas adjacent to the mine.	–	8	–	–	–	6	2
Diplomat coal mine: reclamation since 1950s has involved a variety of methods including ridging, levelling, topsoiling, fertilisation, and seeding to grain, forage, or rangeland crops. Undisturbed soils were from areas adjacent to the mining area.	–	11	–	–	–	8	3
Samples from heavy oil extraction areas							
Disturbed samples represented operational and experimental soil mixes by Suncor Inc. since 1971 for erosion control on sand–covered tailings pond dikes. Reclamation included amendment with peat, mineral overburden, bitumen sludge, and fertilisers and revegetation with trees, grasses, and legumes. Undisturbed soils were taken from forested areas close to the mine.	–	–	–	–	11	9	2
Samples taken following pipeline construction							
Reclamation involved topsoil salvage and replacement and revegetation to the previous land use. Undisturbed samples were taken from areas adjacent to the pipeline.	2	2	2	–	–	3	3
Samples from abandoned oil well drilling sites							
Reclamation involved salvage of topsoil, replacement of topsoil, fertilisation, and revegetation to the previous land use once the drilling operations were concluded.	2	4	4	–	–	5	5
Samples involving hydrocarbon contamination							
An oil spill rehabilitation experiment started in 1973 at the University of Alberta Experimental Farm. Reclamation involved tillage, fertilisation, and revegetation with grain crops.	4	–	–	–	–	3	1
An experimental site set up in 1987 to study the disposal of diesel invert cuttings from oil well drilling by landfarming. Reclamation involved tillage, addition of mineral fertilisers and manure, and reseeding with grasses and legumes.	–	–	–	4	–	3	1
Representative soils not affected by industrial disturbance							
Arable soils producing grains and forages.	3	2	2	4	–	–	11
Soils from native range and pasture.	–	–	2	–	–	–	2
Forested soils.	–	–	–	–	1	–	1

Reproduced from: M.J. Rowell and L.Z. Florence, Soil Biology & Biochemistry, 1993, Vol. 25, p. 1501, with kind permission from Elsevier Science.

* D – Disturbed. UD – Undisturbed.

The conventional (chemical and physical) methods included measurement of pH, electrical conductivity, organic C content, cation-exchange capacity, mineral N (ammonium-N and nitrate-N), soluble sulphate-S, and available phosphate-P contents, saturation percentage and particle size distribution (sand and clay contents).

The biological and conventional data sets were submitted to complex statistical analyses.

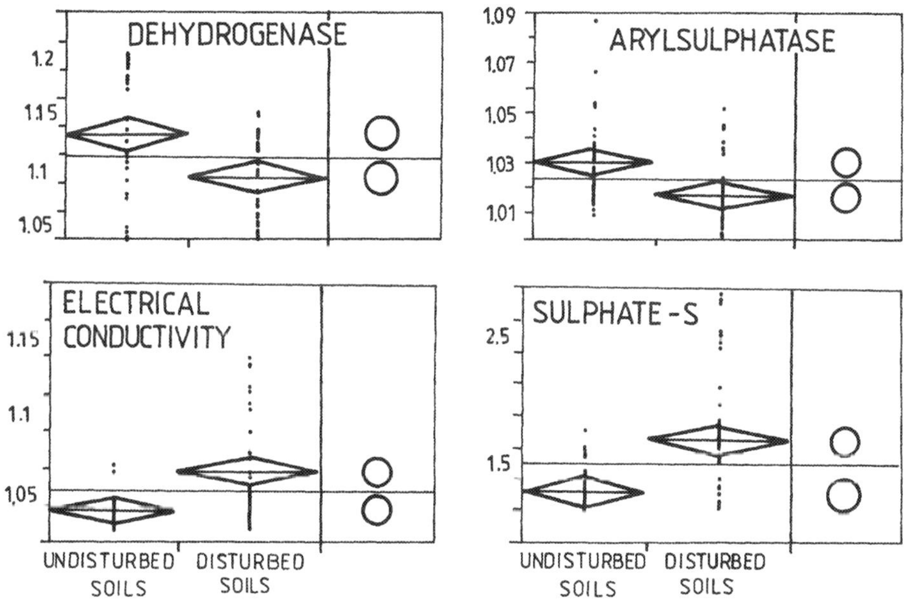

Figure 19. Analysis of variance summaries and comparisons of means (vertical axis. $\log_{10}$) between disturbed and undisturbed soils.

Redrawn from: M.J. Rowell and L.Z. Florence, Soil Biology & Biochemistry, 1993, Vol. 25, p. 1506, with kind permission from Elsevier Science.

The horizontal rule is the overall mean response; soil order means are the lines across and inside each diamond, height of the diamond is the 95% confidence interval of the mean and the diamond width represents the proportional difference in sample size. Circles which are non–intersecting indicate mean differences of $P<0.05$ between disturbed and undisturbed soils.

With the exception of pH (log H^+ concentration), data for the 12 biological and the 11 conventional measurements were transformed to common, base 10 logarithms prior to analysis of variance and multivariate discriminant analysis.

Analysis of variance showed that means of four characteristics were statistically different between disturbed and undisturbed soils; electrical conductivity and soluble sulphate–S content were higher in disturbed soils, while dehydrogenase and arylsulphatase activities were lower (Figure 19).

Multivariate discriminant analysis, where the grouping was done between disturbed and undisturbed soils, revealed that correct classification of soil samples was higher (70% for disturbed soils and 86% for undisturbed soils) when the biological and conventional data sets were combined, than when the biological and conventional data sets were considered separately (68 and 76%, and 64 and 91%, respectively).

The conclusion was drawn that the combination of biological and conventional measurements may provide a practical degree of assurance when determining the extent to which industrially–disturbed sites may have responded to remediation. This conclusion is acceptable by all experts in recultivation of technogenic soils. But usefulness of a combined study of soils suffering from industrial disturbances of very different nature (surface mining of coal, heavy oil extraction, pipeline construction, oil well drilling, hydrocarbon contamination) is questionable. Even Rowell and Florence (1993) mention: "More valuable information often comes from direct comparisons of disturbed and undisturbed soils at the same site".

CONCLUSIONS

The literature reviewed shows that application of enzymological methods makes it possible to indicate the degree of evolution of technogenic soils, the transformation of overburdens and other spoils and wastes into agricultural and forest soils, the efficiency of the recultivation measures applied.

In comparison with microbiological parameters, the enzymes are more synthetic indicators of the evolution of technogenic soils because they reflect *a*) due to their accumulation in form of humic complexes, the past of technogenic soils, and *b*) due to their catalytic activity, which plays a key role in nutrient cycles, the present biological status of these soils.

REFERENCES

Artem'eva, T.I., Nekrasova, K.A., Vertogradskaya, I.A., Zverkov, Yu.V., Borisovich, T.M. and Krylova, L.N. (1980) Biologicheskaya aktivnost' vyrabotannykh torfyanikov na pervykh etapakh ikh okul'turivaniya (Biological activity of exhausted peatlands during the first stages of their cultivation). In *Pochvennaya Fauna i Biologicheskaya Aktivnost' Osushennykh i Rekul'tiviruemykh Torfyanikov* (L.S. Kozlovskaya, Ed.), pp. 83-101. Izdatel'stvo Nauka, Moscow.

Baeva, R.I., Gagarina, E.I., Gorbovskaya, A.D., Dvornikova, L.L., Vinogradova, E.N. and Popov, A.I. (1992) Narushenie prirodnykh ekosistem v raione dobychi fosforitov (Disturbance of natural ecosystems in a phosphorite-mining area). *Pochvovedenie*, No. 5, 86-102.

Balicka, N. and Wegrzyn, T. (1984) Microbial processes in reclaimed power station wastes. In *Soil Biology and Conservation of the Biosphere* (J. Szegi, Ed.), pp. 819-825. Akadémiai Kiadó, Budapest.

Bender, J. (1980) Biologiczna rekultywacja zwalowisk pokopalnianych (Biological recultivation of mine spoil heaps). *Miedzynarodowe Czasopismo Rolnicze*, No. 3, 50-55.

Bender, J. (1989) Pochvoobrazovanie na otval'nykh porodakh v aspekte sovremennykh dostizhenii v rekul'tivatsii (Soil formation on spoil heaps in the light of the contemporary achievements in recultivation). *Sbornik Dokladov, Nauchnyi Seminar "Razrabotka Sposobov Rekul'tivatsii Landshaftov, Narushennykh Promyshlennoi Deyatel'nost'yu* (Konin, 1989), **1**, 1-21.

Bender, J. (1992) Biologicheskaya rekul'tivatsiya v Polshe (Biological recultivation in Poland). In *Ekologiya i Rekul'tivatsiya Tekhnogennykh Landshaftov* (V.M. Kurachev, Ed.), pp. 261-283. Izdatel'stvo Nauka, Sibirskoe Otdelenie, Novosibirsk.

Bender, J. and Gilewska, M. (1980) Biogennyi faktor kak pokazatel' initsial'noi stadii pochvoobrazovaniya industriozema (The biogenic factor as an indicator of the initial stage of pedogenesis in industrial soil). *Sbornik Dokladov, 7. Mezhdunarodnyi Simpozium "Rekul'tivatsiya Landshaftov, Narushennykh Promyshlennoi Deyatel'nost'yu* (Katowice-Zabrze-Konin, 1980), **3**, 178-184.

Bender, J. and Gilewska, M. (1983) The enzymatic activity as an indicator of the initial stage of the soil-forming process. In *Recultivation of Technogenous Areas* (J. Szegi, Ed.), pp. 193-198. Mátraalja Coal Mining Co., Gyöngyös.

Bender, J. and Gilewska, M. (1984a) Aktywność enzymatyczna gruntów zwalowiskowych Konińskiego Zaglebia Wegla Brunatnego. III. Aktywność katalazy w gruntach pogórniczych rekultywowanych rolniczo (Enzymatic activity of spoil heap soils in the Konin Brown Coal Basin. III. Catalase activity in agriculturally recultivated mine soils). *Archiwum Ochrony Środowiska*, No. 1, 163-176.

Bender, J. and Gilewska, M. (1984b) The influence of fertilizers on the enzymatic activity of industrial soils. In *Soil Biology and Conservation of the Biosphere* (J. Szegi, Ed.), pp. 45-52. Akadémiai Kiadó, Budapest.

Bender, J. and Gilewska, M. (1984c) The influence of pioneer forest plantations on the enzymatic activity of industrial soil. In *Soil Biology and Conservation of the Biosphere* (J. Szegi, Ed.), pp. 827-835. Akadémiai Kiadó, Budapest.

Bender, J. and Gilewska, M. (1984d) Rol' antropogennogo faktora v razvitii plodorodiya otval'nogo grunta (Role of the anthropogenic factor in the development of fertility in spoil heap soil). *Sbornik Dokladov, 8. Mezhdunarodnyi Simpozium "Razrabotka Sposobov Rekul'tivatsii Landshaftov, Narushennykh Promyshlennoi Deyatel'nost'yu* (Bucharest-Craiova-Tg.Jiu, 1984), pp. 155-162.

Bender, J. and Gilewska, M. (1984e) Produktsiya CO_2 kak pokazatel' biologicheskoi aktivnosti rekul'tivirovannykh otval'nykh gruntov (CO_2 production as an indicator of the biological activity in recultivated spoil heap soils). *Sbornik Dokladov, 8. Mezhdunarodnyi Simpozium "Razrabotka Sposobov Rekul'tivatsii Landshaftov, Narushennykh Promyshlennoi Deyatel'nost'yu* (Bucharest-Craiova-Tg.Jiu, 1984), pp. 295-303.

Bender, J. and Strzyszcz, Z. (1978) Vozdeistvie rekul'tivatsionnykh meropriyatii na obrazovanie antropogennykh pochv i ikh plodorodie (The effect of recultivation measures on the formation of anthropogenic soils and on their fertility). *Sbornik Dokladov Soveshchaniya "Rekul'tivatsiya Tekhnogennykh Landshaftov"* (Göngyös-Visonta, 1978), pp. 113-127.

Bentham, H., Harris, J.A., Birch, P. and Short, K.C. (1992) Habitat classification and soil restoration assessment using analysis of soil microbiological and physico-chemical characteristics. *Journal of Applied Ecology*, **29**, 711-718.

Bezkorovainaya, I.N. (1994) Eksperimental'noe izuchenie roli dozhdevykh chervei v mineralizatsii podstilki (Experimental study of the role of earthworms in the mineralisation of litter). *Ekologiya*, No. 5-6, 67-73.

Biondini, M.E., Bonham, C.D. and Redente, E.F. (1985) Relationships between induced successional patterns and soil biological activity of reclaimed areas. *Reclamation and Revegetation Research*, **3**, 323-342.

Birch, P., Bentham, H. and Harris, J.A. (1991) Soil microbial ecosystems: importance for the effective restoration of mined lands. *Proceedings of the Meeting of the American Society for Surface Mining and Reclamation "Reclamation 2000: Technologies for Success"* (Princeton, 1991), pp. 631-640.

Blaga, G., Bunescu, V., Nastea, S., Filip, F. and Dumitru, M. (1979) Proprietăţile fizice, chimice şi biologice ale materialului de haldă rezultat prin degradarea solurilor de pe versanţi (Exploatarea Minieră Căpuş – jud. Cluj) (Physical, chemical and biological properties of spoil heap material resulting from degradation of soils on slopes (The Căpuş Mining Enterprise – Cluj county)). *Simpozionul "Exploatarea Intensivă a Terenurilor Agricole în Pantă"* (Cluj, 1979), pp. 87-94.

Blaga, G., Nastea, S., Răuţă, C., Bunescu, V. and Dumitru, M. (1981) Studiul proprietăţilor fizice, chimice şi biologice ale materialelor de haldă nivelate în scopul recultivării (Study of the physical, chemical, and biological properties of spoil heap materials levelled with the aim of recultivation). *Lucrările Conferinţei Naţionale pentru Ştiinţa Solului* (Braşov, 1979), *Publicaţiile Societăţii Naţionale Române pentru Ştiinţa Solului* (Bucharest), No. 19D, 125-132.

Blaga, G., Nastea, S., Răuţă, C., Bunescu, V., Miclăuş, V. and Dobai, R. (1984) Izuchenie fizicheskikh i khimicheskikh svoistv antropogennogo protosola (materialov otvala, nivelirovannykh v tselyakh rekul'tivatsii) (Study of the physical and chemical properties of an anthropogenic protosol (spoil heap materials levelled with the aim of recultivation)). *Sbornik Dokladov, 8. Mezhdunarodnyi Simpozium "Razrabotka Sposobov Rekul'tivatsii Landshaftov, Narushennykh Promyshlennoi Deyatel'nost'yu"* (Bucharest-Craiova-Tg.Jiu, 1984), pp. 169-176.

Blaga, G, Nastea, S., Mihályi, M., Dumitru, M. and Bunescu, V. (1988) Issledovaniya v svyazi s rekul'tivatsiei sel'skokhozyaistvennykh zemel', narushennyk otkrytymi razrabotkami zheleznoi rudi (Investigations related to recultivation of agricultural soils disturbed by strip mining of iron ore). *Sbornik Dokladov, 9. Mezhdunarodnyi Simpozium "Rekul'tivatsiya Landshaftov, Narushennykh Promyshlennoi Deyatel'nost'yu"* (Kompolt-Gyöngyös, 1988), **1**, 213-220.

Blaga, G., Bunescu, H., Ştefănescu, L., Guéron, I., Blaga, L., Lechinţan, T. and Spînu, M. (1991) Îmbunătăţirea fertilităţii protosolului antropic de la Căpuş (jud. Cluj) prin utilizarea nămolului orăşenesc provenit de la staţia de epurare a municipiului Cluj-Napoca (Improving the fertility of the anthropic protosol in Căpuş (Cluj county) by using sludge derived from the sewage treatment plant of the city of Cluj-Napoca). *Buletinul Institutului Agronomic Cluj-Napoca, Seria Agricultură şi Horticultură*, **45** (1), 93-99.

Blanco, M.D., Casas, A., Cives, J., Galán, P., Gil-Sotres, F., González-Sangregorio, M.V., Guitián-Ojea, F., Leirós, M.C., Saá, A., Seoane, S., Trasar-Cepeda, M.C., Varela, M.C. and Vázquez, C. (1991) Suelos sobre estériles de la mina de lignito de Meirama: un ecosistema de rápida evolución (Soils on spoils of the Meirama lignite mine: an ecosystem of rapid evolution). *Resumen, V Encontro Galego-Portugués de Química* (La Coruña, 1991), pp. 98-99.

Bolewski, A. and Skawina, T. (1972) Próba użycia skal montmorillonitowych do rekultywacji piasków (Experimental application of montmorillonitic clays for recultivation of sands). *Prace Mineralogiczne*, No. 30, 1-69.

Bradshaw, A.D. and Chadwick, M.J. (1980) *The Restoration of Land. The Ecology of Reclamation of Derelict and Degraded Land.* Blackwell, Oxford.

Bunescu, V. and Blaga, G. (1980) Activitatea enzimatică a solurilor şi a materialului de haldă de la Exploatarea Minieră de suprafaţă Căpuş (jud. Cluj) (Enzymatic activity in soils and spoil heap material at the surface Mining Enterprise in Căpuş (Cluj county)). *Buletinul Institutului Agronomic Cluj-Napoca, Seria Agricultură*, **34**, 17-20.

Clark, R.K. and Clark, S.C. (1981) Floristic diversity in relation to soil characteristics in a lead mining complex in the Pennines, England. *The New Phytologist*, **87**, 799-815.

Cristea, V., Kiss, S., Paşca, D., Drăgan-Bularda, M., Crişan, R. and Muntean, V. (1995) Dynamics of the vegetation and evolution of the enzymatic potential of technogenic soils submitted to biological recultivation. *Colloques Phytosociologiques* (Camerino), **24**, 169-180.

Cundell, A.M. (1977) The role of microorganisms in the revegetation of strip-mined land in the western United States. *Journal of Range Management*, **30**, 299-305.

Dan'ko, V.N., Keleberda, T.N. and Pertsev, E.V. (1980) Osnovnye rezul'taty issledovatel'skikh rabot UkrNIILKhA po lesnoi rekul'tivatsii na Ukraine (Basic results of the investigations by the Ukrainian Scientific Research Institute for Forestry and Sylvicultural Reclamation on forest recultivation in Ukraine). *Sbornik Dokladov, 7. Mezhdunarodnyi Simpozium "Rekul'tivatsiya Landshaftov, Narushennykh Promyshlennoi Deyatel'nost'yu"* (Katowice-Zabrze-Konin, 1980), **1**, 69-80.

Daraseliya, N.A. (1979) *Biologicheskaya Aktivnost' Osnovnykh Pochv Zapadnoi Gruzii* (Biological Activity in Representative Soils of Western Georgia), pp. 235-267. Metsniereba, Tbilisi.

Daraseliya, N.A. and Kalatozova, G.B. (1973) Fermentativnaya aktivnost' pri okul'turivanii pochv, narushennykh otkrytymi razrabotkami margantsa (Enzymatic activity during the cultivation of soils disturbed by opencast mining of manganese). *Soobshcheniya Akademii Nauk Gruz.SSR*, **70**, 429-432.

Daraseliya, N.A. and Kalatozova, G.B. (1976) Biologicheskie protsessy pri rekul'tivatsii otvalov Chiaturskikh margantsevykh mestorozhdenii (Biological processes during the recultivation of mine spoils in the Chiatura manganese ore zone). *Trudy Nauchno-Issledovatel'skogo Instituta Pochvovedeniya, Agrokhimii i Melioratsii* (Tbilisi), **17**, 195-213.

Daraseliya, N.A., Kalatozova, G.B. and Lapanashvili, E.F. (1978) Nekotorye biologicheskie pokazateli otsenki effektivnosti priemov rekul'tivatsii zemel' pri razrabotkakh bentonitovykh glin (Some biological indicators for evaluation of the efficiency of soil recultivation measures at bentonitic clay mines). *Soobshcheniya Akademii Nauk Gruz.SSR*, **92**, 425-428.

Drăgan-Bularda, M., Kiss, S., Paşca, D. and Olar-Gherghel, V. (1983) Contribuţii la studierea activităţii enzimatice a solurilor tehnogene (Contribution to the study of enzyme activity in technogenic soils). *Lucrările Conferinţei Naţionale pentru Ştiinţa Solului* (Brăila, 1982), *Publicaţiile Societăţii Naţionale Române pentru Ştiinţa Solului* (Bucharest), No. 21B, 109-117.

Drăgan-Bularda, M., Blaga, G., Kiss, S., Paşca, D., Gherasim, V. and Vulcan, R. (1987) Effect of long-term fertilization on the enzyme activities in a technogenic soil resulted from the recultivation of iron strip mine spoils. *Studia Universitatis Babeş-Bolyai, Biologia*, **32** (2), 47-52.

Edgerton, D.L., Harris, J.A., Birch, P. and Bullock, P. (1995) Linear relationship between aggregate stability and microbial biomass in three restored soils. *Soil Biology & Biochemistry*, **27**, 1499-1501.

Eterevskaya, L.V., Lekhtsier, L.V., Mikhnovskaya, A.D. and Lapta, E.I. (1985) Pochvoobrazovanie v tekhnogennykh landshaftakh na lessovykh porodakh (Soil formation in technogenic landscapes on loess spoils). In *Tekhnogennye Ekosistemy: Organizatsiya i Funktsionirovanie* (A.A. Titlyanova, Ed.), pp. 107-135. Izdatel'stvo Nauka, Sibirskoe Otdelenie, Novosibirsk.

Fresquez, P.R. and Lindemann, W.C. (1982) Soil and rhizosphere microorganisms in amended coal mine spoils. *Soil Science Society of America Journal*, **46**, 751-755.

Fresquez, P.R., Aldon, E.F. and Sorensen, D.L. (1985) Microbial and soil enzyme activities in stockpiled topsoil and coal mine spoil materials. *Proceedings of the National Meeting of the American Society for Surface Mining and Reclamation* (Denver, 1985), pp. 340-345.

Fresquez, P.R., Aldon, E.F. and Lindemann, W.C. (1987) Enzyme activities in reclaimed coal mine spoils and soils. *Landscape and Urban Planning*, **14**, 359-367.

Gel'tser, Yu.G. (1990) Pokazateli biologicheskoi aktivnosti v pochvennykh issledovaniyakh (Indicators of biological activity in soil investigations). *Pochvovedenie*, No. 9, 47-60.

Gel'tser, Yu.G. and Tsvetkova, L.A. (1982) Problema biologicheskoi aktivnosti rekul'tiviruemykh pochv zapadnogo Donbassa (Problem of the biological activity of recultivated soils in the western Donets Basin). In *Biogeotsenologicheskie Issledovaniya Stepnykh Lesov, Ikh Okhrana i Ratsional'noe Ispol'zovanie* (A.P. Travleev, Ed.), pp. 103-112. Gosudarstvennyi Universitet, Dnepropetrovsk.

Gel'tser, Yu.G., Tsvetkova, L.A. and Trofimov, S.Ya. (1985) Osobennosti biologicheskoi aktivnosti podstilok v beloakatsievykh nasazhdeniyakh na estestvennykh i rekul'tiviruemykh pochvakh Prisamar'ya (Particularities of the biological activity of litters in black locust plantations on native and recultivated soils along the Samara River). In *Voprosy Stepnogo Lesovedeniya i Nauchnye Osnovy Lesnoi Rekul'tivatsii Zemel'* (A.P. Travleev, Ed.), pp. 63-70. Gosudarstvennyi Universitet, Dnepropetrovsk.

Gel'tser, Yu.G., Travleev, A.P., Tsvetkova, L.A. and Utinova, I.S. (1986) Primenenie biologicheskikh metodov dlya diagnostiki rekul'tiviruemykh pochv zapadnogo Donbassa (Application of biological methods for diagnosis of recultivated soils in the western Donets Basin). *Tezisy Dokladov 2. S'ezda Pochvovedov i Agrokhimikov Ukr.SSR* (Kharkov, 1986), *Agrokhimiya i Plodorodie Pochv, Rekul'tivatsiya Zemel', Bor'ba s Eroziei Pochv, Obrabotka Pochv*, pp. 110-111.

Gel'tser, Yu.G., Bobrov, A.A., Gel'tser, V.Yu. and Chemerovskaya, E.K. (1988) Biologicheskaya aktivnost' dernovo-podzolistykh pochv tekhnogenno-narushennykh territorii (Biological activity of soddy-podzolic soils of technogenically disturbed territories). *Tezisy Dokladov 3. Vsesoyuznogo Simpoziuma "Biodinamika Pochv"* (Harku-Tallinn, 1988), p.62.

Gel'tser, Yu.G., Tsvetkova, L.A. and Utinova, I.S. (1989) Biologicheskaya aktivnost' pochv rekul'tiviruemykh landshaftov zapadnogo Donbassa (Biological activity in soils of the recultivated landscapes in the western Donets Basin). In *Problemy Okhrany, Ratsional'nogo Ispol'zovaniya i Rekul'tivatsii Chernozemov* (V.A. Kovda, Ed.), pp. 217-227. Izdatel'stvo Nauka, Moscow.

Germanova, N.I., Antipina, G.S. and Fedorets, N.G. (1995) Mikrobiologicheskie aspekty rekul'tivatsii zheleznykh otvalov Kostomukshskogo mestorozhdeniya (Microbiological aspects of the recultivation of iron mine spoils at the Kostomuksha deposit). *Ekologiya*, No. 2, 103-109.

Gilewska, M. (1991) Rekultywacja biologiczna gruntów pogórniczych na przykladzie KWB "Konin" (Biological recultivation of mine soils in the case of the "Konin" Brown Coal Enterprise). *Roczniki Akademii Rolniczej* (Poznan), No. 211, 1-59.

Gilewska, M. and Bender, J. (1978) Biologicheskaya aktivnost' antropogennykh pochv (Biological activity of anthropogenic soils). *Sbornik Dokladov Soveshchaniya "Rekul'tivatsiya Tekhnogennykh Landshaftov"* (Gyöngyös-Visonta, 1978), pp. 189-203.

Gilewska, M. and Bender, J. (1979) Znaczenie zabiegów rekultywacyjnych w ksztaltowaniu enzymatycznej aktywności industrioziemu (The importance of recultivation measures for development of enzymatic activity in industrial soil). *Archiwum Ochrony Środowiska*, No. 1, 49-56.

Gilewska, M. and Bender, J. (1983a) Aktywność enzymatyczna industrioziemów Konińskiego Zaglebia Wegla Brunatnego. I. Aktywność celulazy w gruntach pogórniczych rekultywowanych rolniczo (Enzymatic activity of industrial soils in the Konin Brown Coal Basin. I. Cellulase activity in agriculturally recultivated mine soils). *Archiwum Ochrony Środowiska*, No. 3-4, 157-169.

Gilewska, M. and Bender, J. (1983b) Aktywność enzymatyczna industrioziemów Konińskiego Zaglebia Wegla Brunatnego. II. Aktywność beta-fruktofuranozydazy w gruntach pogórniczych rekultywowanych rolniczo (Enzymatic activity of industrial soils in the Konin Brown Coal Basin. II. Beta-fructofuranosidase activity in agriculturally recultivated mine soils). *Archiwum Ochrony Środowiska*, No. 3-4, 171-178.

Gilewska, M. and Bender, J. (1984) Aktywność enzymatyczna gruntów zwalowiskowych Konińskiego Zaglebia Wegla Brunatnego. IV. Aktywność ureazy w gruntach pogórniczych rekultywowanych rolniczo (Enzymatic activity of spoil heap soils in the Konin Brown Coal Basin. IV. Urease activity in agriculturally recultivated mine soils). *Archiwum Ochrony Środowiska*, No. 2, 125-132.

Gilewska, M. and Wójcik, A. (1984) Aktywność enzymatyczna gruntów pogórniczych pod pionierskimi nasadzeniami leśnymi (Enzymatic activity of mine soils under pioneer forest plantations). *Archiwum Ochrony Środowiska*, No. 3-4, 141-156.

Gil-Sotres, F., Trasar-Cepeda, M.C., Ciardi, C., Ceccanti, B. and Leirós, M.C. (1992) Biochemical characterization of biological activity in very young mine soils. *Biology and Fertility of Soils*, **13**, 25-30.

Gil-Sotres, F., Leirós, M.C., Trasar-Cepeda, M.C., Saá, A. and González-Sangregorio, M.V. (1995a) The importance of soil biochemical properties in the reclamation of lignite mining land. In *Environmental Biogeochemistry. Vol. 1. Wastes and Soil Management (1994)* (B. Ceccanti and C. García, Eds.), pp. 131-170. Istituto per la Chimica del Terreno, Pisa.

Gil-Sotres, F., Leirós, M.C., Trasar-Cepeda, M.C. and Benito, E. (1995b) Modificación del riesgo de contaminación en los estériles recuperados (Modification of the risk of contamination in recovered mine spoils). In *Recuperación de las Escombreras de la Mina de Lignitos de Meirama (A Coruña)* (F. Guitián-Ojea, Ed.), pp. 185-211. Univesidade, Santiago de Compostela.

Golebiowska, J. and Bender, J. (1980) Faktory, obuslovlivayushchie obrazovanie peregnoinogo gorizonta v protsesse rekul'tivatsii otvalov (Factors conditioning the formation of humus horizon during recultivation of spoil heaps). *Sbornik Dokladov, 7. Mezhdunarodnyi Simpozium "Rekul'tivatsiya Landshaftov, Narushennykh Promyshlennoi Deyatel'nost'yu"* (Katowice-Zabrze-Konin, 1980), **3**, 249-262.

Golebiowska, J. and Bender, J. (1983) Czynniki warunkujace powstawanie poziomu próchniczego w procesie rekultywacji zwalowisk (Factors conditioning the formation of humus horizon during recultivation of spoil heaps). *Archiwum Ochrony Środowiska*, No. 1-2, 65-75.

González-Sangregorio, M.V., Trasar-Cepeda, M.C., Leirós, M.C., Gil-Sotres, F. and Guitián-Ojea, F. (1991) Early stages of lignite mine soil genesis: changes in biochemical properties. *Soil Biology & Biochemistry*, **23**, 589-595.

González-Sangregorio, M.V., Trasar-Cepeda, M.C., Saá, A., Gil-Sotres, F. and Leirós, M.C. (1993) Characteristics of phosphatase activity in reclaimed opencast lignite mine soils. *Abstracts, XI International Symposium on Environmental Biogeochemistry* (Salamanca, 1993).

Greszta, J. (1973) The recultivation of post-industrial territories. In *Protection of Man's Natural Environment*, pp. 396-417. Polish Academy of Sciences, Warsaw.

Greszta, J. and Olszowski, J. (1974) The effect of fertilization on the biological activity of the soils of former sand opencasts. *Ekologia Polska*, **22**, 339-368.

Harris, J.A. (1991) The biology of soils in urban areas. In *Soils in the Urban Environment* (P. Bullock and P.J. Gregory, Eds.), pp. 139-152. Blackwell, Oxford.

Harris, J.A. and Birch, P. (1989) Soil microbial activity in opencast coal mine restorations. *Soil Use and Management*, **5**, 155-160.

Harris, J.A. and Birch, P. (1992) Land reclamation and restoration. In *Microbial Control of Pollution* (J.C. Fry, G.M. Gadd, R.A. Herbert, C.W. Jones and I.A. Watson-Craik, Eds.), pp. 269-291. Cambridge University Press, Cambridge.

Harris, J.A. and Hill, T.C.J. (1995) Soil biotic communities and new woodland. In *The Ecology of Woodland Creation* (R. Ferris-Kaan, Ed.), pp. 91-112. Wiley, Chichester.

Haubold, M., Henkes, L. and Schröder, D. (1987) Qualität und Entwicklung rekultivierter Böden aus Löß (Quality and evolution of recultivated soils from loess). *Mitteilungen der Deutschen Bodenkundlichen Gesellschaft*, **53**, 173-178.

Hazuk, A. (1967) Gestaltung der enzymatischen Aktivität von Versatzsand-Restlöchern (Development of enzyme activity in filling sand floor drift). *Referaten-Sammlung, 3. Internationales Symposium über Rekultivierung der durch den Bergbau beschädigten Böden* (Prague, 1967). Cited in Bolewski and Skawina (1972).

Hersman, L.E. and Klein, D.A. (1979) Retorted oil shale effects on soil microbiological characteristics. *Journal of Environmental Quality*, **8**, 520-524.

Hersman, L.E. and Temple, K.L. (1979) Comparison of ATP, phosphatase, pectinolyase, and respiration as indicators of microbial activity in reclaimed coal strip mine spoils. *Soil Science*, **127**, 70-73.

Idrisova, Z.N. (1984) Vliyanie rekul'tivatsii zemel', narushennykh pri stroitel'stve magistral'-nykh truboprovodov na plodorodie pochv i urozhainost' sel'skokhozyaistvennykh kul'tur (The effect of the recultivation of soils disturbed by pipeline construction on soil fertility and productivity of agricultural crops). In *Pochvennye Usloviya i Effektivnost' Udobrenii* (F.Sh. Garifullin, Ed.), pp. 43-50. Sel'skokhozyaistvennyi Institut, Ufa.

Idrisova, Z.N. (1990) Vliyanie biologicheskoi rekul'tivatsii na uluchshenie osnovnykh para-metrov plodorodiya tekhnogenno narushennykh pochv (The effect of biological recul-tivation on the improvement of the basic fertility parameters in technogenically disturbed soils). In *Okul'turivanie i Intensivnoe Ispol'zovanie Meliorirovannykh Zemel' Nechernozemnoi Zony*, pp. 14-15. Sel'skokhozyaistvennyi Institut, Gorkii.

Idrisova, Z.N., Garifullin, F.Sh. and Ishem'yarov, A.Sh. (1986a) Fosfatnoe sostoyanie rekul'tiviruemykh pochv (Phosphate status in recultivated soils). *Vestnik Sel'sko-khozyaistvennoi Nauki*, No. 6, 53-59.

Idrisova, Z.N., Garifullin, F.Sh. and Ishem'yarov, A.Sh. (1986b) Azotnoe sostoyanie rekul'tiviruemykh pochv (Nitrogen status in recultivated soils). *Agrokhimiya*, No. 12, 14-19.

Idrisova, Z.N., Garifullin, F.Sh. and Ishem'yarov, A.Sh. (1987) Dopustimyi uroven' razbavleniya gumusogo sloya vyshchelochennogo chernozema vskryshnoi porodoi pri stroitel'stve magistral'nykh truboprovodov (Permissible level for diluting the humus layer of leached chernozem with overburden during pipeline construction). *Pochvovedenie*, No. 6, 82-88.

Ishem'yarov, A.Sh., Garifullin, F.Sh. and Idrisova, Z.N. (1984) Issledovanie pochvennykh kharakteristik putem eksperimental'nogo modelirovaniya protsessa rekul'tivatsii (Investigation of soil characteristics by experimental modelling of the recultivation process). In *Povyshenie Effektivnosti Rekul'tivatsii Zemel', Narushennykh pri Stroitel'stve Truboprovodov* (A.P. Iofinov, Ed.), pp. 26-31. Sel'skokhozyaistvennyi Institut, Ufa.

Katzur, J. and Haubold-Rosar, M. (1996) Amelioration and reforestation of sulfurous mine soils in Lusatia (Eastern Germany). *Water, Air, and Soil Pollution*, **91**, 17-32.

Keleberda, T.N. (1973) K otsenke lesorastitel'nykh svoistv rekul'tiviruemykh zemel' po dannym aktivnosti fermenta katalazy (Evaluation of the forest-growing properties of recultivated soils by values of the activity of the catalase enzyme). *Lesnoi Zhurnal*, No. 3, 158-161.

Keleberda, T.N. (1976) Fitomelioratsiya tekhnogennykh gruntov i invertaznaya aktivnost' (Phytoreclamation of technogenic soils and invertase activity). *Pochvovedenie*, No. 10, 126-131.

Keleberda, T.N. (1977) Fermentativnaya aktivnost' kak bioindikator izmeneniya plodorodiya tekhnogennykh gruntov putem fitomelioratsii (Enzymatic activity as a bioindicator of the change in fertility of technogenic soils by phytoreclamation). *Tezisy Dokladov 5. Delegatskogo S'ezda Vsesoyuznogo Obshchestva Pochvovedov* (Minsk, 1977), No. 2, 271-272.

Keleberda, T.N. (1978) Pochvoobrazovanie na promyshlennykh otvalakh pod lesnoi rastitel'nost'yu (Soil formation on industrial spoil heaps under forest vegetation). *Pochvovedenie*, No. 9, 109-115.

Keleberda, T.N. (1979) Melioratsiya promyshlennykh otvalov listvennymi nasazhdeniyami (Reclamation of industrial spoil heaps by frondous plantations). *Vestnik Sel'skokhozyaistvennoi Nauki*, No. 2, 87-90.

Keleberda, T.N. and Dan'ko, V.N. (1975) Biologicheskaya aktivnost' meliorirovannykh gruntov promyshlennykh razrabotok (Biological activity of reclaimed soils at industrial workings). *Tezisy Dokladov Koordinatsionnogo Soveshchaniya "Rekul'tivatsiya Zemel', Narushennykh pri Dobyche Poleznykh Iskopaemykh"* (Tartu, 1975), pp. 245-249.

Keleberda, T.N. and Drugov, A.N. (1984) Optimizatsiya tekhnogennykh landshaftov otkrytykh vyrabotok Donbassa (Optimisation of the technogenic landscapes at opencast mine workings in the Donets Basin). *Rasteniya i Promyshlennaya Sreda* (Sverdlovsk = Ekaterinburg), No. 10, 92-97.

Keleberda, T.N., Verbin, A.E. and Zharoms'kii, V.Ya. (1974) Polipshennya lisoroslinnikh vlastivostei gruntiv vidvaliv kul'turoyu siderativ (Improving the forest-growing properties of soils of mine spoil heaps by green manure crops). *Visnik Sil's'kogospodars'koi Nauki*, No. 12, 53-56.

Khazanov, M.I. (1975) *Iskusstvennye Grunty, Ikh Obrazovanie i Svoistva* (Artificial Soils, Their Formation and Properties). Izdatel'stvo Nauka, Moscow.

Kiss, S., Drăgan-Bularda, M. and Rădulescu, D. (1975) Biological significance of enzymes accumulated in soil. *Advances in Agronomy*, **27**, 25-87.

Kiss, S., Drăgan-Bularda, M. and Paşca, D. (1985) Enzymological study of the evolution of technogenic soils. *Evolution and Adaptation* (Cluj), **2**, 159-186.

Kiss, S., Drăgan-Bularda, M. and Paşca, D. (1989a) Enzymology of the recultivation of technogenic soils. *Advances in Agronomy*, **42**, 229-278.

Kiss, S., Drăgan-Bularda, M., Paşca, D., Blaga, G., Zborovschi, E., Pintea, H., Crişan, R., Muntean, V. and Mitroescu, S. (1989b) Enzimologia recultivării biologice a solului iazului de decantare de la Exploatarea Minieră Rodna (Enzymology of the biological recultivation of spoils from the decantation pond at the Rodna Mining Enterprise). *Lucrările Conferinţei Naţionale pentru Ştiinţa Solului* (Piteşti, 1988), *Publicaţiile Societăţii Naţionale Române pentru Ştiinţa Solului* (Bucharest), No. 26A, 211-218.

Kiss, S., Paşca, D., Drăgan-Bularda, M., Cristea, V., Blaga, G., Crişan, R., Muntean, V., Zborovschi, E. and Mitroescu, S. (1990) Enzymological analysis of lead and zinc mine spoils submitted to biological recultivation. *Studia Universitatis Babeş-Bolyai, Biologia*, **35** (2), 70-79.

Kiss, S., Drăgan-Bularda, M. and Paşca, D. (1991) Enzymology of the evolution of some technogenic soils. *Evolution and Adaptation* (Cluj), **4**, 125-132.

Kiss, S., Paşca, D., Drăgan-Bularda, M., Crişan, R. and Muntean, V. (1992) Enzymological evaluation of the efficiency of the measures applied for biological recultivation of lead and zinc mine spoils. *Studia Universitatis Babeş-Bolyai, Biologia*, **37** (2), 103-107.

Kiss, S., Drăgan-Bularda, M. and Paşca, D. (1993) *Enzymology of Technogenic Soils*. Casa Cărţii de Ştiinţă, Cluj.

Kiss, S., Paşca, D., Drăgan-Bularda, M., Crişan, R. and Muntean, V. (1994) Evaluarea enzimologică a eficienţei biotehnologiei de recultivare a haldei de steril de la o mină de plumb şi zinc (Enzymological evaluation of the efficiency of the biotechnological recultivation of spoils at a lead and zinc mine). *Proceedings of the 8th National Symposium of Industrial Microbiology and Biotechnology* (Bucharest, 1994), pp. 357-363.

Kiss, S., Drăgan-Bularda, M. and Paşca, D. (1996) Enzymology of technogenic soils. *Studia Universitatis Babeş-Bolyai, Biologia*, **41** (1-2), 3-21.

Klein, D.A., Sorensen, D.L. and Metzger, W. (1982) Soil microorganisms and management of retorted shale reclamation. In *Revegetation Studies on Oil Shale Related Disturbances in Colorado* (E.F. Redente and C.W. Cook, Eds.), pp. 27-44. Department of Range Science, Colorado State University, Fort Collins.

Klein, D.A., Sorensen, D.L. and Redente, E.F. (1985) Soil enzymes: a predictor of reclamation potential and progress. In *Soil Reclamation Processes. Microbiological Analyses and Applications* (R.L. Tate III and D.A. Klein, Eds.), pp. 141-171, Dekker, New York.

Klevenskaya, I.L. (1989) Mikrobo-rastitel'nye assotsiatsii tekhnogennykh ekosistem (Microbe-plant associations of technogenic ecosystems). *Tezisy Dokladov 8. Vsesoyuznogo S'ezda Pochvovedov* (Novosibirsk, 1989), **6**, 122-128.

Klevenskaya, I.L. (1992) Evolutsiya mikrobotsenozov i ikh funktsii (Evolution of the microbial coenoses and their functions). In *Ekologiya i Rekul'tivatsiya Tekhnogennykh Landshaftov* (V.M. Kurachev, Ed.) pp. 149-198. Izdatel'stvo Nauka, Sibirskoe Otdelenie, Novosibirsk.

Klevenskaya, I.L., Trofimov, S.S. and Kandrashin, E.R. (1986) Mikrobo-rastitel'nye assotsiatsii – kak konsortsionnye edinitsy tekhnogennykh ekosistem (Microbe-plant associations as consortium units of technogenic ecosystems). *Tezisy Dokladov 3. Vsesoyuznoi Nauchnoi Konferentsii "Mikroorganizmy v Sel'skom Khozyaistve"* (Moscow, 1986), pp. 89-90.

Klubek, B., Carlson, C.L., Oliver, J. and Adriano, D.C. (1992) Characterization of microbial abundance and activity from three coal ash basins. *Soil Biology & Biochemistry*, **24**, 1119-1125.

Kolk, A. and Hüttl, R.F. (1996) Awendung bodenmikrobiologischer Untersuchungsmethoden zur Charakterisierung der mikrobiellen Biomasse und ihrer Aktivitäten von tertiären, aschemeliorierten Kippensubstraten des Niederlausitzer Braunkohlenreviers (Application of soil microbiological research methods for characterisation of the microbial biomass and of its activities in tertiary, ash-reclaimed spoils of the Niederlausitz (Lower Lusatian) Brown Coal District). *Mitteilungen der Deutschen Bodenkundlichen Gesellschaft*, **81**, 291-294.

Korsunova, T.M. and Shugalei, L.S. (1986) Otsenka sovremennogo sostoyaniya i antropogennykh izmenenii pochv zapadnoi zony KATEKa (Evaluation of the present status and of the anthropogenic changes of soils in the western zone of the Kansk-Achinsk Fuel-Energetic Complex). *Tezisy Dokladov Konferentsii "Zemel'no-Otsenochnye Problemy Sibiri i Dal'nego Vostoka"* (Barnaul, 1986), **2**, 156-157.

Leirós, M.C., Varela, M.C., Gil-Sotres, F., Trasar-Cepeda, M.C. and Seoane, S. (1995) Los suelos de la mina (The mine soils). In *Recuperación de las Escombreras de la Mina de Lignitos de Meirama (A Coruña)* (F. Guitián-Ojea, Ed.), pp. 125-157. Universidade, Santiago de Compostela.

Leirós, M.C., Gil-Sotres, F., Trasar-Cepeda, M.C., Saá, A. and Seoane, S. (1996) Soil recovery at the Meirama opencast lignite mine in northwest Spain: a comparison of the effectiveness of cattle slurry and inorganic fertilizer. *Water, Air, and Soil Pollution*, **91**, 109-124.

Lessmann, U. and Krämer, F. (1985) Mikrobiologische Aktivität eines Rohbodens aus Löß im Rheinischen Braunkohlengebiet (Microbiological activity of a raw soil from loess in the Rhine Brown Coal Region). *Landwirtschaftliche Forschung*, **38**, 110-114.

Lindemann, W.C., Lindsey, D.L. and Fresquez, P.R. (1984) Amendment of mine spoil to increase the number and activity of microorganisms. *Soil Science Society of America Journal*, **48**, 574-578.

Lukina, N.N. (1991) Fermentativnaya aktivnost' rekul'tiviruemykh otvalov v Podmoskovskom basseine (Enzymatic activity of recultivated spoil heaps in the Moscow Basin). *Vestnik Moskovskogo Universiteta, Seriya 17. Pochvovedenie*, No. 1, 63-66.

Machulla, S. and Hickisch, B. (1988) Bodenmikrobiologische Charakterisierung unterschiedlich alter Kippböden (Soil microbiological characterisation of differently-aged tipper soils). *Tagungsberichte, Akademie der Landwirtschaftswissenschaften* (Berlin), No. 269, 333-336.

Mikhnovskaya, A.D. (1981) Mikrobiologicheskie protsessy pri rekul'tivatsii gornykh porod burougol'nykh razrabotok (Microbiological processes during recultivation of spoils at brown coal mines). *Tezisy Dokladov 6. Delegatskogo S'ezda Vsesoyuznogo Obshchestva Pochvovedov* (Tbilisi, 1981), **2**, 190-191.

Mikhnovskaya, A.D. and Lapta, E.I. (1989) Mikrobiologicheskie protsessy transformatsii azota lessovykh otvalov i vozmoshnosti ikh regulirovaniya v khode rekul'tivatsii (Microbiological processes of the transormation of nitrogen in loess spoil heaps and possibilities for their regulation during recultivation). In *Ispol'zovanie Dostizhenii Mikrobiologicheskoi Nauki v Povyshenii Effektivnosti Zemledeliya*, pp. 26-30. Yuzhnoe Otdelenie VASKhNIL, Kiev.

Miller, R.M. (1978) Laboratory research studies on the reestablishment of below-ground ecosystems. In *Land Reclamation Program. Annual Report (July 1976 – October 1977)*, pp. 95-119. Argonne National Laboratory, Argonne, Illinois.

Müller, R., Schneider, R. and Schröder, D. (1988) Physikalische, chemische und biologische Eigenschaften trocken rekultivierter Lößböden unter Wald-, Grünland- und Ackernutzung (Physical, chemical, and biological properties of dry-recultivated loess soils under forest, grassland and arable land use). *Mitteilungen der Deutschen Bodenkundlichen Gesellschaft*, **56**, 387-392.

Naplekova, N.N. and Gorskikh, V.G. (1986) Vliyanie zhidkogo navoza na mikrofloru nizinnykh torfyanikov i urozhai kormovykh kul'tur (Effect of liquid dung on the microflora of low moor peatlands and yield of fodder plants). In *Mikrobiologicheskie Protsessy v*

Pochvakh i Urozhainost' Sel'skokhozyaistvennykh Kul'tur, p. 264. Institut Botaniki Akademii Nauk Lit.SSR, Vilnius.

Naplekova, N.N., Kandrashin, E.R., Trofimov, S.S. and Fatkulin, F.A. (1982) Formirovanie mikrobnykh tsenozov pochv tekhnogennykh landshaftov Kuzbassa (Formation of microbial coenoses in soils of the technogenic landscapes in the Kuznetsk Basin). *Izvestiya Sibirskogo Otdeleniya Akademii Nauk SSSR, Seriya Biologicheskie Nauki*, No. 1/5, 69-73.

Naplekova, N.N., Kandrashin, E.R., Trofimov, S.S. and Fatkulin, F.A. (1983) Microbial coenoses formation of the Kuzbass technogenic landscapes. In *Recultivation of Technogenous Areas* (J. Szegi, Ed.), pp. 177-183. Mátraalja Coal Mining Co., Gyöngyös.

Naplekova, N.N., Trofimov, S.S., Kandrashin, E.R., Fatkulin, F.A. and Barannik, L.P. (1985) Mikrobnye tsenozy tekhnogennykh ekosistem Sibiri (Microbial coenoses in technogenic ecosystems of Siberia). In *Tekhnogennye Ekosistemy: Organizatsiya i Funktsionirovanie* (A.A. Titlyanova, Ed.), pp. 38-69. Izdatel'stvo Nauka, Sibirskoe Otdelenie, Novosibirsk.

Naprasnikova, E.V. (1983) Biokhimicheskaya aktivnost' molodykh tekhnogennykh pochv (Biochemical activity of young technogenic soils). In *Dinamika Veshchestva v Geosistemakh* (V.A. Snytko, Ed.), pp. 55-61. Institut Geografii, Irkutsk.

Naprasnikova, E.V. (1985a) Formirovanie biogennykh svoistv narushennykh zemel' posle otkrytoi dobychi uglya (Development of biogenic properties in disturbed soils after opencast mining of coal). *Transactions of the 8th International Symposium "Humus et Planta"* (Prague, 1983), pp. 484-486.

Naprasnikova, E.V. (1985b) Formirovanie biogennykh svoistv narushennykh zemel' posle otkrytoi dobychi uglya (Development of biogenic properties in disturbed soils after opencast mining of coal). *Tezisy Dokladov 7. Delegatskogo S'ezda Vsesoyuznogo Obshchestva Pochvovedov* (Tashkent, 1985), **2**, 192.

Naprasnikova, E.V. (1986) Biologicheskaya aktivnost' pochv v protsesse rekul'tivatsii posle dobychi uglya (Biological activity in soils during recultivation after coal mining). In *Mikrobiologicheskie Protsessy v Pochvakh i Urozhainost' Sel'skokhozyaistvennykh Kul'tur*, pp. 265-266. Institut Botaniki Akademii Nauk Lit.SSR, Vilnius.

Naprasnikova, E.V. (1987a) Biological activity of the plant rhizosphere in young technogenic soils. *Abstracts, International Symposium "Interrelationships between Microorganisms and Plants in Soil"* (Liblice, 1987), p. 81.

Naprasnikova, E.V. (1987b) Biogennye svoistva narushennykh zemel' posle dobychi uglya (Biogenic properties of disturbed soils after coal mining). *Geografiya i Prirodnye Resursy*, No. 2, 98-102.

Naprasnikova, E.V. (1988) Mikrobotsenozy i biokhimicheskaya aktivnost' narushennykh zemel' v usloviyakh tekhnogeneza (zona KATEKa) (Microbial coenoses and biochemical activity in disturbed soils under conditions of technogenesis (zone of the Kansk-Achinsk Fuel-Energetic Complex)). *Tezisy Dokladov 3. Vsesoyuznogo Simpoziuma "Biodinamika Pochv"* (Harku-Tallinn, 1988), p. 115.

Naprasnikova, E.V. (1993) Biokhimicheskie i mikrobiologicheskie pokazateli ekologicheskikh funktsii pochv (Biochemical and microbiological indices of the ecological functions of soils). In *Pochvy i Povyshenie Ikh Proizvoditel'noi Sposobnosti*, pp. 57-60. Krasnoyarskii Nauchno-Issledovatel'skii Institut Sel'skogo Khozyaistva, Novosibirsk.

Naprasnikova, E.V. (1995) Opredelenie biokhimicheskogo potentsiala pochvy yuzhnoi chasti Leno-Angarskogo plato kak pokazatelya ikh sostoyaniya (Determination of the biochemical potential of soils in the southern part of the Lena-Angarsk Plateau, as an indicator of their status). *Geografiya i Prirodnye Resursy*, No. 3, 195-197.

Naprasnikova, E.V. and Makarova, A.P. (1986) Mikrobotsenozy i biologicheskaya aktivnost' narushennykh zemel' v protsesse rekul'tivatsii (zona KATEKa) (Microbial coenoses and biological activity in disturbed soils during recultivation (zone of the Kansk-Achinsk Fuel-Energetic Complex)). *Tezisy Dokladov 3. Vsesoyuznoi Nauchnoi Konferentsii "Mikroorganizmy v Sel'skom Khozyaistve"* (Moscow, 1986), p. 128.

Naprasnikova, E.V. and Makarova, A.P. (1992) Mikroorganizmy i biokhimicheskie svoistva rizosfery rastenii rekul'tiviruemykh zemel' zony KATEKa (Microorganisms and biochemical properties of the rhizosphere of plants on the recultivated soils in the zone of the Kansk-Achinsk Fuel-Energetic Complex). *Tezisy Dokladov 4. Vsesoyuznoi Nauchnoi Konferentsii "Mikroorganizmy v Sel'skom Khozyaistve"* (Pushchino, 1992), p. 144-145.

Naprasnikova, E.V., Nikitina, Z.L. and Makarova, A.P. (1982) Mikroflora formiruyushchikhsya biogeotsenozov na narushennykh zemlyakh posle ugledobychi (zona

KATEKa) (Microflora of biogeocoenoses developing on disturbed soils after coal mining (zone of the Kansk-Achinsk Fuel-Energetic Complex)). *Materialy Vsesoyuznogo Simpoziuma "Mikroorganizmy kak Komponent Biogeotsenoza"* (Alma-Ata, 1982), p. 177-179.

Nastea, S., Răuţă, C., Marin, N. and Blaja, I. (1973) Degradarea terenurilor agricole ca urmare a exploatărilor miniere la zi şi măsurile de recultivare a lor (Degradation of agricultural lands as a consequence of opencast minings and the measures for their recultivation). *Ştiinţa Solului*, No. 4, 17-26.

Nekrasova, K.A. and Krylova, L.N. (1975) Pochvennye vodorosli kak pokazateli toksichnosti i biologicheskoi aktivnosti pochv (Soil algae as indicators of toxicity and biological activity of soils). *Tezisy Dokladov, 5. S'ezd Vsesoyuznogo Mikrobiologicheskogo Obshchestva, Sektsiya: Sel'skokhozyaistvennaya Mikrobiologiya* (Erevan, 1975), pp. 116-118.

Nekrasova, K.A., Krylova, L.N., Vertogradskaya, I.A., Zverkov, Yu.V., Kormshchikova, L.G., Artem'eva, T.I. and Borisovich, T.M. (1977) Izmenenie biologicheskoi aktivnosti vyrabotannykh torfyanikov na pervykh etapakh ikh okul'turivaniya (Changes in the biological activity of exhausted peatlands during the first stages of their cultivation), In *Razvitie i Znachenie Vodoroslei v Pochvakh Nechernozemnoi Zony* (E.A. Shtina, Ed.), pp. 47-48. Sel'skokhozyaistvennyi Institut, Perm.

Nekrasova, K.A., Artem'eva, T.I. and Krylova, L.N. (1979) Dinamika chislennosti pedobiontov i nekotorykh biokhimicheskikh pokazatelei vyrabotannykh torfyanikov na pervykh etapakh ikh osvoeniya (Dynamics of the number of soil biota and of some biochemical indicators in exhausted peatlands during their cultivation). In *Sezonnaya Dinamika Pochvennykh Protsessov*, pp. 105-107. Akademiya Nauk Est.SSR, Tallinn.

Osmańczyk, D. (1980) Wplyw roślinności zielnej na aktywność enzymatyczna rekultywowanych odpadów górnictwa weglowego (Effect of herbaceous vegetation on the enzymatic activity of recultivated coal mine wastes). *Archiwum Ochrony Środowiska*, No. 3-4, 175-181.

Osmańczyk-Krasa, D. (1984a) Enzyme activity of reclaimed coal mining wastes. In *Soil Biology and Conservation of the Biosphere* (J. Szegi, Ed.), pp. 837-846. Akadémiai Kiadó, Budapest.

Osmańczyk-Krasa, D. (1984b) Aktywność enzymatyczna odpadów wegla kamiennego rekultywowanych roślinnościa drzewiasta (Enzymatic activity of coal mine wastes recultivated with ligneous vegetation). *Archiwum Ochrony Środowiska*, No. 1, 177-182.

Osmańczyk-Krasa, D. (1987) Some factors of the soil forming process in wastes after deep coal mining. In *Soil Biology and Conservation of the Biosphere* (J. Szegi, Ed.), pp. 671-678. Akadémiai Kiadó, Budapest.

Panas, R.N. (1989) *Agroekologicheskie Osnovy Rekul'tivatsii Zemel' (na Primere Mestorozhdenii Sery Predkarpatskogo Basseina)* (Agro-Ecological Bases of the Recultivation of Soils (in the Case of the Sulphur Mines in the Cis-Carpathian Basin)). Izdatel'stvo Universiteta, Lvov.

Pancholy, S.K., Rice, E.L. and Turner, J.A. (1975) Soil factors preventing revegetation of a denuded area near an abandoned zinc smelter in Oklahoma. *Journal of Applied Ecology*, **12**, 337-342.

Paşca, D., Kiss, S., Drăgan-Bularda, M., Crişan, R. and Muntean, V. (1994) Evolution of the enzymatic potential in lead and zinc mine spoils submitted to biological recultivation. *Studia Universitatis Babeş-Bolyai, Biologia*, **39** (1), 95-101.

Paşca, D., Cristea, V., Crişan, R., Muntean, V., Kiss, S. and Drăgan-Bularda, M. (1996) Potenţialul enzimatic al unui sol tehnogen supus revegetării cu cătină albă (*Hippophaë rhamnoides* L.) (Enzymatic potential of a technogenic soil submitted to revegetation with sea buckthorn (*Hippophaë rhamnoides* L.)). *Studia Universitatis Babeş-Bolyai, Biologia*, **41** (1-2), 183-193.

Paşca, D., Crişan, R., Muntean, V., Simihăian, M., Kiss, S. and Drăgan-Bularda, M. (1997a) Evaluarea eficienţei bioremedierii unui sol tehnogen pe baza indicatorilor enzimatici şi proprietăţilor fizico-chimice (Evaluation of the efficiency of bioremediation of a technogenic soil on the basis of enzymatic indicators and physico-chemical properties). *Lucrările Conferinţei Naţionale pentru Ştiinţa Solului* (Bucharest, 1997), *Publicaţiile Societăţii Naţionale Române pentru Ştiinţa Solului* (Bucharest), **29B**, 55-61.

Paşca, D., Crişan, R., Muntean, V., Kiss, S., Popovici, I., Fabian, L., Harşia, T. and Ciobanu, M. (1997b) Monitoringul ecologic al solurilor tehnogene (Ecological monitoring of the technogenic soils). *Studii şi Cercetări. Ştiinţele Naturii* (Bistriţa), **3**, 211-223.

Paşca, D., Crişan, R., Muntean, V., Popovici, I., Kiss, S. and Drăgan-Bularda, M. (1998) Enzymological and microbiological study of the evolution of a technogenic soil submitted to biological recultivation at the lead and zinc mine in Rodna (Romania). *Soil & Tillage Research*, in press.

Păunescu, A.D. and Ştefanic, G. (1989) Cercetări privind microbiologia şi enzimologia haldelor de cenuşă de termocentrală luate în cultură la Işalniţa-Craiova (Investigations concerning microbiology and enzymology of the cultivated power plant ash heaps at Işalniţa-Craiova). *Ştiinţa Solului*, No. 2, 32-37.

Pereverzev, V.N. and Podlesnaya, N.I. (1986) *Biologicheskaya Rekul'tivatsiya Promyshlennykh Otvalov na Krainem Severe* (Biological Recultivation of Industrial Spoil Heaps in the Far North). Izdatel'stvo Kol'skogo Filiala Akademii Nauk SSSR, Apatity.

Persson, T. and Funke, B. (1986) The microbiology of topsoil storage at a North Dakota stripmining site. *Proceedings of the North Dakota Academy of Science*, **40**, 122.

Persson, T.J. and Funke, B.R. (1988) Microbiology of stored topsoil at North Dakota stripmining sites. *Arid Soil Research and Rehabilitation*, **2**, 235-250.

Peterson, N.V., Kurylyak, E.K. and Panas, R.N. (1976) Mikrobnoe naselenie i fermentativnaya aktivnost' v rekul'tiviruemom otval'nom grunte otkrytoi razrabotki sery (Microbial population and enzymatic activity in recultivated spoil heap soil at a sulphur strip mine). *Tezisy Dokladov 4. Vsesoyuznogo Soveshchaniya po Fiziologo-Biokhimicheskim Osnovam Vzaimodeistviya Rastenii v Fitotsenozakh* (Kiev, 1976), pp. 117-118.

Peterson, N.V., Kurylyak, E.K. and Dubkovetskii, S.V. (1979) Degidrogenaznaya aktivnost' rekul'tiviruemykh zemel' (Dehydrogenase activity of recultivated soils). *Mikrobiologicheskii Zhurnal* (Kiev), No. 2, 129-134.

Petrov, M.Yu., Vasilyauskene, V.A. and Shleinis, R.I. (1994) Napravlennost' elementarnykh pochvoobrazovatel'nykh protsessov pri sel'skokhozyaistvennoi rekul'tivatsii graviinykh kar'erov (Trends of the elementary soil-forming processes during agricultural recultivation of gravel quarries). *Pochvovedenie*, No. 1, 140-144.

Polyakov, M.I., Boiko, A.T. and Shvedovskii, P.V. (1987) *Rekul'tivatsiya Zemel' i Okhrana Prirody* (Recultivation of Soils and Protection of Nature), pp. 93-109. Izdatel'stvo Uradzhai, Minsk.

Popovici, I. (1993) The use of soil nematodes to assess the quality of technogenic soil rehabilitation. *Studia Universitatis Babeş-Bolyai, Biologia*, **38** (1-2), 43-53.

Popovici, I. (1995) Soil nematodes used in biomonitoring of technogenic soil rehabilitation in Romania. *Nematologica*, **41**, 331.

Popovici, I., Paşca, D., Kiss, S., Crişan, R., Muntean, V. and Drăgan-Bularda, M. (1995) Enzymatic potential and nematodes as indicators of the bioremediation efficiency of technogenic soils. *Abstracts, 7th European Ecological Congress* (Budapest, 1995), p. 104.

Potapova, S.S. (1989) Biokhimicheskaya aktivnost' vyrabotannykh torfyanikov (Biochemical activity of exhausted peatlands). *Tezisy Dokladov 8. Vsesoyuznogo S'ezda Pochvovedov* (Novosibirsk), **2**, 325.

Ras'kova, N.V., Gel'tser, Yu.G., Tsvetkova, L.A. and Trofimov, S.Ya. (1984) Sravnitel'naya otsenka biologicheskoi aktivnosti rekul'tiviruemykh zemel' zapadnogo Donbassa (Comparative evaluation of the biological activity of recultivated soils in the western Donets Basin). *Tezisy Dokladov, Vsesoyuznaya Shkola "Vliyanie Promyshlennykh Predpriyatii na Okruzhayushchuyu Sredu"* (Zvenigorod, 1984), pp. 157-158.

Răuţă, C. (1988) Particularităţile interacţiunii dintre activitatea antropică şi mediul înconjurător (Particularities of the interaction between anthropic activity and environment). *Ştiinţa Solului*, No. 1, 3-7.

Redente, E.F. and Cook, C.W. (Eds.) (1986) *Structural and Functional Changes in Early Successional Stages of a Semiarid Ecosystem*. Department of Range Science, Colorado State University, Fort Collins.

Ross, D.J., Speir, T.W., Tate, K.R., Cairns, A., Meyrick, K.F. and Pansier, E.A. (1982) Restoration of pasture after topsoil removal: effects on soil carbon and nitrogen mineralization, microbial biomass and enzyme activities. *Soil Biology & Biochemistry*, **14**, 575-581.

Ross, D.J., Speir, T.W., Tate, K.R., Cowling, J.C. and Watts, H.M. (1984) Restoration of pasture after topsoil removal: changes in soil biochemical properties over a 5-year period – a note. *New Zealand Journal of Science*, **27**, 419-422.

Ross, D.J., Speir, T.W., Cowling, J.C. and Feltham, C.W. (1992) Soil restoration under pasture after lignite mining: management effects on soil biochemical properties and their relationships with herbage yields. *Plant and Soil*, **140**, 85-97.

Ross, D.J., McQueen, D.J. and Kettles, H.A. (1994) Land rehabilitation under pasture on volcanic parent materials: changes in soil microbial biomass and C and N metabolism. *Australian Journal of Soil Research*, **32**, 1321-1337.

Rowell, M.J. and Florence, L.Z. (1993) Characteristics associated with differences between undisturbed and industrially-disturbed soils. *Soil Biology & Biochemistry*, **25**, 1499-1511.

Rtskhiladze, T.G., Rostiashvili, K.A. and Lapanashvili, E.F. (1981) Biologicheskaya aktivnost' rekul'tiviruemykh ploshadei bentonitovykh glin i kvartsevykh peskov (Biological activity in recultivated areas of bentonitic clays and quartz sands). *Tezisy Dokladov 6. Delegatskogo S'ezda Vsesoyuznogo Obshchestva Pochvovedov* (Tbilisi, 1981), **2**, 189.

Saá, A., García, C., Cives, J., Gil-Sotres, F., Trasar-Cepeda, M.C., Leirós, M.C. and Ceccanti, B. (1994) Biochemical properties, development and humification in minesoils amended with cattle slurry. *Transactions, 15th World Congress of Soil Science* (Acapulco, 1994), **3a**, 210-220.

Schneider, R. and Schröder, D. (1995) Development of soil structure and physical, mechanical, chemical and biological properties of reclaimed soils derived from loess and harbour sediments. In *Soil Structure. Its Development and Function* (K.H. Hartge and B.A. Stewart, Eds.), pp. 135-158. Lewis, Boca Raton.

Schneider, R., Schneider, H. and Schröder, D. (1995) Beziehung zwischen Bodenorganismen und Gefügeentwicklung auf Löß-Neuland unter Acker- und Waldnutzung (Relationship between soil organisms and soil structure in new reclaimed loess-derived soil under agricultural and forestal use). *Zeitschrift für Pflanzenernährung und Bodenkunde*, **158**, 197-204.

Schröder, D. (1986) Probleme der Bodenschätzung bei rekultivierten Böden aus Löß (Problems of soil assessment of recultivated soils from loess). *Zeitschrift für Kulturtechnik und Flurbereinigung*, **27**, 318-325.

Schröder, D. (1988a) Initiale Bodenentwicklung in aufgespültem Löß und Hafenschlick (Initial soil development in wet-deposited loess and harbour mud). *Zeitschrift für Pflanzenernährung und Bodenkunde*, **151**, 3-8.

Schröder, D. (1988b) Bodenschonende Rekultivierung von Lößböden in Braunkohlentagebauen (Soil-sparing recultivation of loess soils at brown coal strip mines). In

Bodenschutz: ergänzbares Handbuch der Maßnahmen und Empfehlungen für Schutz, Pflege und Sanierung von Böden, Landschaft und Grundwasser (D. Rosenkranz, Ed.), pp. 1-22. Schmidt Verlag, Berlin.

Schröder, D. and Schneider, R. (1992) Bodenschätzung von Neukulturen (Soil assessment of new reclaimed disturbed lands). *Mitteilungen der Deutschen Bodenkundlichen Gesellschaft*, **67**, 265-268.

Schröder, D., Stephan, S. and Schulte-Karring, H. (1985) Eigenschaften, Entwiklung und Wert rekultivierter Böden aus Löß im Gebiet des Rheinischen Braunkohlen-Tagebaues (Properties, development, and value of the recultivated soils from loess in the Rhine Brown Coal Strip Mining Region). *Zeitschrift für Pflanzenernährung und Bodenkunde*, **148**, 131-146.

Schröder, D., Haubold, M. and Henkes, L. (1987a) Humusgehalt und bodenbiologische Eigenschaften rekultivierter Böden aus Löß (Humus content and soil biological properties of recultivated soils from loess). *Landwirtschaftliche Zeitschrift* (Bonn), No. 21, 1466-1469.

Schröder, D., Oeß, M. and Köpping, R. (1987b) Qualität und Entwicklung aufgespülter Böden aus Hafenschlick (Quality and development of wet-deposited soils from harbour mud). *Mitteilungen der Deutschen Bodenkundlichen Gesellschaft*, **53**, 281-287.

Schulz, V.M., Vollmer, G. and Mathes, K. (1985a) Rekultivierung als ökologisches Problem. 10. Zur ökologischen Bedeutung der Protease-Aktivität von Böden (Recultivation as an ecological problem. 10. Ecological significance of protease activity in soils). *Verhandlungen der Gesellschaft für Ökologie* (Bremen, 1983), **13**, 805-808.

Schulz, V.M., Bölling, C. and Mathes, K. (1985b) Rekultivierung als ökologisches Problem. 11. Der Einfluß von Rekultivierungsmaßnahmen auf Stickstoffmineralisation und Stickstoffretention (Recultivation as an ecological problem. 11. The influence of recultivation measures on nitrogen mineralisation and nitrogen retention). *Verhandlungen der Gesellschaft für Ökologie* (Bremen, 1983), **13**, 809-815.

Schumacher, B., Kutsch, H. and Schröder, D. (1989) Beziehungen zwischen mikrobieller Aktivität und Humusgehalt sowie Humuszusammensetzung in Neulandböden unterschiedlichen Alters (Relationships between microbial activity and humus content as well as humus composition in new reclaimed soils of different age). *VDLUFA-Schriftenreihe*, **30**, 473-478.

Schumacher, B., Kutsch, H. and Schröder, D. (1993) Huminstoffsysteme in Rekultivierungsböden des Rheinischen Braunkohlenreviers bei Erstnutzung als Acker, Grünland oder Wald (Systems of humic substances in recultivation soils of the Rhine Brown Coal District during the first use as arable land, grassland or forest). *Mitteilungen der Deutschen Bodenkundlichen Gesellschaft*, **71**, 209-212.

Seoane, S., Benito, E., Leirós, M.C., Gil-Sotres, F. and Guitián-Ojea, F. (1995) Los materiales estériles de la mina (The mine soils). In *Recuperación de las Escombreras de la Mina de Lignitos de Meirama (A Coruña)* (F. Guitián-Ojea, Ed.), pp. 69-96. Universidade, Santiago de Compostela.

Shcherbakov, A.P., Uskov, B.V. and Kosolapova, A.V. (1991) Biologicheskaya aktivnost' pochvogruntov gidrootvala pri sel'skokhozyaistvennoi rekul'tivatsii (Biological activity of hydraulic-fill spoil grounds during agricultural recultivation). *Doklady VASKhNIL*, No. 11, 2-4.

Shirey, J.J. and Sexstone, A.J. (1989) Denitrification and nitrate-reducing bacterial populations in abandoned and reclaimed minesoils. *FEMS Microbiology Ecology*, **62**, 59-69.

Shirokikh, A.A. (1988) Mikrobnye tsenozy vyrabotannykh torfyanikov pri rekul'tivatsii (Microbial coenoses of exhausted peatlands during recultivation). *Tezisy Dokladov 3. Vsesoyuznogo Simpoziuma "Biodinamika Pochv"* (Harku-Tallinn, 1988), p. 166.

Shirokikh, I.G. (1993) Aktinomitsety kak komponent bioty lesnoi podstilki (Actinomycetes as a component of the forest litter biota). In *Na Torfyanykh Pochvakh, Chast' 1* (I.A. Vertogradskaya, Ed.), pp. 79-81. Lugobolotnaya Opytnaya Stantsiya, Kirov, formerly Vyatka.

Shirokikh, A.A. and Shirokikh, I.G. (1993) Pochvenno-mikrobiologicheskie issledovaniya osushennykh torfyanikov (Soil microbiological investigations of the dried peatlands). In *Na Torfyanykh Pochvakh, Chast' 1* (I.A. Vertogradskaya, Ed.), pp. 16-20. Lugobolotnaya Opytnaya Stantsiya, Kirov, formerly Vyatka.

Shirokikh, A.A. and Ulanov, A.N. (1986) Vliyanie peskovaniya vyrabotannykh nizinnykh torfyanikov na produktivnost' seyanykh trav i biologicheskuyu aktivnost' pochvy (Effect of the sand treatment of exhausted low moor peatlands on the yield of sown grasses and on the biological activity of soil). *Tezisy Dokladov 3. Vsesoyuznoi Nauchnoi Konferentsii "Mikroorganizmy v Sel'skom Khozyaistve"* (Moscow, 1986), p. 151.

Shirokikh, A.A. and Vertogradskaya, I.A. (1992) Biologicheskie aspekty transformatsii organicheskogo veshchestva vyrabotannykh torfyanikov (Biological aspects of the transformation of organic matter in exhausted peatlands). *Torfyanaya Promyshlennost'*, No. 2, 26-27.

Shirokikh, A.A. and Zimenka, T.G. (1989) Zmyanenne mikrobabiyakhimichnai aktyvnastsi vypratsavanaga tarfyaniku pad uplyvam ugnaennyav (Changes in the microbial biochemical activity of exhausted peatland under the influence of fertilisers). *Vestsi Akademii Navuk BSSR, Seryya Biyalagichnykh Navuk*, No. 2, 48-51.

Shirokikh, A.A., Shirokikh, I.G., Vertogradskaya, I.A. and Zimenka, T.G. (1987) Biyalagichnaya aktyvnasts' vypratsavanykh tarfyanikav roznykh terminav sel'skagaspadarchaga vykarystannya (Biological activity of exhausted peatlands in agricultural use for different times). *Vestsi Akademii Navuk BSSR, Seryya Biyalagichnykh Navuk*, No. 3, 52-56.

Shugalei, L.S. and Yashikhin, G.I. (1985) Biologicheskaya aktivnost' rekul'tivirovannykh zemel' KATEKa (Biological activity in recultivated soils of the Kansk-Achinsk Fuel-Energetic Complex). In *Biologicheskaya Aktivnost' Lesnykh Pochv* (V.M. Korsunov, Ed.), pp. 80-88. Institut Lesa i Drevesiny, Sibirskoe Otdelenie Akademii Nauk SSSR, Krasnoyarsk.

Shugalei, L.S., Yashikhin, G.I. and Korsunova, T.M. (1984) Vliyanie ugledobychi na sel'skokhozyaistvennye zemli KATEKa (The effect of coal mining on the agricultural soils of the Kansk-Achinsk Fuel-Energetic Complex). *Tezisy Dokladov, Vsesoyuznaya Shkola "Vliyanie Promyshlennykh Predpriyatii na Okruzhayushchuyu Sredu"* (Zvenigorod, 1984), pp. 224-226.

Shugalei, L.S., Korsunova, T.M. and Yashikhin, G.I. (1985) Gumusnoe sostoyanie i biologicheskaya aktivnost' rekul'tiviruemykh zemel' KATEKa (Humus status and biological activity in recultivated soils of the Kansk-Achinsk Fuel-Energetic Complex). *Izvestiya Sibirskogo Otdeleniya Akademii Nauk SSSR, Seriya Biologicheskie Nauki*, No. 13/2, 70-76.

Šíša, R. (1985; 1988) Výzkum změn biologické acktivity půd v procesu rekultivací v SHR Most (Investigation of the changes in biological activity of soils during the process of recultivation in the North Czech Brown Coal District Most), I. pp. 1-32; II. pp. 1-24. Vysoká Škola Zemědělská, Prague.

Šíša, R. (1993) Enzymová aktivita půdy jako ukazatel její biologické aktivity (Enzymatic activity of soil as an indicator of its biological activity). *Rostlinná Výroba*, **39**, 817-825.

Šíša, R. and Belatková, P. (1991) Effect of recultivation on enzymatic activity and respiration of soil. *Poster, 10th International Symposium "Humus et Planta"* (Prague, 1991).

Šíša, R., Vořišek, K., Růžek, L., Randová, D. and Strnadová, S. (1997) Biologická aktivita neporušených a antropogenních půd v Severočeském hnědouholném revíru (Biological activity in undisturbed and anthropogenic soils in the North Czech Brown Coal District). *Papers, Seminář "Život v Půde"* (Bratislava, 1997).

Smith, J.L. and Paul, E.A. (1990) The significance of soil microbial biomass estimations. In *Soil Biochemistry, Vol. 6* (J.-M. Bollag and G. Stotzky, Eds.), pp. 357-396. Dekker, New York.

Snytko, V.A., Bazhenova, O.I., Bessolitsyna, E.P., Dubynina, S.S., Naprasnikova, E.V. and Nefed'eva, L.G. (1988) Kompleksnyi analiz tekhnogenno izmenennykh geosistem KATEKa (Complex analysis of the technogenically changed geosystems of the Kansk-Achinsk Fuel-Energetic Complex). In *Geografiya Pochv i Geokhimiya Landshaftov Sibiri* (E.G. Nechaeva and V.A. Snytko, Eds.), pp. 114-134. Institut Geografii, Irkutsk.

Soreanu, I. (1983) Activitatea biologică în iazul de la E.P. Săsar în urma aplicării unor tratamente agrochimice (Biological activity in spoils following application of some agrochemical treatments at the Săsar Lead-Mining Enterprise). *Buletin Ştiinţific, Institutul de Învăţămînt Superior Baia-Mare, Seria B*, **6**, 93-98.

Sorensen, D.L., Klein, D.A., Ruzzo, W.J. and Hersman, L.E. (1981) Enzyme activities in revegetated surface soil overlying spent Paraho process oil shale. *Journal of Environmental Quality*, **10**, 369-371.

Speir, T.W., Ross, D.J. and West, A.W. (1986) The role of soil biochemical properties in soil restoration studies in New Zealand. *Proceedings, New Zealand Society of Soil Science – Australian Society of Soil Science Inc. Joint Conference "Surface Soil Management"* (Rotorua, New Zealand, 1986), pp. 19-25.

Starodumova, E.V. and Shirokikh, I.G. (1993) Effektivnost' mineral'nykh udobrenii na mnogoletnykh zlakovykh travostoyakh vyrabotannogo nizinnogo torfyanika (Effectiveness of mineral fertilisers on the perennial grass stands of an exhausted low moor peatland). In *Na Torfyanykh Pochvakh, Chast' 1* (I.A. Vertogradskaya, Ed.), pp. 21-25. Lugobolotnaya Opytnaya Stantsiya, Kirov, formerly Vyatka.

Stefurak, V.P. (1989) Pochvennaya mikroflora antropogenno narushennykh nazemnykh ekosistem na trassakh gazoprovodov v Karpatakh (Soil microflora of anthropogenically disturbed terrestrial ecosystems along the pipelines in the Carpathians). *Mikrobiologicheskii Zhurnal* (Kiev), No. 2, 12-17.

Stroo, H.F. and Jencks, E.M. (1982) Enzyme activity and respiration in minesoils. *Soil Science Society of America Journal*, **46**, 548-553.

Stroo, H.F. and Jencks, E.M. (1985) Effect of sewage sludge on microbial activity in an old, abandoned minesoil. *Journal of Environmental Quality*, **14**, 301-304.

Strzyszcz, Z. (1996) Recultivation and landscaping in areas after brown-coal mining in Middle-East European countries. *Water, Air, and Soil Polluton*, **91**, 145-157.

Sulyok, L. and Vörös, I. (1983) Relationship of saccharase activity and other soil biological characteristics in the refuse soils of Gyöngyös-Visonta. In *Recultivation of Technogenous Areas* (J. Szegi, Ed.), pp. 207-212. Mátraalja Coal Mining Co., Gyöngyös.

Sutton, P. and Dick, W.A. (1987) Reclamation of acidic mined lands in humid areas. *Advances in Agronomy*, **41**, 377-405.

Sviridova, I.K. and Panozishvili, K.P. (1979) Puti povysheniya biologicheskoi aktivnosti otval'nykh zemel' tekhnogennykh landshaftov KMA (Ways for increasing the biological activity of spoil heap soils of the technogenic landscapes in the Kursk Magnetic Anomaly region). *Tezisy Dokladov Vsesoyuznogo Soveshchaniya "Biologicheskaya Produktivnost' Pochv i Ee Uvelichenie v Interesakh Narodnogo Khozyaistva"* (Moscow, 1979), pp. 141-142.

Szegi, J. (1983) Soil biological processes as affected by recultivation in the spoil heaps of mines. In *Recultivation of Technogenous Areas* (J. Szegi, Ed.), pp. 105-111. Mátraalja Coal Mining Co., Gyöngyös.

Tate, K.R., Ross, D.J., Speir, T.W. and Hart, P.B.S. (1985) Soil – the fundamental life support system and its disturbance by man. *Proceedings of the Soil Dynamics and Land Use Seminar* (Blenheim, New Zealand, 1985), pp. 243-256.

Tóth, S. (1988) *A Rekultiváció Helyzete és Eredményei a Magyar Népköztársaságban 1983-1988 között* (Position and Results of the Recultivation in Hungary between 1983 and 1988). Gödöllői Agrártudományi Egyetem Kutató Intézete, Kompolt.

Trasar-Cepeda, M.C., Ceccanti, B., Ciardi, C., Gil-Sotres, F. and Leirós, M.C. (1990a) Recuperación de escombreras de minas de lignito: evolución en el tiempo de varias actividades enzimáticas, ATP y sustancias húmicas (Recovery of lignite mine spoils: evolution in time of various enzymatic activities, ATP, and humic substances). *Actas, I Congreso Internacional de Química de la Asociación Nacional de Químicos de España (ANQE) "Residuos Solidos y Liquidos: Su Mejor Destino"* (Puerto de la Cruz-Tenerife, 1990), **3**, 409-415.

Trasar-Cepeda, M.C., Gil-Sotres, F., Ciardi, C. and Ceccanti, B. (1990b) Evoluzione della fertilità biochimica in terreni di miniera di lignite (Evolution of the biochemical fertility in lignite mine soils). *Atti, VIII Convegno Nazionale della Società Italiana di Chimica Agraria* (Bari, 1990), pp. 241-244.

Trasar-Cepeda, M.C., de Blas, E., Garcia-Rodeja, I., Leirós, M.C. and Gil-Sotres, F. (1993) Modifications of the phosphorus cycle with development of lignite mine soils. *Land Degradation and Rehabilitation*, **4** (3), 131-142.

Trofimov, S.S., Naplekova, N.N., Kandrashin, E.R., Fatkulin, F.A. and Stebaeva, S.K. (1986) *Gumusoobrazovanie v Tekhnogennykh Ekosistemakh* (Humus Formation in Technogenic Ecosystems). Izdatel'stvo Nauka, Sibirskoe Otdelenie, Novosibirsk.

Tsvetkova, L.A., Ras'kova, N.V. and Gel'tser, Yu.G. (1982) K otsenke lesorastitel'nykh svoistv rekul'tiviruemykh zemel' po ikh biokhimicheskoi aktivnosti (Evaluation of the forest-growing properties of recultivated soils by their biochemical activity). *Materialy Vsesoyuznogo Simpoziuma "Mikroorganizmy kak Komponent Biogeotsenoza"* (Alma-Ata, 1982), pp. 109-110.

Uzbek, I.Kh. (1986) Ob intensivnosti pochvoobrazovaniya rekul'tivirovannykh zemel' (Intensity of pedogenesis in recultivated soils). *Tezisy Dokladov 2. S'ezda Pochvovedov i Agrokhimikov Ukr.SSR* (Kharkov, 1986), *Agrokhimiya i Plodorodie Pochv, Rekul'tivatsiya Zemel', Bor'ba s Eroziei Pochv, Obrabotka Pochv*, p. 112.

Uzbek, I.Kh. (1989) O stepeni obogashchennosti fermentami tekhnogennykh landshaftov (Degree of enrichment of technogenic landscapes with enzymes). *Tezisy Dokladov 8. Vsesoyuznogo S'ezda Pochvovedov* (Novosibirsk, 1989), **2**, 242.

Uzbek, I.Kh. (1991) Osobennosti fermentativnoi aktivnosti rekul'tivirovannykh pochv (Particularities of the enzymatic activity in recultivated soils). *Pochvovedenie*, No. 3, 91-96.

Uzbek, I.Kh., Yaroshevich, I.V., Barskaya, L.M. and Golinskii, A.G. (1988) Aktivnost' nekotorykh gidroliticheskikh fermentov v vskryshnykh gruntakh Nikopol'skogo margantsevorudnogo basseina (Activity of some hydrolytic enzymes in overburdens of the Nikopol manganiferous basin). *Tezisy Dokladov 5. Ural'skogo Soveshchaniya "Problemy Rekul'tivatsii Narushennykh Zemel'"* (Sverdlovsk=Ekaterinburg, 1988), pp. 34-35.

Verbin, A.E. and Keleberda, T.N. (1974) Opyt rekul'tivatsii porod razvedeniem lesnykh kul'tur (A trial for recultivation of mine spoils through plantation of forest species). *Pochvovedenie*, No. 2, 116-120.

Vörös, I. and Antal, M. (1988) Formirovanie tsellyulozorazrushayushchei aktivnosti i enzimnoi aktivnosti sakharazy v Vishontskikh otval'nykh zemlyakh v protsesse ikh rekul'tivatsii (Development of cellulose-decomposing and saccharase enzyme activities in the Visonta refuse soils during their recultivation). *Sbornik Dokladov, 9. Mezhdunarodnyi Simpozium "Rekul'tivatsiya Landshaftov, Narushennykh Promyshlennoi Deyatel'nost'yu"* (Kompolt-Gyöngyös, 1988), **3**, 109-117.

Weidemann, G. (1985) Rekultivierung als ökologisches Problem. 1. Konzept und Probeflächen (Recultivation as an ecological problem. 1. Concept and sampling sites). *Verhandlungen der Gesellschaft für Ökologie* (Bremen, 1983), **13**, 751-758.

Wigfull, S. and Birch, P. (1987) The microbiology of landfill soils. *Land and Minerals Surveying*, **5**, 349-353.

Wigfull, S.D., Harris, J.A. and Birch, P. (1987a) The activity of micro-organisms in landfill soils. *Poster, Conference on "Reclamation and Restoration of Soils"* (London, 1987).

Wigfull, S.D., Harris, J.A. and Birch, P. (1987b) The activity of micro-organisms in landfill soils. *Journal of the Science of Food and Agriculture*, **40**, 230-231.

Williams, S.T., McNeilly, T. and Wellington, E.M.H. (1977) The decomposition of vegetation growing on metal mine waste. *Soil Biology & Biochemistry*, **9**, 271-275.

Zaitsev, G.A., Motorina, L.V. and Dan'ko, V.N. (1977) *Lesnaya Rekul'tivatsiya* (Forest Recultivation), pp. 64-69. Izdatel'stvo Lesnaya Promyshlennost', Moscow.

Zasorina, E.V. (1985a) Biokhimicheskaya aktivnost' molodykh tekhnogennykh pochv (na primere Stoilenskogo GOKa) (Biochemical activity of young technogenic soils (in the case of the Stoilensk iron strip mine)). In *Rekul'tivatsiya Zemel', Narushennykh Gornymi Rabotami na KMA* (A.M. Burykin, Ed.), pp. 74-84. Sel'skokhozyaistvennyi Institut, Voronezh.

Zasorina, E.V. (1985b) Fermentativnaya aktivnost' molodykh tekhnogennykh pochv (Enzymatic activity of young technogenic soils). *Tezisy Dokladov 7. Delegatskogo S'ezda Vsesoyuznogo Obshchestva Pochvovedov* (Tashkent, 1985), **2**, 190.

Żukowska-Wieszczek, D., Stuszewska, J. and Żurawska-Olszewska, J. (1985) Aktywność enzymatyczna gleb rekultywowanych w doświadczeniu modelowym (Enzymatic activity of recultivated soils in a model experiment). *Czlowiek i Środowisko*, **9**, 407-421.

Zverkov, Yu.V. (1982) *Vtoraya Zhizn' Torfyanikov* (The Second Life of Peatlands). Volgo-Vyatskoe Izdatel'stvo, Kirovskoe Otdelenie, Kirov, formerly Vyatka.

SUBJECT INDEX

Acetylene reduction 89, 109, 177, 217-220, 222-224
-, *see also* N_2 fixation
N-Acetylglucosaminidase 113
Acid
-, benzoic 24
-, hydrochloric 65, 106
-, hydrofluoric 106
-, nitric 65
-, palmitic 24
-, salicylic 24
-, sulphuric 41, 65, 93, 192
Acidification/acid precipitation 41, 67, 93, 94, 104, 121, 122, 124, 127, 192
Acids
-, amino 19, 89, 96, 184, 214
-, fulvic 41, 235
-, humic 7, 41, 235
Actinomycetes, number of 6, 7, 14, 17, 22, 32, 33, 43, 44, 51, 77, 78, 81, 82, 89, 96, 103, 109, 118, 133, 134, 173, 174, 192, 209, 217, 221, 235, 249, 264, 272
Activation energy 199
- enthalpy 199
Albic luvisol 287
Alder 133, 134, 248
-, black 157, 158, 160, 179-181, 231, 256
-, white 270
Alfalfa 159, 165, 166, 176, 177, 186, 188, 192, 208, 229, 232, 233, 259, 260, 267
Algae, number of 8, 211
Alkali sacaton 167
n-Alkanes 6, 20-23, 33
Allergens 109
Aluminium
- content 91, 108, 127, 205
- oxide 79, 111
Aminopeptidase 120
Ammonia 97, 108
Ammonification 8, 81, 96, 102, 109, 121, 124, 200, 206, 214, 272
Ammonium
- availability 50
- content 19, 71, 108, 157, 183, 200, 236, 279, 289
Ammonium (*continued*)
- oxidisers 164, 166, 167
- oxidising capacity 77, 78
Amylase 74, 77, 104, 121, 133, 134, 159, 165, 168, 169, 171, 172, 184, 204, 205, 276, 277
Andesite 245
Andesitic tuff 186
Apple trees 129
Arginine deaminase 287
Arsenic content 74-76, 113, 114
Arylsulphatase 93, 99-101, 121-123, 126, 165, 168, 287, 289, 290
-, *see also* Sulphatase
Ascorbate oxidase 98
Ash
-, European 179, 180
-, green 256, 257
-, mountain 120, 131
Ash, for soil/spoil reclamation 106, 108, 166, 175, 180, 181, 192, 193, 263-265
Askanite 259
Asparaginase 19, 79, 80,87, 91, 118, 121, 179, 180, 263, 264
Aspen 179, 183, 248
Asphalt 24
ATP 164, 196-198, 200, 217, 275, 282, 283
ATPase 18

Bacteria
-, inorganic N-assimilating/-utilising 8, 96, 109, 214
-, iron-reducing 243
-, N_2-fixing 14, 15, 109, 118, 173, 174
-, nitrate-reducing 175
-, number of 6, 8, 10, 14, 35, 42, 51, 75, 77, 78, 81-83, 89, 102, 109, 118, 121, 133, 134, 173-175, 188, 192, 207, 217, 221, 235, 249, 264, 265, 272
-, organic N-utilising 6, 8
-, spore-forming/endosporogenic 7, 34, 82, 89, 102, 175, 272
-, sulphate-reducing 243
Bacterial biomass 109, 114, 182, 272

Barley 29, 31-34, 46-48, 105, 175, 207, 215, 251, 285
-, two-row 32, 33
Bean
-, field 207
-, kidney 208
Beech 120, 123, 124, 183
Bentonite 118, 205, 263, 264
Benzene 24, 26, 27, 39
-, ethyl 39
Benzine 24, 26
Benzo[a]pyrene 15, 27
Birch 104, 120, 131, 184, 248
-, weeping 159, 256
Blackberry 170
Black cherry 170, 178
Blackgum 209
Black locust 158-162, 169, 170, 178, 228, 255-257
Blue grama grass 165
Bluegrass 98, 199
Bog soils 7, 34
Box elder 209
Bromegrass 159, 212, 229
-, awnless 211, 215, 247, 248
Broomsedge 175
Brown earths 99, 102, 120
Bryophytes 203
-, *see also* Mosses
BTEX 39
Bulk density 189, 196, 197

Cadmium
- content 27, 69-71, 74, 75, 77, 78, 83, 86, 87, 91, 93, 94, 104, 112-114, 117, 118, 120, 124, 130, 134, 237
- oxide 71, 79
Calcium
- content 11, 94, 96, 110, 117, 189, 247, 286
- carbonate 91, 111, 286
- hydrocarbonate 111
- oxide 79, 111
Cambisols 287
Capillary barrier 217-222

Carbon
- content 22, 41, 49, 87, 104, 130, 132, 133, 169, 189, 191-193, 196, 197, 199, 200, 202, 203, 206, 209, 210, 246, 247, 261, 275, 282, 286, 287, 289
- -, *see also* Humus and Organic matter contents
- decomposition/mineralisation 52, 71, 72, 200, 203, 217, 287
- -, *see also* Respiration
- disulphide 102
- monoxide 108, 127
Carboxymethylcellulase 101
-, *see also* Cellulase
Carcinogens 27
Catalase 5-7, 13, 14, 17, 19-24, 26, 29-33, 35, 36, 40, 42-44, 51, 78, 90, 96, 98, 103-109, 112, 118, 119, 130, 131, 134, 157-159, 161, 162, 175-180, 183-185, 191-193, 204-208, 211-215, 225, 226, 228-233, 239, 251, 253, 256, 263, 264, 269, 270, 272, 273, 286
Cation-exchange capacity 189, 286, 289
Cattail 209
Cellulase 7, 16, 34, 77, 78, 81, 82, 90, 93, 109, 127, 128, 131, 134, 163, 168, 214, 276, 277
Cellulolysis/cellulolytic activity (capacity) 102, 106, 111, 175, 176, 270, 272, 273
Cellulose decomposition 8, 9, 81, 89, 96, 109, 130, 184, 186-190, 192, 207, 214, 249, 285
Cesium-137 content 112
Chernozems 5, 6, 22, 24, 26, 28, 31, 41, 43-45, 48, 49, 52, 91, 97, 110, 161, 162, 183, 184, 186, 204, 229, 232, 269, 287
Chromium content 69, 74, 94-96, 104, 113, 114, 117, 119
Clay
- content 39, 169, 196, 198, 289
- minerals 177, 259, 264
- -, *see also* individual clay minerals
- treatment of exhausted peatlands 213
- - - sandy spoils 263, 265

Clover 194, 229, 248, 275, 277
-, foxtail 215
-, meadow 238, 239
-, red 182
-, white 199
Cobalt content 87, 88, 94, 117
Common coltsfoot 248
- lilac 97, 98
Compaction 169, 170, 188, 190, 194, 197, 198, 282
Copper
- content 11, 27, 28, 69, 71-75, 78, 83-88, 91, 93-95, 104, 113, 114, 117, 118, 130, 211, 237
- oxide 70
Cotton 133
Covering of spoils with (top)soil 163, 181, 183, 185, 186, 192, 204, 205, 207, 217, 219, 220-222, 224, 238-240, 243, 245, 246, 267, 268, 282
-, *see also* Topsoiling
Crop rotation 34, 227, 270, 279
Cryolite 105
C-to-N ratio 39, 40, 104, 169, 177, 210, 282
C-to-organic P ratio 202
Cucumber 110
Cycloalkanes 23
Cyclohexane 23, 26, 27, 104

Dark-brown soils 5, 272
Dehydrogenase 5-11, 14, 17, 19-24, 26-35, 39-45, 51, 69, 70, 77, 79-83, 91-94, 98, 101, 102, 104-108, 112, 114, 115, 117-120, 124-131, 161-165, 167, 168, 173, 174, 179, 180, 184, 185, 188-192, 195-198, 202, 207-209, 212, 213, 215, 217-228, 231, 232, 236, 237, 239, 249, 251, 260, 261, 265, 271, 272, 279-283, 285-287, 289, 290
Denitrification 15, 105, 126
- enzyme 174, 175
Diesel fuel/oil 10, 23, 36-39, 49, 50
Dimethylsulphoxide reductase 126
- reduction 25
Dimethyl terephthalate 102
DNase 18
Drainage 132, 188, 208, 211, 285
Dry redeposition of spoil (loess)/topsoil 188-190, 192, 229
Dry remediation 39, 40
Duff-humus 84, 85
Dusts 78, 79, 81, 82, 95, 109-111, 127, 132, 133

Electric conductivity 286, 289, 290
Elm 256, 257
Emulsifier of pellicular oil 41, 42
Endocellulase 15, 16, 113, 114, 132, 133
-, *see also* Cellulase
Enzymatic indicator 208, 227, 239, 240-243
Enzyme reactivation 72, 73
Ergosterol 196
Erosion 155, 173, 229
Ester-sulphate 99, 122
Ethylenediamine 108
Ethylene glycol 102
Exocellulase 15, 16, 132, 133
-, *see also* Cellulase

Fallow 19, 34, 40, 192
Fauna 40, 93, 235, 236
-, collembolans 240
-, earthworms 32, 131, 183, 184, 190, 191
-, enchytraeids 128, 131
-, invertebrates 182, 191
-, mites 240
-, nematodes 128, 240
-, snails 131
Fertilisation/fertilisers 46, 93, 105, 163, 170, 188, 195, 207
-, Biohumus 31-33
-, cattle slurry 202, 203
-, coal 185, 186
-, compost 31, 32, 52, 265
-, farmyard manure 29-31, 34, 51, 89, 98, 185, 186, 190, 191, 208, 213, 227, 238, 239, 267-270
-, grape marc 35, 45
-, green manure 31, 157-159, 185, 187, 213, 255
-, hay 165-168
-, lignite 187, 207
-, liquid dung 192, 214

Fertilisation/fertilisers (*continued*)
-, microelements (trace elements) 97
- -, copper 211
- -, molybdenum 214
-, municipal refuse 265
-, NPK 29-31, 34, 35, 37, 41, 42, 45, 46, 50, 51, 81-83, 98, 99, 106, 108, 118, 119, 130, 164-168, 171, 172, 175-177, 179, 180, 185-187, 192-194, 198, 199, 202, 203, 207, 208, 211-215, 219-222, 224, 227, 228, 237-240, 245, 248, 251, 259, 260, 263, 264, 267, 269, 270, 275
-, peat 131, 248, 263-265
-, sapropel 268
-, sawdust 31, 33, 35, 42, 52
-, sludge
- -, activated 31, 33, 42, 52
- -, sewage 106, 108, 165-167, 171, 172, 187, 213, 228, 265
-, straw 186
-, urea 37, 46, 49, 50, 194, 245, 275
-, wastewater 31, 35, 42, 43
-, whey 35
Fescue 229
-, meadow 182, 215, 247, 248
-, red 199, 267
-, tall 170-172, 175
Fire cherry 170
Fireweed 248
Flax 207
Flooding 197, 198
Fluoranthene 27, 28
Fluorescein diacetate (FDA)-hydrolysing enzymes 37-39, 48, 50, 51
Fluoride 65, 82, 90, 91, 105
Forecrops 215
Forest soils 5, 6, 13, 17, 18, 22, 26, 27, 30-32, 42-44, 51, 87, 111, 134, 181, 233
Forest-steppe 5, 34, 157, 181, 184, 269
Formaldehyde 104, 108
Fourwing saltbush 163, 165, 167
Fungal biomass 89, 109, 114, 182
- genera, distribution of 164-167
- mycelial length 51, 76, 109, 114, 128
Fungi/microfungi
-, mycorrhizal 128, 166, 167, 173, 174

Fungi/microfungi (*continued*)
-, number of 6, 8, 10, 14, 17, 20, 22, 35, 42, 43, 51, 74, 77, 78, 81, 82, 96, 103, 109, 110, 118, 121, 133, 134, 166, 167, 173-175, 192, 209, 217, 221, 235, 249, 264, 265, 272

Galena 95
Gasoline
-, leaded 23, 49
-, light 51, 52
Gelatin decomposition/gelatinolytic activity (capacity) 102, 111, 184
Gley/gleyic soils 7, 8, 26, 119
β-Glucosidase 71, 80, 91, 93, 113, 118, 120, 124, 179, 180, 202, 264
Glutaminase 19
Golden elder 179
Goldenrod 170
Grading of spoils 164, 166
Grapevine 233, 265
Grazing 163, 190, 272, 275
Grey-brown soils 13, 14, 23, 34, 35, 45
Gumbrine 259
Gypsum 41

Heating oil 4, 8, 9, 19, 20, 25, 37-39, 52
Heavy metals 27, 28, 65, 67, 70-72, 74, 75, 83, 86-90, 92, 93, 95-97, 105, 108, 113, 114, 117, 119, 120, 124, 127, 130, 228, 237
- -, *see also* individual heavy metals
n-Hexadecane 23, 24, 26, 27
1-Hexadecanol 24
Hornbeam 190
Horse chestnut 97, 98
Humus
- accumulation 176, 188, 190, 213, 228, 229, 249, 268
- content 26, 158-160, 182, 183, 187, 190, 204, 205, 229, 231, 233, 236, 240, 253, 255, 256, 259, 260, 267, 269, 272
- -, *see also* Carbon and Organic matter contents
Hydrocarbon hydroxylase 15

Hydrocarbons/oil
-, aliphatic 3, 25, 35
-, aromatic 3, 23, 25, 27
-, chlorinated 108
- content 5-7, 10, 13-15, 17-20, 22-32, 34-37, 39, 41-46, 48-52
- -, residual 6, 14, 24, 26, 32, 34-37, 39, 42-46, 52
-, polycyclic aromatic 27, 39, 40
Hydrogen sulphide 97, 102, 127
Hydroxylamine reductase 19, 22, 28

Illite 264
INT-dehydrogenase 25
-, *see also* Dehydrogenase
Invertase 4-7, 13, 14, 17, 22-26, 28-33, 35, 40, 42, 44, 51, 77, 78, 80, 83, 87-89, 98, 103-108, 112, 118, 119, 121, 124, 125, 128, 133, 134, 157-162, 165, 168, 175-182, 186, 187, 190, 191, 194, 195, 204-207, 211, 212, 225-229, 231-233, 237, 246, 248, 255-257, 259, 260, 263-265, 268, 269, 275-277, 286, 287
Iron
- content 69, 110, 117, 119, 205
- disulphide 192
- oxides 79, 84, 111
Irrigation 27, 164, 168, 208

Jerusalem artichoke 208
Jet fuel 37, 38, 50, 51
Joe-pye weed 170

Kaolin 263
Kastanozems 287
Kerosene 23, 49, 50

Larch 181, 248
-, Siberian 183
Lead
- content 69, 70, 74, 75, 78, 83, 86, 87, 91, 93-96, 104, 113, 114, 117, 119, 120, 124, 127, 130, 134, 235-237
- oxide 71, 79
-sulphide 95, 96
Levelling of spoils 164, 166, 182, 183, 188, 225, 227, 228, 230, 245, 248, 279
Lime/liming 34, 37, 50, 81-83, 93, 94, 118, 119, 170-172, 175, 185, 186, 192, 193, 198, 200, 245, 264, 275
Loess 157, 159, 161, 162, 188-192, 204, 205
Lupine 215
-, perennial 157-159

Magnesium content 117, 188, 189, 286
Maize 170, 227, 228, 259, 260, 270
Manganese content 11, 94, 117, 119
Maple 190
-, red 209
Marcasite 192
Maximum velocity 199
Meadow grass 231
Meadow-swampy soils 7
Mercury content 74, 108
Methanol 102, 103
5-Methyl-3-heptanone 26
Methyl mercaptan 102
Michaelis constant 199
Microbial biomass 35, 46, 48, 105, 124, 127, 128, 130, 131, 177, 183, 189, 190, 193, 194, 196, 198, 213, 261, 275, 282, 285-287
- diversity 168, 230
Microorganisms
-, ammonifying 14, 15, 17, 79, 96, 109, 118, 243, 272
-, cellulose-decomposing/-degrading (cellulolytic) 6, 17, 22, 26, 32, 33, 42-44, 83, 96, 109, 118, 134, 192, 213
-, denitrifying 14, 15, 17, 118, 272
-, diesel fuel-degrading 10
-, heterotrophic 20, 22, 26, 32, 33, 42, 44, 45, 82, 110, 164, 166, 167, 209, 243, 251
-, hydrocarbon-degrading/-oxidising 3, 6-8, 10, 14, 31, 33-36, 42-45, 50
-, nitrifying 6, 15, 17, 22, 26, 33, 42, 43, 118, 173, 174

Microorganisms (*continued*)
-, number of 6, 7, 10, 11, 13, 14, 19, 31, 34, 35, 42, 51, 77, 79, 82, 87, 90, 91, 101, 102, 108, 119, 130, 165, 167, 182-184, 208, 209, 213, 214, 230, 231, 233, 235, 248, 251, 259, 265, 283
-, oligonitrophilic 96, 109, 134, 272
-, oligotrophic 96, 251
-, phenol-oxidising 97-99
-, phosphate-mobilising 192
-, proteolytic 192
-, species diversity of 89, 168, 209
Mixing of upper soil layers
- with each other 34, 193, 194
- with overburdens 194, 269, 270
Moistening 40, 42, 43, 51, 110, 184, 202, 238, 239
-, *see also* Watering
Moisture content 24, 25, 50, 74, 99, 132, 177, 196, 202, 221, 237, 246, 282
- -, *see also* Water content
Molybdenum content 94, 95, 214
Monosaccharides 202, 203
Montmorillonite 263
Mors 69-75, 94-96, 110, 111, 117, 127, 128, 131
Mosses 132
-, *see also* Bryophytes
Motor oil 23, 49, 50
Mulch/mulching 40, 52, 164, 168, 170, 199
Mycorrhyza 16, 121, 128, 167

Naphthalene 26
Nature reserves 196
Nepheline 247, 248
N_2 fixation 89, 105, 109, 124, 159, 177, 214, 217, 222, 268
- -, *see also* Acetylene reduction and Nitrogenase
Nickel content 69, 70, 74, 84-88, 93-95, 110, 113, 114, 117, 118
Nitrate
- accumulation 249
- content 18, 19, 71, 108, 157, 183, 200, 211, 236, 279, 280, 289

Nitrate (*continued*)
- reductase 14, 15, 19, 22, 28, 98, 108, 185, 186
Nitrification 19-21, 81, 82, 87-89, 93, 102, 109, 120, 121, 124, 175, 200, 206, 207, 214, 229, 230, 246, 272, 279, 280, 287
Nitrite
- content 71, 108
- reductase 19, 22, 28, 98, 108
Nitrogen
- accumulation 169, 190, 192, 193, 199, 202, 228, 249
- availability 194, 209
- content 87, 108, 127, 128, 134, 157-160, 169, 171, 177, 179, 182, 183, 189, 190, 196, 197, 199, 200, 202, 203, 207, 209-211, 225, 229-231, 240, 246, 255, 256, 261, 267-269, 271, 275, 282, 286
- -, *see also* Ammonium, Nitrate, and Nitrite contents
- flush 194, 195, 275
- immobilisation 280
- mineralisation 26, 71, 82, 93-95, 104, 126, 127, 191, 194, 200, 246, 279, 280, 286
- oxides 108, 127
Nitrogenase 165, 168, 185, 186, 213, 214
Nonenzymatic catalysts 36, 221, 227, 231, 239, 243, 253
N-to-organic P ratio 202

Oak 79
-, blackjack 76, 77
-, pedunculated 159
-, post 76, 77
-, red 120, 170, 179, 180
-, willow 209
Oats 5, 17, 30-33, 105, 110, 133, 185, 214, 215, 227, 251, 269, 270
Oleaster 228, 231, 255, 256
On-site bioremediation 36, 39
Open burning 113
- detonation 113-115
Orchard grass 98, 170, 176, 177, 225, 226
Organic matter
- - accumulation 169, 202

Organic matter (*continued*)
- - content 36, 74, 105, 179, 183, 199, 202, 222-224, 237, 248, 271
- - -, *see also* Humus and Carbon contents

Paprin 31, 33, 34
Paraffin-degrading mycobacteria 4
Pea 34, 215, 251
-, field 46-48
Pectinase 124, 125
Pectinolyase 164
Perchlorate 114
Permafrost 131
Peroxidase 7, 24, 25, 32, 97, 99, 109, 181, 182, 184, 213
pH 22, 26, 37, 49, 72-75, 77, 91, 93-96, 104, 110, 111, 124-128, 161, 168, 169, 175, 179, 181, 182, 188, 192, 196, 197, 199, 200-202, 204, 205, 208, 211, 215, 225, 233, 236, 240, 246-248, 251, 260, 261, 263, 267, 268, 271, 282, 289, 290
Phenol-degrading capacity 98, 99
Phenol oxidase 97-99
Phenols 65, 90, 97, 98, 102, 202, 203
Phosphatase 6, 13, 25, 27, 29-32, 43, 71, 74-76, 79-81, 83, 91, 94, 95, 101, 103-105, 118, 125, 126, 164, 165, 168, 169, 171, 172, 182, 184, 199-202, 206, 208, 217, 219-227, 231-233, 237, 239, 259, 260, 269, 271, 272, 276
-, acid 16, 47, 48, 72, 73, 82, 84-87, 96, 104, 110, 111, 113, 120, 121, 132, 133, 181, 199, 201, 202, 205, 236
-, alkaline 26, 81, 82, 93, 104, 113, 114, 124, 125, 133, 173, 174, 181, 191, 193, 199, 201, 202, 286, 287
-, neutral 104, 111, 181
Phosphate
- content 18, 19, 71, 105, 169, 171, 177, 179, 182, 199, 202, 208, 211, 231, 236, 240, 245-248, 256, 263, 267-270, 289
-, solubilisation of 102
Phosphodiesterase 120, 132, 203
Phosphomonoesterase 18, 19, 132, 203
-, *see also* Phosphatase
Phosphorite 105, 185, 186
Phosphorus
- accumulation 199, 202, 228
- anhydride 105
- mineralisation 71, 72, 169
Phytase 18, 42
Phytomass 35
-, *see also* Plant biomass
Pine 5, 69, 74, 77, 79, 90, 105, 118-120, 127, 131, 133, 134, 180, 181, 193, 248, 270
-, jack 159
-, loblolly 209
-, longleaf 209
-, Scotch 157-160, 183, 231
-, Siberian 183
Plant
- biomass 31, 32, 41, 113, 119, 177, 182, 248, 265
- cover 39-41, 132, 159, 168, 173, 175, 190, 203, 209, 220-224, 238, 243, 248, 271, 272, 282
- growth 17, 22, 37, 39, 46, 48, 99, 104, 157, 159, 161, 230, 231, 243, 245, 246, 256, 264
- residues 264
- -, accumulation of 270
- -, decomposition/degradation of 15, 89, 109, 164
- -, humification of 177
- species composition 221-224, 246
- succession 218, 221-223
Ploughing 170, 175, 176, 190, 198, 269, 270, 286
Podzolic soils 5, 17, 71, 87, 88, 96, 103, 105, 108, 110, 112, 117-119, 131, 180, 184, 230, 248, 287
Polychlorinated biphenyls 27
Poly(ethylene terephthalate) 102
Polyphenol oxidase 7, 13, 32, 90, 103, 109, 181, 182, 184, 213
Polysaccharides 41
Poplar 193, 196
- hybrid 209
Potassium content 117, 169, 171, 179, 188, 208, 211, 231, 236, 240, 247, 248, 256, 263, 267-270, 286
Potato 207, 208, 215
Prairie 173

Protease/proteinase 6, 7, 11, 13-15, 19, 22, 24, 25, 28, 34, 35, 41, 81, 93, 96, 98, 103, 108-110, 124, 125, 127, 134, 159-162, 165, 181-184, 191, 200, 202, 212-214, 231, 255-257, 268, 270, 279, 280, 286, 287
Proteins 33, 41, 109
Proteolysis/proteolytic activity 162, 163, 272, 273
Purging buckthorn 179
Pyridine 97
Pyrite 168, 174, 192, 198

Quartz sand 265

Radioactivity 112
Rape 176, 285
Regosols 26, 27
Removal of contaminated soil layer 16
Rendzina 26, 225, 287
Respiration
-, basal 6, 8, 9, 14-17, 19, 20, 23, 25, 26, 32, 33, 35, 45, 47, 48, 51, 52, 69, 71, 74, 75, 81, 82, 84-87, 89, 91, 93-96, 102, 104, 106, 109, 118, 121, 124, 128, 130, 159, 161, 164, 169, 171, 172, 177, 183, 184, 186-189, 200, 204, 205, 209, 213, 214, 217, 229, 236, 237, 246, 249, 251, 256, 264, 265, 269, 272, 273, 275, 287
-, substrate-induced 25, 26, 40, 74, 93, 114, 126, 190, 191, 193, 204-206, 246
Rhizosphere 97, 104, 157, 165-167, 181, 182, 184, 229
Rhodanese 100, 101, 121
RNase 18
Root
- biomass 15, 46, 268
- nodules 157, 181
- zones 157, 236, 242, 243
Roses 129
Rye 6, 34, 208, 269, 270
Ryegrass 37, 194, 233, 259, 260
-, Italian 238, 239
-, perennial 199

Sainfoin 159, 225-227, 229, 232
-, Hungarian 267
Sagebrush 231
Salinisation 11
Sand 161
- content 196, 198, 261, 289
- treatment of exhausted peatlands 213
Sandstone 163, 168, 179
Sea buckthorn 160, 228, 229, 243
Seed germination 37, 39, 110
Selenium content 74
Serozems 91, 105
Shale 163, 179
-, alun 103
-, carbolithic 174
-, clay 161, 184
-, oil 217-221
Shoot mass 46
Silicon dioxide 79, 111
Silt content 196
Silver content 95
Slurry poldering 188, 189, 285
Small-leaf lime (linden) 97, 98, 256, 257
Smolnitsa soil 134
Sodium
- content 11, 110, 188, 189, 286
- EDTA 72, 73
- hexafluoroaluminate 105
- humate 98, 99
Soil aggregates 48, 191, 198
Soot 97, 99, 100, 124
Sorghum 166, 187
Soybean 37, 208
Spreading of overburdens on soil 269, 270
Spruce 69, 71-74, 91, 94, 103, 106, 110, 117, 120, 121, 248, 270
-, Siberian 183
Stemflow 124, 126
Steppe 5, 157, 159, 161, 184
Streptomycetes, number of 20, 164, 166, 167
Subsidence 285
Sugarbeet 208
Sugar content 134
Sulphatase 131, 132, 194, 195, 275-277
-, *see also* Arylsulphatase

Sulphate 99, 100
- availability 122, 123
- content 11, 84, 90, 289, 290
- reductase 90
-, *see also* Ester-sulphate
Sulphoxide oxidase 119
Sulphur
- compounds 90, 96, 105
- content 103, 122, 127, 128
-, *see also* Sulphate content
- dioxide 65, 67, 74, 75, 82, 84, 87, 97, 100, 103, 106, 120, 127
-, elemental 99, 121
- mineralisation 99
-, oxidation of 99, 121
-, reduced forms of 99, 100
- trioxide 106, 108
Sunflower 237
Surfactant 31, 44
Sweetclover 31, 32, 181, 182, 248, 251, 255, 256
Sweet gum 209
Sycamore 99-101, 179, 180, 209

Taiga 87, 184, 230
Tar 23, 24, 39, 51, 52
Temperature coefficient 199
n-Tetradecane 26
Tetrathionate 99
Texture
-, clay 168, 170, 186, 187, 204, 225, 231-233, 251
- -, silty 170, 285
-, loam 50, 174, 182, 188, 191, 197, 198, 228, 231, 232, 269
- -, clay 168, 170
- -, fine 195, 218
- -, heavy 6, 18, 30, 41, 43
- -, light 17
- -, medium 17, 181, 231
- -, medium-heavy 175, 178
- -, sandy 157, 168, 170, 178, 198, 225, 251, 253, 267
- -, sandy clay 168, 170, 184
- -, silt 170, 189, 193, 197, 198, 245
- -, silty clay 133
-, sand 182, 186, 187, 205, 211, 212, 228, 229, 232, 247

Texture (*continued*)
- -, fine 215
- -, loamy 19, 20, 157, 168, 238, 255, 261
- -, loose 106, 118, 248, 263
- -, mellow 78
- -, slightly clayey 78
- -, slightly loamy 106
-, silt 161, 248
Thiocyanates 97, 98
Thiosulphate 99-101
-, oxidation of 99
Tillage 37, 50, 188
Timothy 105, 215
Toluene 39
Tomato 129
Topsoil
- inoculant 165-167
- (stock)pile 45, 46, 162-167, 173, 183, 195
Topsoiling 167, 168, 170, 205, 206, 221, 232, 233, 246, 282
-, *see also* Covering of spoils with (top)soil
Toxic/toxicity 11, 14, 15, 23, 24, 37-39, 49, 51, 53, 110, 127, 205, 207, 236
Trefoil 251
-, birdsfoot 170, 175, 267
TTC-dehydrogenase 25, 26, 52
-, *see also* Dehydrogenase
Tundra 15, 131, 132

Urease 6, 7, 11, 14, 17, 19, 22-24, 26-35, 41, 44, 49, 50, 71-80, 83, 87-89, 91, 93-96, 98, 101-108, 111, 118, 119, 121, 124-126, 128-131, 133, 134, 157-163, 165, 168, 169, 171, 172, 175-180, 182-186, 188, 200, 203, 208, 212, 213, 215, 227, 229-232, 235-238, 247, 248, 255-257, 263-265, 269, 270, 273
Ureolytic activity 102, 270

Vanadium content 69, 70, 74, 94, 110, 117
Vetch 31, 33, 34, 185, 251, 269, 270
Vitamins 33

Water
- content 71, 72, 118, 217, 230

Water (*continued*)
- -, *see also* Moisture content
- holding capacity 20, 26, 36, 41-43, 48, 51, 71, 72, 110, 177
- logging 24, 25, 198
Watering 37, 98, 238, 239
-, *see also* Moistening
Wax myrtle 209
Wet redeposition of spoil (loess)/topsoil 188, 189, 229
-, *see also* Slurry poldering
- remediation 39, 40
Wheat 34, 91, 110, 176, 177, 187, 251, 270
Wheatgrass 159, 163, 229
Willow 181, 196, 209, 248

Xylanase 93, 94, 124-126, 130, 131, 276, 277
Xylenes 39, 102

Yeasts, number of 10, 14, 35, 45
Yellow soil 259, 260
Yield
-, crop 34, 171, 172, 186, 187, 207, 215, 227-229, 236, 269, 270
-, grain 17, 29, 31-33, 176
-, green matter 203, 214
-, hay 211, 213
-, herbage 194, 195, 214, 238-240, 246, 264, 268, 275-277

Zinc
- content 11, 27, 28, 71-74, 77, 78, 83, 86, 87, 91, 93-95, 104, 112-114, 117-119, 130, 134, 235-237
- oxide 70,79

Index of the scientific names of microorganisms, plants, and animals

Acer negundo 129, 209
- *platanoides* 129
- *pseudoplatanus* 99, 129, 179
- *rubrum* 209
Achillea millefolium 279
Aesculus hippocastanum 97, 98
Agropyron 163
- *desertorum* 222
- *elongatum* 222
- *inerme* 223
- *intermedium* 222
- *riparium* 223
- *sibericum* 222
- *smithii* 223
- *trichophorum* 222
Agrostis tenuis 235, 236, 279
Alnus glutinosa 157, 179, 231, 256
- *nepalensis* 133
Andropogon virginicus 175
Anthyllis vulneraria 264
Arrhenatherum elatius 179
Artemisia 163
- *austriaca* 231
Arthrobacter 43
Astragalus cicer 222
Atriplex canescens 163, 167, 223
- *confertifolia* 163
- *gardneri* 163
Azotobacter 77, 109, 164-166
- *chroococcum* 192

Bacillus 174, 272
Betula fruticosa 183
-*nana* 16
- *verrucosa* 130, 159, 256
Bouteloua gracilis 165
Bromus biebersteinii 222
- *erectus* 279
- *inermis* 179, 180, 211, 222, 248, 264

Calamagrostis villosa 120
Candida 14, 33, 43
-*guilliermondii* 35, 45
-*maltosa* 43
Caragana arborescens 222
Carex aquatilis 15
Ceratoides lanata 223
Chenopodium album 279
Clostridium 90, 174
- *pasteurianum* 272
Colutea arborescens 243
Cowania mexicana stansburiana 223

Dactylis glomerata 98, 176, 179, 225, 245
Dupontia fischeri 15

Eisenia 184
- *foetida* 32
Elaeagnus angustifolia 222
Elymus junceus 222
Ephedra viridis 223
Eriophorum vaginatum 15, 16, 132, 133
Escherichia coli 110

Festuca 203
- *arundinacea* 171, 175
- *capillata* 179
- *heterophylla* 179
- *ovina* 235, 236, 264, 279
- - *duriuscula* 223
- *pratensis* 179, 180, 248
- *rubra* 179, 180, 199
Fraxinus excelsior 130, 179
- *viridis* 256
Fusarium solani 99, 100

Glomus fasciculatum 166
- *mosseae* 166
Gossypium hirsutum 133

Halogeton glomeratus 163
Hedysarum boreale 223
Hippophaë rhamnoides 242, 243
Hordeum vulgare 46

Larix sibirica 183
Lezula confusa 15
Liquidambar styraciflua 209
Lolium x hybridum hausslen 245
- *multiflorum* 179, 238
- *perenne* 179, 180, 199, 245, 279
- *westerwolicum* 179
Lotus corniculatus 175, 179, 180, 264, 279
- *pedunculatus* 245
Lupinus luteus 264
- *polyphyllus* 264

Medicago sativa 222
Melilotus albus 264
Melilotus officinalis 222
- *volgicus* 255
Minuartia verna 236
Myrica cerifera 209

Nitrobacter 77
Nitrosomonas 77
Nyssa sylvatica 209

Onobrychis viciaefolia 222, 225
Oryzopsis hymenoides 223

Penstemon palmeri 223
Photobacterium phosphoreum 37
Picea abies 69, 71, 117, 121
-*obovata* 183, 184
Pinus banksiana 159
- *kesiya* 133
- *nigra* 243
- *palustris* 209
- *sibirica* 183
- *sylvestris* 69, 77, 79, 118, 127, 157, 183, 193, 231
- *taeda* 209
Pisum arvense 46
Platanus occidentalis 209
Poa ampla 223
- *angustifolia* 231
- *pratensis* 98, 199, 279
Populus 209
- *nigra* 130, 193
- *tremula* 179, 183
Prunus serotina 178, 179
Pseudobacterium lacticum 98
Pseudomonas 43
- *aeruginosa* 35
- *fluorescens* 192
- *liquefaciens* 98
Purshia tridentata 223

Quercus marilandica 77
- *pedunculata* 159
- *phellos* 209
Quercus (*continued*)
- *robur* 130
- *rubra* 179, 193
- *stellata* 76, 77

Rhamnus catharctica 179
Rhizobium 264
Rhodotorula 33
Robinia pseudacacia 158, 243

Salix 209
- *alba* 130
- *gracilis* 243
Salsola kali 163
Sambucus nigra 179
Sanguisorba minor 222
Saponaria officinalis 222
Sinapis arvensis 279
Sorghum bicolor 187
Sphaeralcea munroana 223
Sphagnum 122
Sporobolus aëroides 167, 223
Stipa viridula 223
Syringa vulgaris 97, 98

Thiobacillus ferrooxydans 192
Tilia cordata 97, 98, 130, 193, 256
Torulopsis 14
Trifolium pratense 238
- *repens* 179, 199, 245, 264
Typha latifolia 209

Ulmus pinnato-ramosa 256

Index of geographic names

Aberdeen 112, 113
Achinsk 111, 181
Alaska 15, 131, 132
Alberta 45, 48, 287
Albinovo námešti 121
Aleksandrov 231
Altamont 49, 50
Angarsk 271
Appalachian Mountains 122
Apsheron Peninsula 13, 14, 23, 45
Arlan 5
Austria 25, 52, 92, 106, 123, 124, 135
Avesta 94, 95
Azerbaijan 13, 23, 34, 45

Baia Mare 83, 237
Baidakov 157, 159, 161
Baikal Lake 102
Bald Knob 174, 175
Barnwell 208
Barrow 15
Bashkiria 5, 17, 18, 22, 24, 26, 29, 31, 269
Baukol-Noonan 172
Belorussia 17, 102, 135, 215
Berezovo 182
Bergheim 190
Berlin-Buch 27
Berrenrath 188
Berwyn Mountains 121-123
Big Horn 163
Big Run Bog 122
Binagadin 14, 45
Bistriţa-Năsăud 238
Bitkov 22
Blagoslovo 90
Bohemia 120
Böhmerwald 124
Bonn-Bad Godesberg 130, 131
Bothnia Bay 127
Bremen 279
British Isles 236
Brixlegg 92, 93
Bryansk 112
Buckhead 195
Bulgaria 134, 135

California 10, 22, 49
Canada 45, 83, 135, 287
Căpuş 225-228
Carpathians 108, 272
Center 172
Čepirohy 206
Chapeltown 99
Chasov Yar 255, 257
Cherepovets 119
Cherkassy 157
Chernobyl 112
Chiatura 232, 233, 265
Chimkent 91
China 24
Choszczówka 265
Cluj 225
Colchester 281
Colchis Lowland 7
Colorado 217, 221, 271
Colstrip 164
Coniston 84
Copper Cliff 84, 85
Copşa Mică 83
Coromandel Range 245
Cottbus 179, 192
Cracow 79, 263
Czech Republic 120, 135, 204

Dalton 132
Denby 196, 197
Derbyshire 196, 197
Dnepropetrovsk 231
Dneprov 255
Dnieper-Donets Depression 22, 28
Donbass 161
Donets Basin 161
Donetsk 253, 255
Dorsten 131

Eagle Creek 15
East Khasi Hills 133
Elizavetgrad 157
Emden 285
Ems River 285
England 99, 100, 121, 195, 236, 282
Erin 196, 197
Erskine 45, 48
Espenhain 191
Essex 282
Exelberg 124, 125

Fairbanks 132
Falconbridge 84
Farmington 164
Finland 127, 131, 135
Finspång 117
Fliseryd 69, 70
Frankfurt am Main 192

Gadovo 211
Galicia 198, 201, 202
Georgia 7, 232, 259, 265
Germany 8, 19, 27, 39, 120, 130, 135, 179, 187, 261, 279, 285

Gießen 8, 19
Glenharold 173
Goslawice 176, 178, 179
Göttinger Wald 120
Grassington Moor 236, 237
Grefrath 188
Grevenbroich 188
Grodno 215
Gumbra 259, 260
Gunpowder Falls 113
Gustorf 188
Gusum 70-73
Gyöngyös-Visonta 186, 207

Haard 120
Halemba 207
Halle 191
Hallein-Gamp 126
Harkey 133
Henryetta 76
Hidden Water Creek 163
Holme Moss 121-123
Hungary 186, 207

Ibinogorks 247
Illinois 163
India 133, 135
Inntal Lowland 92
Işalniţa-Craiova 208
Ishambai 5
Italy 200
Ivano-Frankovsk 272, 273

Jowai 133
Judgeford 275, 277

Kaltasin 5, 29, 33, 42, 43
Kalush 108
Kama River 6
Kansk 111, 181
Karelia 96, 230
Karlshamn 110
Katowice 77
Kavykta 271
Kazakhstan 91, 135
Kimovsk 185
Kingisepp 248
Kirchhoven 188
Kirk 195
Kirov 211
Kirovograd 157
Kirovsk 247
Kitchen Creek 49, 50
Klaj 78
Kola Peninsula 87, 247
Köln 190, 191
Konin 175, 178, 179
Kostomuksha 96, 230
Kuda River 271
Kulata 134
Kulenga River 271
Kursk 228, 229
Kuzbass 184
Kuznetsk Basin 184
Kvarntorp 103, 104
Kvirila Basin 232

Laisvall 95
Lebedin 228, 229
Leipzig 191
Lich 8
Lida 215
Lithuania 267
Little Bledowska Desert 263
Loxley 100
Lyudvikov 272, 273

Macoupin 163
Maramureş 83, 237
Mariánské Radčice 204
Maryland 112, 113
Meghalaya 133
Meirama 198, 199, 201-204
Meshchera 119
Miasteczko Slaskie 77
Migneint 121-123
Mineral 174
Mogilev 102
Monchegorsk 87
Monongalia 175
Montana 164
Moravian-Silesian Beskids 121
Morrells 195, 196
Moscow 119, 131, 185, 270
Murmansk 247
Mürztal 106, 107

Nazarovo Basin 111, 181, 183
Nefteyugansk 44
Neuwid Basin 261
New Jersey 36, 50
New Mexico 164
New Zealand 193, 245, 275
Niederlausitz 192
Niederrheinische Bucht 189
Niepolomice Forest 78, 79, 81
Nieslusz 178, 179
Nikopol 231, 232
Nordrhein-Westfalien 120
Norrland 75
North Dakota 162, 172, 173
North Island 245
Nová Ves 120
Novosibirsk 214
Novotroitsk 253

Ob River 7
Oklahoma 76, 162
Okmulgee 76
Olt 82
Ontario 83
Ore Mountains 120
Osage 175
Östergötland 70, 72
Oulu 127, 131

Patnów 175-177
Pennines 121, 236
Piceance Basin 217, 218, 221, 271
Pico 49, 50
Podolsk 270
Podorozhnen 251
Poland 77, 95, 106, 111, 118, 128, 135, 175, 179, 207, 263
Polevskoi 105
Powys 235
Preston 168, 170-172
Pre-Urals 269
Primorye 87
Prudhoe Bay 15, 131, 132

Québec 84

Rechitsa 17
Řečice 121
Reichwalde 193
Rhine 188-191
Riepe 285
Riepster Hammrich 285
Rock Springs 163
Rodna 238, 240, 243
Romania 82, 135, 208, 225, 237
Romashkino 6
Rönnskär 74, 75
Rouyin-Noranda 84, 87
Rozdol 251
Ruhr 120, 131
Russian Federation 5, 11, 17, 22, 24, 26, 29, 41, 51, 87, 96, 102, 104, 109, 111, 112, 114, 119, 131, 133, 135, 181, 211, 228, 247, 269
Ryefield 195

Saalfelden-Ramseiden 126
Saarbrücken 39
Saar River 39
Salajka 121
Salterwood 195
Salzburg 126
- Freisaal 126
Samara River 161
San Bernardino 10, 36
Saneli 133
Sankt-Petersburg 248
Săsar 237
Savanna River 208
Schöneben 124
Sheffield 100
Sheridan 163
Shillong 133
Shilo 196
Siberia 7, 41, 102, 111, 181, 184, 214, 271
Sibiu 83
Siedenburg 279
Silesia 179
Skellefteå 74
Slatina 82
Slivnitsa 134
Slope Mountains 131
Smolnica 179
Smrčiny 121
Sofia 134
Someşul Mare River 238
South Carolina 208
Southland 193
South Yorks 99
Spain 198, 200
Sredne-Uralsk 90
Stanton 173
Starooskol 228
Staunton 163
Steadmill 195
St.Koloman 126
Stoilensk 229
St.Osyth 281
Střimice 205
Strizhev 157-159
Sudbury 83, 84
Sweden 69, 74, 75, 94, 95, 103, 104, 110, 117, 135
Switzerland 91, 135
Szczakowa 263

Tataria 6, 11, 34
Terfens 93
Tirol 92, 93, 127
Tolmachevo-Krivodanov 214
Toolik Lake 132
Toronto 72
Torun 106
Trier 130
Trollhätten 94, 95
Tskhaltubo 259, 260
Tutura River 271
Tyumen 7, 8, 31, 34, 44

Ukraine 22, 28, 97, 108, 135, 157, 159, 231, 251, 253, 255, 272
United Kingdom 99, 100, 121, 135, 195, 235, 281
United States of America 10, 15, 22, 36, 49, 50, 76, 112, 122, 131, 135, 162, 208, 217, 271
Upper Silesia 118
Upper Styria 106
Užin 204, 206

Vienna 123, 125
Vilnius 267
Visonta 207
Volgo-Akhtubinsk 114
Vologda 119
Vršany 204
Vyatka 211
Vygonichi 112
Vyshkov 272, 273

Waimumu 193
Waitekauri Valley 245
Wales 121, 235
Wallis (Ronetal) 91
Warsaw 128
Wellington 275
Welzow-Süd 193
West Virginia 122, 168, 170-172, 174
Wetterau 192
Wienerwald 123, 124
Wschodnie 176
Wyoming 163

Y Fan 235
Yorkshire 236, 237
Yurkov 157-160

Zap 162
Zhitomir 157
Zlynka 112

www.ingramcontent.com/pod-product-compliance
Ingram Content Group UK Ltd.
Pitfield, Milton Keynes, MK11 3LW, UK
UKHW020236290926
15470UKWH00002B/575

9 780444 500571